Achieving sustainable cultivation of maize

Volume 1: From improved varieties to local applications

It is widely recognised that agriculture is a significant contributor to global warming and climate change. Agriculture needs to reduce its environmental impact and adapt to current climate change whilst still feeding a growing population, i.e. become more 'climate-smart'. Burleigh Dodds Science Publishing is playing its part in achieving this by bringing together key research on making the production of the world's most important crops and livestock products more sustainable.

Based on extensive research, our publications specifically target the challenge of climate-smart agriculture. In this way we are using 'smart publishing' to help achieve climate-smart agriculture.

Burleigh Dodds Science Publishing is an independent and innovative publisher delivering high quality customer-focused agricultural science content in both print and online formats for the academic and research communities. Our aim is to build a foundation of knowledge on which researchers can build to meet the challenge of climate-smart agriculture.

For more information about Burleigh Dodds Science Publishing simply call us on +44 (0) 1223 839365, email info@bdspublishing.com or alternatively please visit our website at www.bdspublishing.com.

Related titles:

Achieving sustainable cultivation of maize Volume 2: Cultivation techniques, pest and disease control
Print (ISBN 978-1-78676-012-8); Online (ISBN 978-1-78676-014-2, 978-1-78676-015-9)

Achieving sustainable cultivation of wheat Volume 1: Breeding, quality traits, pests and diseases
Print (ISBN 978-1-78676-016-6); Online (ISBN 978-1-78676-018-0, 978-1-78676-019-7)

Achieving sustainable cultivation of wheat Volume 2: Cultivation techniques
Print (ISBN 978-1-78676-020-3); Online (ISBN 978-1-78676-022-7, 978-1-78676-023-4)

Achieving sustainable cultivation of rice Volume 1: Breeding for higher yield and quality
Print (ISBN 978-1-78676-024-1); Online (ISBN 978-1-78676-026-5, 978-1-78676-027-2)

Achieving sustainable cultivation of rice Volume 2: Cultivation, pest and disease management
Print (ISBN 978-1-78676-028-9); Online (ISBN 978-1-78676-030-2, 978-1-78676-031-9)

Achieving sustainable cultivation of sorghum Volume 1: Genetics, breeding and production techniques
Print (ISBN 978-1-78676-120-0); Online (ISBN 978-1-78676-122-4, 978-1-78676-123-1)

Achieving sustainable cultivation of sorghum Volume 2: Sorghum utilisation around the world
Print (ISBN 978-1-78676-124-8); Online (ISBN 978-1-78676-126-2, 978-1-78676-127-9)

Chapters are available individually from our online bookshop: https://shop.bdspublishing.com

climate•SMART•publishing
IN AGRICULTURAL SCIENCE

BURLEIGH DODDS SERIES IN AGRICULTURAL SCIENCE

NUMBER 01

Achieving sustainable cultivation of maize

Volume 1: From improved varieties to local applications

Edited by Dr Dave Watson, CGIAR Maize Research Program Manager, CIMMYT, Mexico

burleigh dodds
SCIENCE PUBLISHING

Published by Burleigh Dodds Science Publishing Limited
82 High Street, Sawston, Cambridge CB22 3HJ, UK
www.bdspublishing.com

Burleigh Dodds Science Publishing, 1518 Walnut Street, Suite 900, Philadelphia, PA 19102-3406, USA

First published 2017 by Burleigh Dodds Science Publishing Limited

Library of Congress Control Number: 2017932196

British Library Cataloguing in Publication Data
A catalogue record for this book is available from the British Library

ISBN 978-1-78676-008-1 (print)
ISBN 978-1-78676-010-4 (online)
ISBN 978-1-78676-011-1 (online)
ISSN 2059-6936 (print)
ISSN 2059-6944 (online)

Typeset by Deanta Global Publishing Services, Chennai, India
Printed by Lightning Source

Contents

Series list

Title	Series number
Achieving sustainable cultivation of maize - Vol 1 From improved varieties to local applications Edited by: Dr Dave Watson, CGIAR Maize Research Program Manager, CIMMYT, Mexico	001
Achieving sustainable cultivation of maize - Vol 2 Cultivation techniques, pest and disease control Edited by: Dr Dave Watson, CGIAR Maize Research Program Manager, CIMMYT, Mexico	002
Achieving sustainable cultivation of rice - Vol 1 Breeding for higher yield and quality Edited by: Prof. Takuji Sasaki, Tokyo University of Agriculture, Japan	003
Achieving sustainable cultivation of rice - Vol 2 Cultivation, pest and disease management Edited by: Prof. Takuji Sasaki, Tokyo University of Agriculture, Japan	004
Achieving sustainable cultivation of wheat - Vol 1 Breeding, quality traits, pests and diseases Edited by: Prof. Peter Langridge, The University of Adelaide, Australia	005
Achieving sustainable cultivation of wheat - Vol 2 Cultivation techniques Edited by: Prof. Peter Langridge, The University of Adelaide, Australia	006
Achieving sustainable cultivation of tomatoes Edited by: Dr Autar Mattoo, USDA-ARS, USA & Prof. Avtar Handa, Purdue University, USA	007
Achieving sustainable production of milk - Vol 1 Milk composition, genetics and breeding Edited by: Dr Nico van Belzen, International Dairy Federation (IDF), Belgium	008
Achieving sustainable production of milk - Vol 2 Safety, quality and sustainability Edited by: Dr Nico van Belzen, International Dairy Federation (IDF), Belgium	009
Achieving sustainable production of milk - Vol 3 Dairy herd management and welfare Edited by: Prof. John Webster, University of Bristol, UK	010
Ensuring safety and quality in the production of beef - Vol 1 Safety Edited by: Prof. Gary Acuff, Texas A&M University, USA & Prof.James Dickson, Iowa State University, USA	011
Ensuring safety and quality in the production of beef - Vol 2 Quality Edited by: Prof. Michael Dikeman, Kansas State University, USA	012
Achieving sustainable production of poultry meat - Vol 1 Safety, quality and sustainability Edited by: Prof. Steven C. Ricke, University of Arkansas, USA	013
Achieving sustainable production of poultry meat - Vol 2 Breeding and nutrition Edited by: Prof. Todd Applegate, University of Georgia, USA	014
Achieving sustainable production of poultry meat - Vol 3 Health and welfare Edited by: Prof. Todd Applegate, University of Georgia, USA	015

Acknowledgements

We wish to acknowledge the following for their help in reviewing particular chapters:

- Chapter 1: *Dr Joost van Heerwaarden, Wageningen University, The Netherlands*
- Chapter 7: *Dr Supan Langyun Singh, Indian Institute of Maize Research*
- Chapter 10: *Professor Narpinder Singh, Duru Nanak Dev University, India*

Introduction

Maize is one of the most widely grown crops in the world, both for food and livestock feed, biofuels and other uses. Cultivated on 184 million hectares globally, maize is cultivated in 160 countries throughout the continents of America, Europe, Oceania, Africa and Asia. Maize is fundamental to global food and economic security, providing up to 30% of total calorie intake in some developing countries. It has been estimated that maize yields need to increase by 60% by 2050 in the context of increasing competition for land, water, energy and labour. Maize production is currently held back by factors such as: lack of available improved cultivars or failure to take up new improved varieties, inadequate crop management and storage, poor soil quality, the impact of pests and diseases and more extreme weather related to climate change. Research needs to focus on supporting more productive, sustainable and nutritionally valuable maize cultivation, particularly for smallholders in the developing world.

The two volumes of *Achieving sustainable cultivation of maize* summarise key research addressing these challenges. This volume focuses on breeding, improving nutritional quality and ways of supporting smallholders. Volume 2 looks at ways of improving maize cultivation as well as the management of pests and diseases.

Part 1 Genetic diversity and breeding

It is widely accepted that developing improved varieties which are more resilient to pests, diseases and extreme weather depends on preserving and exploiting genetic diversity in maize. As Chapter 1 points out, genebanks and *ex situ* collections are essential to conserving genetic diversity and enabling global access to those resources. The chapter provides a comprehensive and authoritative review of the key issues facing both *in situ* and *ex situ* collections. These issues include better conservation and exploitation of wild varieties and landraces, as well as identifying broader genetic and allelic diversity in *Zea* and *Tripsacum*. The chapter includes case studies of the use of exotic germplasm to support breeding of more resilient varieties. As Chapter 1 argues, the future role of genebanks must continue to evolve from the initial concept as an emergency and long-term repository of genetic diversity which otherwise would be lost. Genebanks should occupy a pivotal point in the global flow of germplasm, facilitating both conservation and providing access for use, as well as providing characterisation data on accessions to support breeding programmes.

Building on Chapter 1, Chapter 2 provides a review of key challenges in breeding programmes. The chapter focuses specifically on the challenges facing breeding programmes in sub-Saharan Africa which are critical to meeting the need for new varieties suited to local conditions. An example is screening new varieties for drought tolerance season under carefully monitored conditions of water stress. However, most national agricultural research system (NARS) research stations in African countries do not have the necessary weather stations and irrigation systems to conduct high-quality research in order to develop drought-resistant varieties and boost maize production in the region. Similarly, for national programmes to make full use of methods such as doubled haploid (DH) techniques or marker-assisted selection (MAS), adequately equipped facilities are required and appropriately trained scientist/personnel must be available.

As well as looking at what is needed to improve infrastructure and personnel, the chapter explores developments such as the improved use of broader germplasm resources in African breeding programmes. As an example, and echoing Chapter 1, it shows how *Striga* tolerance in maize germplasm has been increased by introgression of favourable genes from teosinte (*Zea diploperennis*), a wild relative of maize. It also discusses advances in hybrid development in the areas of heterotic grouping and identification of testers in maize germplasm and the potential of participatory plant breeding research, including innovations such as mother-baby trials (MBTs). An MBT involves evaluating a relatively large number of potential new varieties in a trial (mother trial) and subsets of the varieties (baby trials) in satellite farmers' fields. MBTs have been shown to encourage better interaction between researchers and farmers, resulting in better targeted and adopted varieties for local conditions. They illustrate the potential of African breeding programmes given appropriate support.

Picking up on themes in both Chapters 1 and 2, Chapter 3 looks at key challenges facing breeders in an era of climate change. As the chapter points out, inbred-hybrid techniques are still widely used in maize breeding but have resulted in a decline in genetic diversity and have sometimes focused too much on breeding for ideal environmental conditions. Only maize breeding programmes utilising large samples of genetically broad-based germplasm have the potential to contribute useful and unique alleles to improve abiotic and biotic stress tolerance. There is also a need for breeders to improve techniques for developing traits which are genetically complex, difficult to screen and largely influenced by the environment. The chapter includes case studies addressing the challenge of using exotic germplasm and improved techniques to develop more resilient varieties such as short-season quality maize hybrids, as well as cold and drought-resistant varieties.

Chapter 4 provides a case study of cutting-edge developments in maize breeding, focusing on improving protein content in maize. The chapter reviews the mechanisms that limit the supply and quality of protein, including transcriptional and post-transcriptional regulation of gene expression. It looks in particular at the regulation of storage protein genes, particularly the synthesis and deposition of zein proteins. It also describes mutants in maize that interfere with these mechanisms and how genetic approaches can improve protein quality traits, including ways of maintaining sulphur and nitrogen storage in maize seeds.

The following two chapters illustrate how developments in breeding can be used to target and improve particular traits in maize. Depending on the growing environment, maize grains can be infected by one or more ear rot fungi which may decrease grain yield and quality. Many of them also produce secondary metabolites, known as mycotoxins, which can have serious detrimental effects on humans and animals that consume the infected grain. The successful creation of maize germplasm with resistance to aflatoxin infection requires the identification of genetic variation for resistance in maize germplasm, techniques to identify and quantify resistance, and breeding methods or procedures to develop resistant lines. Chapter 5 looks at these challenges in developing new varieties, and how they have been addressed in practice in creating *Aspergillus flavus*-resistant maize breeding lines.

In the light of more extreme weather conditions associated with climate change, Chapter 6 looks at research on developing cold-resistant maize varieties. This chapter discusses the physiological and genetic background of cold tolerance, the methodology required for its analysis, and ways of improving cold tolerance by breeding. It begins by reviewing research on the effects of temperature on maize growth, as well as factors affecting seed quality and germination at low temperatures. It goes on to discuss germplasm evaluation and the Inheritance of cold tolerance before showing how chilling-tolerant hybrids have been developed by selecting for inbred lines with good early or seedling chilling tolerance. The increasingly widespread use of molecular breeding methods, MAS and the discovery of quantitative trait loci (QTLs) linked to chilling tolerance are helping to shorten the breeding process.

Part 2 Understanding and improving maize nutritional and processing quality

As indicated at the beginning of the Introduction, maize is a critical part of the diet in many developing countries. Improving its nutritional quality is an important target in improving health. Chapter 7 reviews what we know about the protein content of maize, including a discussion of how protein quality can be enhanced by the incorporation of the *opaque-2* gene into elite maize germplasm in order to enhance lysine content. Subsequent sections review the carbohydrate profile of maize, the qualities and uses of corn oil and the micronutrients present in maize.

Building on Chapter 7, Chapter 8 looks at ways of enhancing the micronutrient density of maize through biofortification. The chapter assesses options for biofortification of maize with provitamin A carotenoids, and looks at methods for breeding of provitamin A biofortified maize and the ways it can be delivered to consumers. The chapter includes a case study of the maize biofortification programme in Zambia.

Developing themes in Chapter 4, Chapter 9 looks at ways of improving the starch and protein content of maize. After reviewing maize kernel composition, it provides an overview of methods for improving the protein and starch content of maize kernels. The chapter then goes on to summarise methods for improving the protein and starch content of maize using mutant lines. It also discusses ways of improving the digestibility of maize protein and starch, as well as assessing and reducing the anti-nutritional properties of maize.

Advances in the technologies and methods for analysing the quality characteristics of maize grains can help us to determine the contribution of maize products to human health, as well as helping in the development of maize cultivars with improved characteristics. Chapter 10 reviews the current range of methods for determining important characteristics of maize grains, with sections covering starch content, protein content, phenolic compounds (in particular flavonoids such as carotenoids), kernel hardness/texture and levels of myotoxins.

Part 3 Translating research into practice: improving maize cultivation in the developing world

The final group of chapters focus on ways of translating the results of research into practical outcomes that can benefit farmers, particularly smallholders in developing countries. The adoption of improved technology is essential for increasing productivity in the production of maize in Africa. Chapter 11 reviews the importance of increasing the adoption rate of improved maize technologies, especially among smallholder farming households. It reviews current research on understanding farmers' decisions on whether or not to adopt new technology. It then assesses economic and institutional factors which make adoption difficult, such as land availability, lack of capital, price fluctuations, poor seed supply and infrastructure. It also looks at farmer characteristics and attitudes, which affect adoption such as objectives and goals as well as perceptions of costs and risks associated with adopting new methods and technology. The chapter concludes by looking at practical strategies for improving the adoption of improved maize technologies.

One of the reasons that adoption of new varieties, cultivation techniques or technologies can be poor is the nature of implementation. Paradigms of agricultural innovation have shifted considerably from linear Transfer-of-Technology (ToT) models to 'Farmer First' and Farming Systems Research and, most recently, to the use of Agricultural Innovation Systems (AIS). An innovation system can be defined as a network of organisations, enterprises and individuals focused on bringing new products, new processes and new forms of organisation into use, together with the institutions and policies that affect their behaviour and performance. The agriculture innovation systems framework maintains that improved interaction and joint work between stakeholders result in better information exchange, and more ideas and opportunities for sector improvement. This, in turn, leads to further collaboration and more opportunities for better and faster innovation. Chapter 12 presents an AIS approach to improving innovation in maize production. The chapter considers three important stages of an innovation process: the identification of problems and targeting of interventions, local experimentation, and bringing new ideas into routine use. The chapter discusses the role of research at each stage, and then argue that an AIS approach mandates increased focus on the personal, organisation and institutional capacities of researchers and research bodies.

The final chapter in the book looks at the role of women in maize cultivation. Woman farmers are less likely to use improved technologies for maize production than are their male counterparts. Gender analysis is therefore essential in guiding the design and focus of research to ensure that women participate in and benefit from growth in the maize sector. Chapter 13 begins by analysing the way farming households manage maize cultivation. The following section analyses the ways in which gender impacts agricultural production systems, including access to resources and decisions about inputs. The next three sections offer a detailed analysis of the impact of gender on three vital aspects of maize production: labour, processing and storage and varietal choice. The preferences of women as urban consumers are also considered. Finally, a brief discussion of the Women's Empowerment in Agriculture Index highlights one tool for monitoring the impact of research and agricultural change on women's empowerment.

Part 1

Genetic diversity and breeding

Chapter 1

Ensuring the genetic diversity of maize and its wild relatives

J. Stephen Smith, Iowa State University, USA; Candice A. C. Gardner, USDA-ARS/ Iowa State University, USA; and Denise E. Costich, CIMMYT Germplasm Bank (Maize Collection), Mexico

1 Introduction

'Maize is a sensitive mirror of the people who grow it.' (Edgar Anderson, 1947)

Economic activity and the quality of life for current and future generations of humankind, many other species, and the health and sustainability of the natural environment are inextricably linked with the management of plant genetic resources for food and agriculture (PGRFA).

Management of crop genetic diversity provides the potential upon which all other farm management decisions and the environment interact to annually determine harvestable yield and nutrient quality. The abilities to access, characterize, successfully recombine and evaluate the genetic potential for traits of agronomic importance are crucial elements of successful germplasm management. Conservation of genetic resources is an imperative to allow their continued use, both now and in the future.

Of major importance in this regard is maize (*Zea mays* L.). Maize is the world's most multipurpose crop, a food staple for hundreds of millions of people in the developing

http://dx.doi.org/10.19103/AS.2016.0001.02

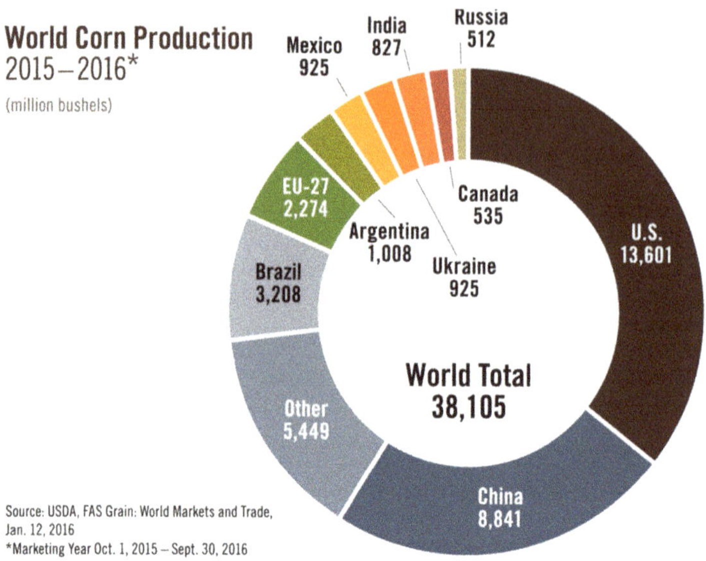

Figure 1 Global maize production (from World of Corn, 2016).

world, feed for billions of livestock and raw material for an increasing number of industrial and biofuel uses. Cultivated on 184 million hectares (M ha) globally (FAOSTAT, 2013), maize is fundamental to global food and economic security, and of immense historical and current cultural value, providing 20% of the total calories in human diets in 21 low-income countries, and over 30% in the 12 countries that are home to a total of more than 310 million people (Anon. Maize Phase II, CGIAR-Research Progam on Maize, 2016) (Fig. 1). On average, 22% additional land globally could be transitioned into productive cropland. Of this additional cropland, greatest additional percentages compared to existing cropland are situated in sub-Saharan Africa (48%), Latin America and the Caribbean (43%), Australia (36%), least additional cropland is available in East Asia and Central Asia (17%), East and South Asia (3%) and the Middle East and North Africa (3%) (Fischer et al., 2014).

To date, most commercial production of maize has been based on relatively few of the ancient races that were developed by farmers during the original domestication process, some 8–10 000 years ago, or even on the additional races that were developed during subsequent millennia as a result of hybridization, transport to and selection in new locations, genetic drift and mutation. Improved management of maize diversity on farms and in breeding, whether informal or formal, is required to better meet the needs of all farmers and of society.

DNA-based taxonomic and phylogenetic evidence have served to revise the scope of genetic resources, through the identification of the wild relatives that may serve as further sources of useful genetic diversity for future maize improvement. Historical data and experience provide abundant evidence that hybridization among maize races and effective selection can markedly increase the potential for maize to be productively adapted to a wide range of production locations. Six compelling reasons to conserve crop diversity include ensuring food security, adapting to climate change, safeguarding biodiversity, protecting nutritional security, reducing poverty and ensuring sustainable agriculture (Global Crop Diversity Trust, 2016, https://www.croptrust.org/our-mission/crop-diversity-why-it-matters/). Genetic resource diversity must be conserved to be potentially available and so complementary approaches to conservation (*in situ* compared to *ex situ*) are reviewed.

The genetic basis for improving potential maize yields through breeding has been largely through selection upon anonymous multigenic factors supporting increasing abilities to yield under stress conditions imposed as a result of higher planting densities. The reallocation of photosynthates to seed production has made an additional but much smaller contribution in maize (Smith et al., 2014) compared to classic examples such as major effects of the reallocation of photosynthates achieved through the introgression of dwarfing genes from the Japanese Daruma wheat variety into high-yielding American varieties or through the use of Norin 10 by Norman Borlaug during the Green Revolution (Dalrymple, 1978). Nonetheless, single native genes and modifiers have been used to achieve pest resistances in maize. More recently, biotechnology has enabled use of hitherto unavailable single genes for pest resistance in combination with herbicide resistance genes from very exotic sources to help preserve and achieve yield potential, as well as realization of environmental benefits resulting from use of conservation tillage and reduced application of pesticides.

Plant breeders face increasing challenges to achieve sustainable yield increases in agricultural systems that must also produce more nutritious food while also contributing more positively to the environment in the face of increasingly unpredictable weather and associated abiotic and biotic pressures. Management of genetic resources fundamentally determines the outcome of plant breeding and agriculture. Consequently, both challenges and imperative opportunities abound regarding the future use of genetic resources. Genetic diversity inevitably declines as a result of pedigree breeding in a system that is closed to the introduction of fresh diversity. Consequently, it will be imperative to broaden the base of useful maize germplasm from that which breeders have become accustomed to using. Conservation of available genetic resource diversity including germplasm arrayed in crop wild relatives is a fundamental prerequisite to enabling their potential use in the future. *However, achieving the goal of conservation alone* is insufficient. Decades of practical experience have demonstrated that identifying and incorporating useful maize genetic diversity is challenging, especially when exotic germplasm is initially un-adapted to target production environments. Nonetheless, new opportunities exist to identify useful new germplasm that potentially resides within international breeding programmes, facilitated through the increased use of more efficient procedures to identify and incorporate useful new diversity.

Factors impacting the ability to sustain advances in maize genetic gain include 1) the dynamic nature of the maize genome itself, 2) access (legal and technical) to diverse and useful genetic resources, 3) the use of both traditional and molecular plant breeding approaches and 4) the application of bioinformatics. There is also an abiding imperative to close yield gaps between those achieved in breeders' trials and those attained on farms. With regard to the germplasm that can be available, we provide a review of past and current conservation and utilization strategies. The necessary tools to enable full realization of the inherent usefulness of genetic resource diversity in both cultivated and wild species are in a period of continuous and innovative development (Yu et al., 2016). We conclude by proposing a more dynamic and central role for genebanks to play in the future as a pivot point connecting farmers, researchers and plant breeders with facilitated access to germplasm that is both fully in compliance with access and benefit-sharing regimes and more thoroughly documented with regard to genetic content and potential utility.

2 Global maize cultivation and yield trends

2.1 Global cultivation

Maize is now cultivated in 160 countries throughout the continents of America, Europe, Oceania, Africa and Asia (Upfront, 2006; Ben-Ari and Makowski, 2014). Predicted yields in metric tonnes/ha (t/ha) for the 2017 harvest range from 0.64 t/ha (Zimbabwe), 3.6 t/ha (Mexico), 5.97 t/ha (China), 8.37 t/ha (Argentina), to a historical high of 10.96 (t/ha) (the United States), with a global mean of 5.78 t/ha (USDA-FAS, 2017). Together, China and the United States accounted for 60% of global maize production in 2014 (FAOSTAT, 2014). Typically, the United States produces approximately 35% of the global corn crop (World of Corn, 2016). Population growth and growing demand for animal products, particularly in developing countries, have contributed to increased global demands on maize production for food and feed.

Per cent of global production (in parentheses) by the top 14 maize producing countries during 2014 was the United States (35), China (25), Brazil (7.7), Argentina (3.2), Ukraine (2.7), Mexico (2.5), India (2.2), Indonesia (1.8), France (1.8), South Africa (1.4), Romania (1.1), Canada (1.1), Russian Federation (1.1) and Nigeria (1.0) (FAOSTAT, 2014). Maize is particularly important to food security in Central and South America and most regions in Africa (FAO, 1997). The introduction of new improved maize varieties into West and Central Africa has moved more than one million people per year out of poverty since the mid-1990s (Alene et al., 2009). The range of environments across which maize can be cultivated is enormous, from 58°N (Canada and the Russian Federation) to 40°S (e.g. Chile) in tropical, subtropical and temperate environments, and from sea level up to 3800 m with growing seasons between 42 and 400 days (Paliwal, 2000; Farnham et al., 2003).

2.2 Global yield trends

Total world production of maize in 2014 was 1040 million metric tonnes achieved by an annual increase of approx. 13.2 million tonnes/yr from the production level of 205 million metric tonnes achieved in 1961. During this period mean global maize yields increased from 1.9 to 6.5 t/ha. During these years, global area harvested increased from 106 million ha to 185 million ha at an annual rate of approx. 1.2 million ha/yr. However, global maize production during 2007–2014 also increased due to a much higher (3.9 million ha/yr) increase in area planted to the crop (FAOSTAT, 2017). Rates of yield increase for Eastern, Central and Western Africa are low (6–19.7 kg/ha/yr). In contrast, rates of yield gain in Southern Africa more than doubled post 1991 from 50.8 to 125 kg/ha/yr. Maize yields rose fourfold in Eastern Europe from 50.7 to 210.6 kg/ha/yr post 2010 and showed a fivefold increase in Southeast Asia from 30 to 105.8 kg/ha/yr post 1991 (Table 1). The fastest rate of genetic gain recorded is for Northern Europe (6.5%) and reflects a relatively short-term catch-up phase as hybrids have been developed for more northern maturities.

Rates of yield gain during the period 1961–2014 are also presented in Table 1 (FAOSTAT, 2017). Ray et al. (2012, 2013) reported trends in maize yields in greater detail for regions, countries and for some countries, to the level of individual states. Globally, maize yields were increasing in 70% of the planted area (103 million hectares), stagnating in 26% and decreasing in 3%. Almost 29% of global area planted to maize has not experienced increased yield during a 47-year period between 1961 and 2008, as summarized in Fig. 2 and 3 (from Ray et al. (2012, 2013)). The area over which maize yields have stagnated is

Table 1 Mean maize yields achieved in 2014 (t/ha) and (annual rates of yield gain) (kg/ha/yr) achieved during 1961–2014 (except where stated), regional groupings and data from FAOSTAT (2017)

Africa 3.2 (37.8)	Asia 5.3 (74.9)	Europe 7.8 (74.5)	America
E. Africa 1.8 (11.8)	Central Asia (1992–2014) 6.6 (215.4)	E. Europe (1961–2009) (50.7)	N. America 10.7 (112.2)
Middle Africa 1.2 (6.2)	E. Asia 5.8 (90.2)	E. Europe (2010–2014) 5.7 (210.6)	S. America 5.3 (67.4)
N. Africa 6.7 (100.5)	S. Asia 33.4	N. Europe (2000–2014) 6.5 (391)	
S. Africa (1961–1992) (50.8)	S.E. Asia (1961–1991) (30)	S. Europe 9.0 (105.9)	
S. Africa (1993–2014) 4.8 (125)	S.E. Asia (1991–2014) 4.2 (105.8)	W. Europe 10.3 (129.5)	
W. Africa 1.8 (19.7)	W. Asia 7.1 (92.6)		

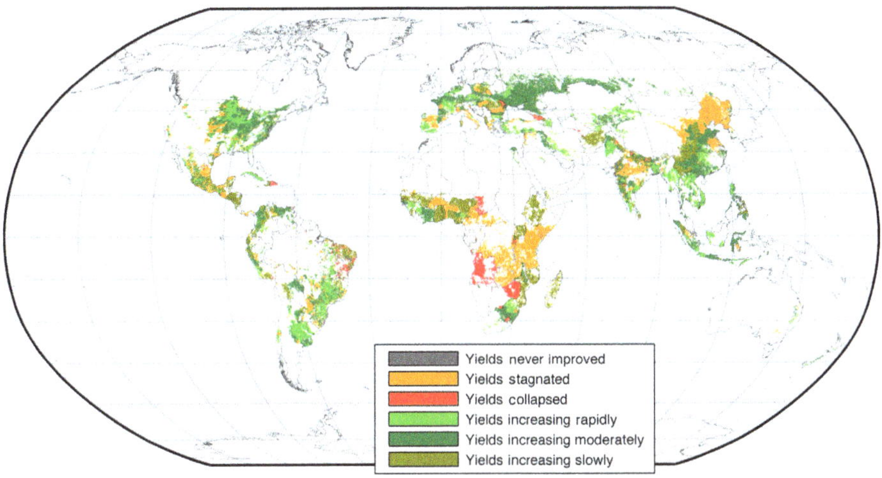

Figure 2 Changes in global maize yields.

less than for wheat (37%) and rice (35%), but more than for soybean (23%). Maize yields are stagnating in Eastern and Southern Mexico, other areas in Latin America, Spain, Italy, Greece and across large areas of Africa. India and China have the largest areas of stagnating maize yields at 31 and 52%, respectively.

Causes of maize yield stagnation include inferior agronomic practices and non-adoption of improved varieties (Mexico), intensified and higher risk crop rotations in drought-sensitive areas (US Great Plains). Lack of water management, low nutrient usage and unavailability of higher-yielding varieties contributed to reduced yields in Africa. Nonetheless, practice shows that fairly small increases in inputs can double maize yields in Africa (Ray et al., 2012). The area of most rapid maize yield increase is the US Corn Belt, although with stagnant yields in fringe areas. Other isolated areas of rapid yield increase exist within Portugal,

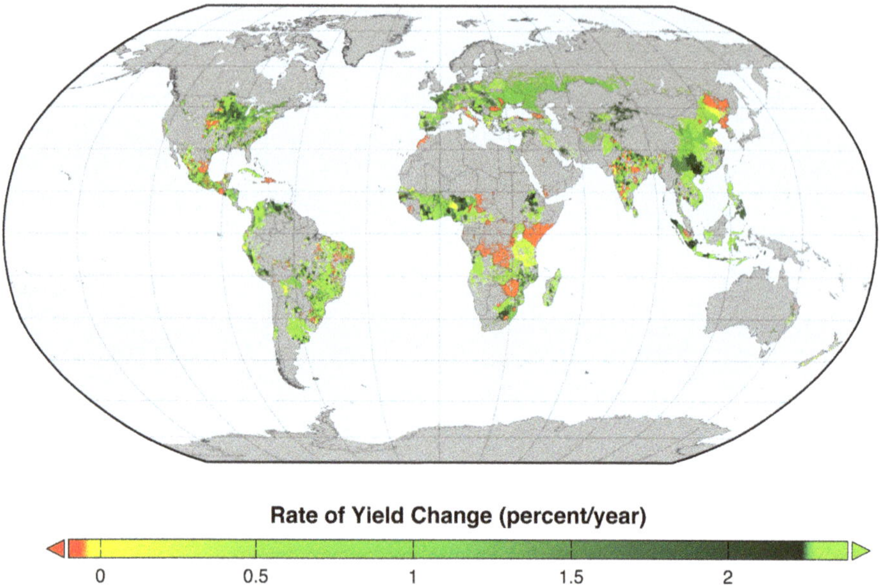

Figure 3 Rate of change of global maize yields.

France, Eastern Europe, Turkey, Nigeria, South Africa, Pakistan, India, China, Vietnam and Indonesia. Brazil and Argentina together account for 90% of maize production in Southern America with harvests of 81 and 34 million tonnes, respectively, in 2016.

During recent decades the US maize harvest has contributed annually approx. 35% of global maize production (FAOSTAT, 2014), hence the use of maize germplasm north of the US–Mexican border warrants particular attention. Yields of maize on farms cultivated by native Mexican farmers have been estimated at 700–1200 kg/ha (11–19 bu/ac) (Cook and Borah, 1979) and might thus approximate yields achieved by Native Americans (in the United States) during the last 3–4 millennia. Maize yields achieved on Native American Indian farms in New England during the 1600s have been estimated at 2500 kg/ha (40 bu/ac) (Bennett, 1955). Centuries later, Kroeber (1947) noted mean US yields of 1567–1880 kg/ha (25–30 bu/ac), significantly below those achieved by Native American Indians some three centuries earlier. Understandably, Kroeber (1947) lamented: 'Maize notoriously increases its yields but little under improved methods of farming. The improvements which we have made over Indian methods have been mainly in the direction of reducing production costs, especially in labor.' Improved farm management and the development of varieties with improved genetic potential have together brought about yields that were unforeseeable by many in the late 1940s, even as hybrids had begun to outyield and thus to replace open-pollinated varieties (OPVs) on farms. The introduction of hybrids offered improved standability, reducing manual labour required to harvest and bring in the crop. This enabled mechanization, allowing farmers to increase production acreage, which in turn drove the demand for improved varieties and quality seed.

The percentage of global maize exports contributed by South America is now 10%, having increased from 2% in 2004 (South American Maize Market, Mordorintelligence.com). Agricultural production has increased in Brazil and Argentina due to the use of new techniques and technologies to bring formally unusable and barren lands into cultivation. However, the rapid rates of increase in Brazilian maize yields may be a one-time event due to the development of improved management of acidic soils and the introduction of hybrids (Grassini et al., 2013). Maize yields in China have the potential to make a one-time step change due to the introduction of hybrids that are better adapted to denser planting conditions, mechanization and improvements in seed technology (Li et al., 2011), and is in progress.

There is a great diversity of maize yields among countries within sub-Saharan Africa. During 2007, the top five countries had maize yields ranging from 2.4 to 7.7 t/ha (mean 4.5 t/ha) whilst the bottom five maize yields ranged from 0.2 to 0.5 t/ha (mean 04 t/ha) (Fact 6, www.gapminder.org) (http://global-growing.org/en/content/fact-6-sub-saharan-african-agricultural-productivity-yield-hectare-about-50-cent-its). Gibbon et al. (2007) attributed percentile causations of low maize yields in sub-Saharan Africa to poor soil quality with an annual net loss of nutrients (44%), competition from weeds (19%) and drought (18%). Lack of infrastructure and effective institutions are also contributing factors (Fischer et al., 2014). There is much evidence that maize yields in this region can be improved, with conservative estimates of achievable yields from 6.1 to 7.1 t/ha by remedial attention to soil nutrition, weed control, better adapted varieties with increased stress resistance and improved crop water supply (Fischer et al., 2014).

Providing stable production in the face of increasing weather variability is a major challenge to sustainable agriculture. Ray et al. (2014) noted that across the nine major grain belts of the globe, 41% of inter-year yield variation was due to climate variability, and the percentage effect of climate variability on yield variability increased to 60–75% for the United States and China. An analysis by Burke et al. (2009) on the impact of expected growing season temperatures in Africa on its three primary cereal crops – maize, millet and sorghum – indicated that the majority of African countries will have novel climates over at least half of their current crop area by 2050. Of these countries, 75% will have novel climates that are analogous to those of five other African countries, suggesting that international movement of germplasm will be necessary for adaptation. The hotter Sahelian countries' climates will have few analogues for any crops. They also identified key African countries whose current crop areas are analogous to future predicted climates, but their germplasm is poorly represented in major genebanks, suggesting a need to increase focus on conservation of their genetic resources.

2.3 Need to increase global maize production

Several studies have concluded that there is a need to double agricultural production by 2050 (FAO, 2009a,b; Pingali, 2006; FAOSTAT, 2009; Godfray et al., 2010, Foley et al., 2011; Tilman et al., 2011; OECD-FAO, 2012). However, a mean annual increase in global maize production of 1.6% will result in a 33% shortfall in relation to this goal (Ray et al., 2013). Grassini et al. (2013) cautioned that rates of yield increase are generally linear leading to an upper yield plateau defined by interactions of the genetic potential of available varieties with the biophysical environment, as well as by policies affecting fertilizer and pesticide applications. Yield projections that use compound, exponential rates of yield increase 'do not exist in the real world' (Grassini et al., 2013). Duvick and Cassman (1999) reported increasing research expenditures for maize during the 1970s–1990s, not only as a result of the new wave of biotechnology, but also by an increase in the numbers of maize breeders, testing locations and numbers of yield test plots. Smith et al. (2014) reported a continuation of the trend in increased research costs per unit of genetic gain for maize into the second decade of the twenty-first century. Thus, 'the marginal yield increase per unit of research investment has decreased substantially over time' (Grassini et al., 2013). Nonetheless, civilized society depends upon agriculture being successful both in the short and long term, with the latter timescale extending, not only over decades but across centuries and millennia.

3 Domestication and global spread of maize

3.1 Domestication and global spread of maize diversity in the Americas

Maize (*Z. mays* L.) was domesticated in the central Balsas River drainage, Mexico, approx. 8700 years before present (BP) from its wild ancestor teosinte *Z. mays* subp. *parviglumis* H. H. Iltis & Doebley (Matsuoka et al., 2002; Piperno et al., 2009) in a lower altitude and moister region of the Río Balsas with subsequent introgression from *Z. mays* subsp. *mexicana* (Schrad.) H. H. Iltis into highland Mexican maize (van Heerwaarden et al., 2011). The domestication of maize, as with other crop species, occurred many hundreds of thousands to millions of years after the origin of the direct ancestral and related primary gene pool wild species. Considering that maize was domesticated 8700 years BP and the genus *Zea* diverged 5 million years BP, it would be as if domesticated maize first appeared during the last two minutes of a 24 hr day. These comparisons help explain why considerable diversity resides in the wild relatives of maize, including both sister genera, *Zea* and *Tripsacum*, as well as in modern cultivars.

Maize diversity developed as grain was spread through the Americas, reaching Ecuador (6400–5000 BP), Peru (4800–4600 BP), Colombia (4745–4380 BP), the Caribbean (1140–1350 BP) and Puerto Rico (3295–2890 BP), as evidenced by archaeological remains of maize cobs, pollen, starch grains, kernels, phytoliths and ceramics (Bonzani and Oyuela-Caycedo, 2006).

Molecular and archaeological cob evidence support that following diversification in the Mexican highlands, maize was carried via the Mexican highlands to the Southwestern US by 4000 BP, with gene flow starting around 2000 BP from lowland maize which had been carried northwards via a Pacific coastal route (da Fonseca et al., 2015). Bracco et al. (2016), comparing simple sequence repeat (SSR) data, confirmed that maize landraces situated in the Andean region encompass much diversity and collectively represent a discrete entity from maize landraces of highland Mexico and the United States. They further identify three distinct gene pools for lowland South America: tropical lowland, Northeastern Argentinean floury and Northeastern Argentinean popcorn. Their data indicate that tropical lowland maize represents genetic diversity that is doubly significant both quantitatively and qualitatively, simply representative of a contact zone between Andean and tropical lowland maize, as previously considered.

Some 12 000 maize collections that were made during 1945–1960 under the guidance of the US National Academy of Sciences with funding from the Rockefeller Foundation provide the main basis upon which maize diversity present in the Americas has, and continues to be, characterized and classified, including to help understand comparative germplasm constitution and phylogenies (Brown and Goodman, 1977). Exercises to classify maize date back to the 1890s and have been successively undertaken using comparative morphological data resulting in contrasting schemes, as summarized by Brown and Goodman (1977). Relatively few racial complexes have attained global importance (Mexican dents, Corn Belt dents (CBD), Tusons, Caribbean Flints, Northern Flints (NF) and Flours, and the Catetos or Argentine Flints), although several additional races are important regionally (Goodman, 1978). Native US maize racial descriptions were made as early as the eighteenth century (Beverly, 1705). Nine US racial groups (excluding sweet and popcorn) have been described, including NF, Great Plains Flints and Flours, Pima-Papago, Southwestern Semidents, Southwestern 12 row, Southern Dents, Derived Southern Dents, Southeastern Flints and CBD (Brown and Goodman, 1977).

Maize was present in the subtropical zone now comprising the Southwestern US by 3500–4200 BP or possibly, earlier (Hard and Roney, 1998). The earliest evidence for the introduction of maize into the temperate northeast of the continent is 275 AD (Ohio) and 460 AD (southern Ontario) (Hart and Means, 2002). Little (2002) found that maize had been introduced into the lower Hudson River valley around 1000 AD, using tree ring calibrated radiocarbon dating. Maize cultivation thus spread from west to east, evidenced by archaeological samples recovered more frequently from New England that date to the period of 900–1500 AD (Hart and Means, 2002; Little, 2002).

Among the races of maize that arose in North America, two that were genetically and morphologically very different, the NF and Southern Dents (SD), achieved primary commercial and economic importance (Brown and Anderson, 1947, 1948; Doebley et al., 1988). Further dispersal and development of maize in North America began in the 1800s as European farmer-settlers spread westward. Maize became a more important commodity, and as cultivated areas of NF and SD OPVs or farmer-developed landraces started to overlap, their contrasting agronomic features were noted by the farmers. Labate et al. (2003) summarized important milestones in this process. These include:

- 1814, only five varieties were known
- 1840s, as many as 250 varieties were known
- 1847, Reid Yellow Dent was created serendipitously by crossing a Southern Dent with a NF which would become the most widely cultivated and well-known OPV of the CBD (Wallace and Brown, 1956; Troyer, 1999)
- 1870–1890, more attention was paid to improving maize production, including the establishment of the US Agricultural Experiment Stations
- 1899, approx. 800 OPVs described, although with no documentation of pedigrees; many were synonyms
- 1900, Illinois Seed Corn Breeders' Association founded, including a registry to record pedigrees
- 1920s, inbred–hybrid breeding and production method began
- 1921, first commercial sale of hybrid seed
- 1933, <1% of the US Corn Belt planted to hybrids
- 1961, 96% of US maize area planted to hybrids

US maize agriculture is fortunate to be largely founded upon germplasm developed following the hybridization during the mid-nineteenth century of two genetically and morphologically distinctive races, the NF and SD, which collectively encompass a relatively broad and well-adapted germplasm base. However, analyses of pedigree backgrounds of widely used US maize inbred lines demonstrate that the breadth of diversity that constitutes the US maize germplasm base today is contributed primarily by a relatively small number of inbreds. The vast majority of these inbred lines were initially developed either by the public sector or trace to a single commercial breeding programme. Germplasm that was presumably developed by several other commercial breeding programmes during the 1930s–2000s has disappeared from the record. Several US hybrids have small percentages (2–6% by pedigree) of temperate, exotic germplasm from the insect-resistant, Argentine cultivar, Maíz Amargo introduced in the 1980s. Flint germplasm sourced from the French inbreds F2 and F7 can contribute 12–25% germplasm in some hybrids. Goodman (1999) describes three geographic types of exotic maize germplasm used in temperate areas. First, those from other temperate areas, including Argentina, Europe and South Africa. Second, those from the lowland tropics, including Cuban Flint, Suwan, Tusón and Tuxpeño races or varieties. Third, low usage germplasm from the highland tropics including races Chalqueño, Cuzco, Sabanero or San Geronimo. Widely sold US hybrids can contain 1–5% tropical germplasm with a few less widely used hybrids containing 25–50% tropical germplasm (Goodman, 1999). Maize germplasm (CBD, flints, iodent) developed in the United States underpins European and Chinese maize agriculture and in other temperate regions of the world, including Argentina and South Africa.

3.2 Spread of maize into other continents

During the late 1500s AD maize spread by various routes into Europe, Africa and Asia (Dubreuil et al., 2006; Rebourg et al., 2003) and during the 1600s AD into Africa (McCann, 2005). The spread of maize along a north-south axis, as opposed to a west-east direction post domestication was relatively slow due to the need to adapt to local day length to initiate flowering (Hung et al., 2012).

Knowledge of distribution routes sheds light on phylogenetic relationships which, in turn can inform conservation and use strategies, for example in helping to determine assignments of germplasm representing complementary germplasm including potential heterotic groupings. However, subsequent admixtures of germplasm and the general paucity of the archaeological record make the comparison of molecular marker data a more comprehensive source of evidence for developing informative taxonomies for conservationists and plant breeders. For example, Arteaga et al. (2016) associated 46 Mexican landraces into seven groups according to their comparative profiles using 36 931 single-nucleotide polymorphisms (SNPs). Likewise, Warburton et al. (2008) showed founder contributions from each of 23 Mexican maize landraces representing maize diversity in the area of initial domestication to each of 261 CIMMYT developed inbred lines.

Showing meaningful germplasm associations is particularly challenging for landraces that were transferred to other continents with subsequent hybridizations. Nonetheless, Mir et al. (2013) have provided at least a preliminary overview of global maize germplasm by using 17 unlinked SSR molecular markers. Having defined maize germplasm of the Americas into seven groups (Northern US flints, Mexican highland, Tropical lowlands, Andes, Middle North American, South American and Middle South American), these categories of germplasm were then associated according to their level of similarity with maize races sourced from other continents. Mir et al. (2013) thereby revealed a global migration pattern for maize (Fig. 4). In a similar manner, Westengen et al. (2012) showed associations among maize races of the Americas and maize cultivated in Africa. Three African clusters of maize germplasm were defined: 1) Sahelian, representing original introductions from Meso-American landraces via Europe with more modern introduction of temperate germplasm; 2) Western, reflecting the introduction of coastal Brazilian landraces

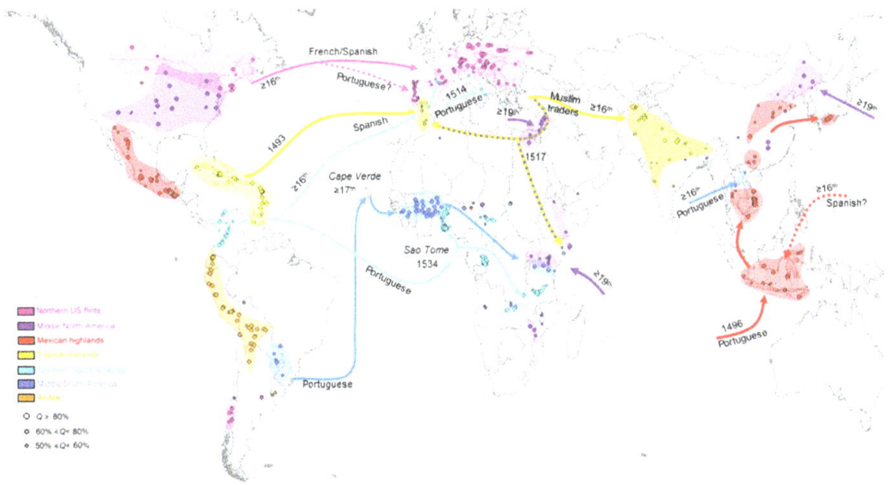

Figure 4 Global migration routes and potential germplasm associations for maize worldwide.

coupled with maize moving NW along Arabic trade routes; and 3) Eastern, reflecting gene flow from modern tropical varieties. Prior to this study there was a paucity of knowledge about the genetic make-up of maize germplasm used in Africa which thereby limited both its conservation and further use in breeding (Westengen et al., 2012). Similarly, Zhang et al. (2016) used evidence from 56 110 SNPs to characterize maize germplasm used in China into multiple groupings (Tropical, Reid, Iowa Stiff Stalk Synthetic, North, PA – group A germplasm derived from modern US hybrids, and PB – group B germplasm derived from modern US hybrids).

4 Genetic diversity in *Zea*: maize and its wild relatives

4.1 Taxonomy and phylogeny of maize and its wild relatives and their divergence over time

Maize is a member of the genus *Zea* which is comprised of four species native to Mexico and Central America. *Zea* includes cultivated maize and, its closest wild relatives, the teosintes. The currently accepted taxonomy of the genus was originally proposed in Doebley and Iltis (1980) and Iltis and Doebley (1980), and was recently summarized by Hufford et al. (2012). The teosintes represent the primary gene pool of maize. The estimated time of divergence, based upon comparative DNA sequence data, between *Z. luxurians* and *Z. mays* subsp. *parviglumis* occurred at least 100 000 years ago (Hilton and Gaut, 1998).

Comparison of nucleotide sequence for genes under neutral selection given mutation rate per nucleotide allows the age of divergence from a common sequence to be estimated. White and Doebley (1999) provided a cautionary note that estimates may be 'off by a factor of two or more' and depend upon improved knowledge of mutation rates for specific genes and ultimately, validation from the fossil record. Nonetheless, as data accumulate from greater numbers of genes, including from additional taxa, the estimated divergence times will become more meaningful.

White and Doebley (1999) provided a summary of divergence times as follows:

- The genera *Zea* and *Tripsacum* diverged between 2.3 and 2.6 m yr and 4.5 and 4.8 m yr.
- The age of the *Zea* gene pool is between 0.7 and 4.7 m yr, possibly between 1.2 m yr and 1.4 m yr.
- *Z. mays* and *Z. luxurians* represent potentially the first phylogenetic divergence within *Zea* and occurred approx. 700 000 yr ago.

The genus *Tripsacum* is the closest wild relative of *Zea*. However, 'close' is a relative term, phylogenetically speaking, for the genera have been estimated to have diverged between 4.5 and 5.2 million years ago, based on a constant mutation rate and comparisons of DNA sequence data (Hilton and Gaut, 1998; Buckler and Stevens, 2005). *Tripsacum* comprises 15–16 perennial species. Members of the genus are widely distributed in the Americas, from Northeastern and North Central US to Mexico and Central America, the West Indies and in South America to Bolivia and Paraguay (Doebley, 1983; Blakey et al., 2007) Comparative DNA sequence data provide no evidence to support gene flow between *Zea* and *Tripsacum*, in their natural habitats (Hilton and Gaut, 1998; Bennetzen et al., 2001). Gene transfer from *Tripsacum* to *Zea* is possible, but only by using special methods, such as embryo rescue (Manglesdorf and Reeves, 1931). The only exception to this is the case of *T. andersonii* or Guatemala grass, which was the result of a *Tripsacum* x teosinte cross; however, this 'species' is mostly sterile and can only reproduce vegetatively (Doebley, 1983; Blakey et al., 2007). Much cytogenetic research was done to examine maize-*Tripsacum* hybrids and even three-way hybrids between maize, teosinte and *Tripsacum* (reviewed in Blakey et al. (2007)), and the main barrier to gene transfer was the difference in chromosome number between *Zea* ($n = 10$) and *Tripsacum* ($n = 18$). For example, unpaired univalent *Tripsacum* chromosomes created unstable genomes in subsequent generations after the initial hybrid crosses (Harlan and de Wet, 1977).

Recent research examining the genomic relationship between *Zea* and *Tripsacum* has revealed the differences in the composition and abundance of the transposable element families between the two, in contrast with their very similar gene contents (Chia, et. al., 2012). Access to the abundant, potentially useful, diversity in *Tripsacum* for maize improvement may thus be attainable directly in the future via gene editing technologies.

4.2 Revised perspectives on the array of genetic diversity in *Zea* and *Tripsacum*

Pioneering genetic studies by Barbara McClintock (McClintock, 1956; McClintock, 1984) led to revolutionary insights into the diversity of the maize genome. For example, transposable elements discovered by McClintock were initially little understood and thought by many to comprise 'junk DNA' (Biemont and Vieira, 2006). It has taken more than 50 years to appreciate their role in regulating genes (Biemont and Vieira, 2006). Further insights into the complex genomic diversity of maize and its closest wild relatives, the teosintes, have accelerated, particularly during the last decade. Detailed genomic and transcriptional analyses have become possible due to advances in genome sequencing, transcriptomics, metabolomics, epigenetic analyses and information management systems.

Hilton and Gaut (1998) estimated that maize contains 60% of the diversity present within its progenitor (*Z. mays* ssp. *parviglumis*). The notion of genetic co-linearity was explored by Fu and Dooner (2002), who discovered that four of ten genes present in inbred McC at the bz genomic region were completely absent in inbred B73; four classes of genic diversity were observed at this region across just nine US maize inbred lines. Only 50% of the genome was found to be of common sequence across four genic regions for inbred lines B73 and Mo17 (Brunner et al., 2005; Morgante et al., 2007). More comprehensive genic and transcriptional diversity, both qualitative and quantitative, has been found as a result of both copy number variation and presence or absence of genes (Gore et al., 2009; Springer et al., 2009; Hansey et al., 2012). A further source of diversity in patterns of gene expression in maize is a result of epigenetic or methylation processes which can be stably inherited (Swanson-Wagner et al., 2012; Eichten et al., 2013). By exploring hitherto underutilized maize landraces and their crop wild relatives, maize researchers are very likely to encounter novel genomic features in terms of genes, allelic variants and haplotypes.

It is now clear that there was a significant genetic bottleneck during the maize domestication process (Eyre-Walker et al., 1998; Beissinger et al., 2014). Sequence comparisons indicate that maize retains 60% of the diversity present in *Z. mays* ssp. *parviglumis* at the *glb1* locus, a non-essential seed storage protein (Hilton and Gaut, 1998) and 75% of the variation at the *Adh1* locus (Eyre-Walker et al., 1998). As a consequence, significant amounts of genetic, physiological and morphological diversity within the genus *Zea* can only be found in the teosintes (Flint-Garcia, 2009, 2013). Further loss of diversity presumably occurred in what are now categorized as specific races of maize, as a result of genetic drift, due to the limited sampling of diversity that occurred as grain was transported to new localities, first in the Americas, and subsequently throughout much of the globe.

Comparative sequence and chromosome map data across species also show synteny among diverse genera, coupled with diversity in haplotypes between grass species (Bennetzen et al., 1990). Consequently, sequence diversity and associated agronomic data provide potentially useful information and resources, not only for a single crop genus, but for many. Sequence, molecular marker and associated agronomic performance data thus need to be in a universally understood format that is readily available via appropriate information systems technology. The capacities to generate DNA sequence data, and to transfer genes across species boundaries, including via gene editing, may lead to a much broader concept of what constitutes practically available and useful gene pools or genetic resources in terms of current botanical classification systems.

5 The importance of genetic diversity in improving maize productivity

5.1 Opportunities and challenges faced by breeders in using exotic germplasm

Maize encompasses great genetic diversity which allows adaptation to a wide range of environmental conditions resulting in great phenotypic variability as a result of genotype x environmental interaction (Gore et al., 2009: Hansey et al., 2012; Makarevitch et al., 2015). Current studies on genetic gain for US maize (Smith et al., 2014) showed 1) an increased rate of genetic gain compared to the previous survey conducted a decade earlier and

2) an increase in the proportion of yield gain due to improved genetics from approx. 50% (Duvick and Cassman, 1999) to 75%. These data imply an approaching plateau for achieving increased yields due to improved farm management, thereby placing a greater emphasis on the need to achieve continued genetic gain (Cassman et al., 2003; Lobell et al., 2009). Reliance upon genetic contributions to improve agricultural productivity will also further increase as needs to reduce negative environmental footprints increase.

Considerable numbers of well-adapted and highly characterized maize inbred lines have been developed within US public and private breeding programmes. Their demonstrated high productivity can be attributed to selection for stress resistances associated with higher planting densities. They were, and are, used directly in hybrids and as parents for further breeding in other regions of the world. These include, with approximate contributions by pedigree in parentheses, temperate climatic zones in Europe (50–100%), China, (50%), and subtropical and tropical climatic zones, including Argentina, Brazil and India (20–35%). Nonetheless, use of temperate germplasm in tropical breeding programmes is poorly documented in the public domain (Goodman, 1999), an exception being tropical maize streak resistant lines (Kim et al., 1987). Surveys have also indicated marginal use of temperate exotic germplasm in US hybrids (usually 2–6%, occasionally 12–25%) and minimal use of tropical germplasm (0.1–5%) in just a few US hybrids (Goodman, 1999).

Special programmes are required to adapt exotic maize germplasm in order to allow initial generations of characterization and evaluation to begin (Goodman, 1999; Wen et al., 2012; Hallauer and Carena, 2014). Other challenges include a potential lack of characterization and pedigree information, the need to alter photoperiod requirements to enable appropriate timing of flowering, susceptibility to insects and disease, and poor grain quality, especially for temperate germplasm introduced into the tropics, where there is additionally a lack of clear arrangement of tropical germplasm according to temperate heterotic groups (Wen et al., 2012). In addition, tropical germplasm has generally been subjected to less inbreeding than highly productive temperate germplasm and thus carries deleterious alleles that further undermine breeding progress and prolong evaluation for their potential utility. This disadvantage can be ameliorated by using tropical hybrids as initial sources of tropical germplasm (Goodman et al., 2000) or through the use of doubled haploids (Strigens et al., 2013). Recent efforts (Teixeira et al., 2015; Jamann et al., 2017; Romero Navarro et al., 2017) have sought to decipher the complexity of photoperiod and flowering time; multiple genes and gene networks influence these traits, with both large and small effects. Their expression may be influenced by day length, latitude, temperature, altitude and even circadian factors.

Identification and incorporation of exotic germplasm to improve traits that are under control of single or few genes such as resistance for specific races of pests or disease, or specific grain quality characteristics with well-understood physiologies and genetic control, will always be far simpler and less demanding of resources compared to traits under multigenic control, such as drought tolerance or yield. Successful searches for useful exotic germplasm in maize include 1) from maize – southern corn leaf blight resistant germplasm developed in response to the race T cytoplasm (Lim and Hooker, 1971; Horner, 1990), resistances to maize lethal necrosis (MLN) disease (Gowda et al., 2015), to *Fusarium* ear rot and reduced toxin production, to ear rot caused by Diplodia, fall army worm and other insects, improved silage quality, starch quality and stress resistance including drought (Blanco and Gardner, 2014); 2) from teosinte – resistances to grey leaf spot from *Z. mays* ssp. *mexicana* (Lennon et al., 2016), a range of pests and diseases (de Lange et al., 2014), to the parasitic weed *Striga* from *Z. diploperennis* (Rich and Ejeta,

2008) and flooding tolerance potential from *Z. nicaraguensis* (Mano and Omori, 2007; Mano et al., 2013, 2016); and 3) from *Tripsacum dactyloides*, resistance to *Striga* (Gurney et al., 2003), resistance to western corn rootworm (Prischmann et al., 2009) and resistance to common rust (Bergquist, 1981), and from *Tripsacum floridanum* resistance to northern corn leaf blight (Hooker, 1981).

Further progress in sourcing useful diversity can also be possible for multigenic traits including drought resistance and higher yields. Data from the NCSU programme showed that significant amounts of tropical germplasm add diversity to US maize without undermining yield and thus with continued selection can potentially contribute to sustained genetic gain (Goodman et al., 2000). Wang et al. (2008) reported that germplasm contributed by teosinte (*Z. mays* ssp. *mexicana*) was associated with higher yields in maize when evaluated at Jinan and Weifang, China.

Plant breeding is, in effect, a cyclical process of germplasm evaluation, introduction of the best germplasm into new breeding populations and selection. Choice of breeding strategies and germplasm critically determine the chances of success for both short and long term. Most important traits have complex inheritance patterns, and their expression is influenced by the environment and other factors. Consequently, it would be a false assumption that all genetic, genetic x environment and genetic x epigenetic interactions that collectively determine genetic potential and agronomic performance can be identified, measured and understood. Similarly, a pedigree breeding programme that is highly productive in achieving genetic gain but which is closed to the introduction of new diversity will inevitably result in a reduction of genetic diversity in space and time, thereby imposing limits to future genetic gain. For example, while use of genomic selection (GS) is expected to maintain or even increase the rate of genetic gain in the short-term, it could also exacerbate the rate at which diversity is diminished within a breeding programme. From the dairy cattle literature, we know that GS can result in dramatic fixation of haplotypes and concurrent loss of diversity within five generations (Kristensen et al., 2015; Ponsart et al., 2014; García-Ruiz, 2016). Consequently, unless efforts are made to constantly introgress new diversity, or use sufficient population sizes to preclude loss of diversity in breeding populations, GS may result in fixation of haplotypes within pools and reduce genetic diversity. However, adoption and adaptation of breeding approaches to characterize exotic germplasm also offer new opportunities to identify useful germplasm that can contribute further to improving genetic gain (Goodman, 1999; Wen et al., 2012).

Technical components include increasingly detailed genotyping, improved phenotyping, doubled haploids, multiple generations per year using off-season nurseries and sophisticated information management to achieve an increased power of marker-assisted selection through the use of successive cycles of modelling, development of training populations, evaluation of performance data with the integration of agronomic performance and environmental data. For traits that are under single-gene control it is also feasible to sequence the controlling loci. This information could then serve to 1) allow a DNA-based screen of germplasm accessions to identify new diversity and 2) provide a basis to generate diversity *de novo* using gene editing technologies (Yu et al., 2016). Several programmes have reported the potential that new breeding methods can provide to facilitate use of multigenic traits from exotic germplasm (Prada, 2009; Yang et al., 2011; Sood et al., 2014; Gorjanc et al., 2016), thereby offering new opportunities to achieve the more effective use of a broader germplasm base (Yu et al., 2016; García-Ruiz, 2016).

5.2 Political challenges to achieving multilateral access to maize genetic diversity

All countries, including those with centres of crop diversity are dependent upon genetic resources that were initially accessed from other countries or regions, thereby exemplifying multilateral dependences upon useful genetic diversity for continued crop improvement (Table 2) (Brown, 1984; Kloppenburg and Kleinman, 1988; Palacious, 1998; Fowler et al., 2000, 2001; Cassaday et al., 2001; Falcon and Fowler, 2002; Gepts,

Table 2 Recorded numbers of maize accessions stored *ex situ*, genebanks individually holding 1% or more of maize accessions are reported (Genesys, 2014 data)

Country of holding institute	No of accessions	% of maize holdings
United States of America*	30,899	24
Mexico, CIMMYT**	27,267	21
Russia	14,233	11
Ukraine	9,375	7
Romania	7,448	6
Italy	6,373	5
Yugoslavia, Socialist Federal Republic of	5,540	4
Serbia	5,475	4
Bulgaria	4,700	4
Spain	2,936	2
Hungary	2,764	2
Portugal	2,140	2
Nigeria ITA	1,561	1
Germany	1,523	1
Poland	957	1
Czech Republic	791	1
Slovak Republic	749	1
Albania	643	1
France	567	0
Azerbaijan	550	0
Other	1,975	2
Total	**128,466**	**100**

* 7,510 from the Maize Genetics Cooperative Stock Center, 20,685 from the North Central Regional Plant Introduction Station, 974 from the Seed Savers Exchange.
** Does not include INIFAP holdings. Only a portion of CIMMYT accessions are from Mexico.
Source 2014 data: Plant genetic resources accession level data provided by: Genesys. All intellectual property rights (including copyright) in the Data are owned and retained by the said institution(s). Data accessed through GENESYS Global Portal on Plant Genetic Resources, http://www.genesys-pgr.org, 1952-09-08.

2004; Brush, 2005; Frison and Halewood, 2006; Ramirez et al., 2013; Wang, 2013; Khoury et al., 2016). These interdependencies are likely to increase as agriculture relies more upon using genetic resource-based solutions provided by plant breeders to allow production of healthy food, fibre and fuel in ways that are environmentally sustainable (Fujisaka et al., 2011; Ramirez et al., 2013). With specific regard to genebanks, Mahmoud Sohl, Director General of the International Center for Agricultural Research in Dry Areas (ICARDA), previously located in Aleppo, Syria, recently stated: 'genebanks are not isolated treasure troves and shouldn't be treated as such. Their power comes from the connections between them, and the worldwide network of genetic resources those connections create' (GCDT, 2015).

Khoury et al. (2016) have analysed interdependencies of countries, not only based on their use of plant genetic resources from distant centres of origin, but also in terms of their dietary food sources. It is ironic, given global interdependency on access to genetic resources, that political challenges have created barriers to access domesticated genetic diversity, whether stored in a genebank or used on farms, and even to crop wild relatives. As a result, it is imperative to help increase understanding among policy makers that multilateral access to germplasm benefits all countries, including those where crops were originally domesticated, and especially for developing countries that lack resources to undertake thousands of bilateral agreements. Discussions on access and benefit sharing have continued in the decades following the publication and widespread agreement by governments of the 20-point action plan embodied in the FAO Global Plan of Action (GPA) (FAOGPA, 1996) and subsequent passage of the FAO ITPGRFA (http://www.fao.org/3/a-i0510e.pdf) which was adopted by the Thirty-First Session of the Conference of the Food and Agriculture Organization of the United Nations on 3 November 2001 and entered into force on 29 June 2004 (http://www.fao.org/plant-treaty/overview/en/). The ITPGRFA is fully concordant with the conservation and sustainable use goals of the CBD (http://www.fao.org/plant-treaty/en/). Political challenges remain to be resolved to allow the goals of the ITPGRFA to be fully realized. Issues include how to improve funding of activities included in the remit of the Treaty; exploration of relationships between WIPO, UPOV and the Nagoya protocol of the CBD; and how traditional knowledge can be respected and recognized.

In our view, progress in achieving the sustained and increasingly beneficial use of PGRFA is essential to the well-being of humankind and the furtherance of civilization. An aspirational goal would be to provide every farmer, regardless of country, region or farm size, with the best possible choice of seed from which to plant the next harvest. This choice could therefore include landrace, improved landrace, OPV or hybrid. Improved breeding support and quality of seed systems are likewise needed, and may include farmer-participatory breeding, farmer breeding of landraces, creolized varieties, including either hybrids that are available via the UPOV breeder exception, or parental lines of hybrids that are publicly available following expiration of PVP and patents and placed into the public domain by the USDA-ARS (Kurtz et al., 2016).

However, partly as a result of false dichotomies, positive progress is being stymied. We recommend review of what is now the 19-point FAO GPA (originally a 20-point plan), and recognition of the positive progress that has been accomplished. Such a reset in thinking could enable stakeholders to focus on common areas of agreement and collaboration. It is vital to improve prospects for achieving increased and sustainable crop production through the more effective use of a broader base of PGRFA by plant breeders and ultimately by all farmers.

6 Case studies of the use of maize genetic diversity in breeding

6.1 Case studies: some valuable contributions of exotic germplasm to maize breeding

Kitale Maize breeding programme: Two notable publications by M.N. Harrison of CIMMYT (1969, 1970) discussed major issues affecting productivity of maize and described the research programme initiated at Kitale, Kenya, in 1955. Over time, collaborations involved the Kenya Ministry of Agriculture, the East African Agriculture and Forestry Research Organization of the East African Community, USDA-AID-EAAFRO and USDA-ARS (Darrah, pers. comm.). Harrison stressed the importance of use of biosystematics and understanding of the ecology of the area as a starting point, wide evaluation of introductions and the importance of genetic diversity found in the centre of origin materials, especially if the ecological conditions at the origin match the development environment. A critical mass of appropriate specialists to tackle objectives and team organization were also considered to be key. This approach essentially was an early precursor of focused identification of germplasm strategy (FIGS). The Kenya Flat White complex used successfully in this programme resulted from the mixing of varieties introduced from South Africa (local farmer's strains); its parentage included lines from the Southern US, with origins tracing to Mexico.

Kitale Synthetic I was not successful and was never released, Kitale Synthetic II was a first-cycle synthetic produced by one cycle of recurrent selection and released in 1961, its yield exceeded that of the parent variety by 15–20%. Kitale III was the result of a second cycle of selection and increased yields by an additional 6–8% (Eberhart, 1967). The breeding system developed and used by the Kenya Maize Research Section had four main phases: 1) evaluation of local and exotic varieties so that the breeding material is the best available; 2) compositing the selected breeding material into two or more populations or varieties in such a manner that each population has considerable genetic variation for the traits requiring improvement and that the crosses of these populations will show heterosis; 3) recurrent selection in each population to increase the frequency of favourable genes so that the populations and population crosses are improved with each cycle of selection; and 4) release of a commercial variety of one of the following forms: (a) the cross of two populations as a variety cross hybrid; (b) single, three-way or double cross hybrids from inbred lines developed from the elite material after each cycle of selection; or (c) a synthetic variety derived from the advanced generation of the population cross in areas where hybrid production is not yet feasible. The authors stressed that two or three years spent introducing and evaluating exotic materials in addition to local varieties could produce greater increases in mean performance than a ten-year breeding programme utilizing only readily available local material, and these principles extend to improvement of other crop species (Eberhart, 1967). Darrah et al. (1972) compared different methods of recurrent selection under East African conditions using local and introduced populations of maize, using Kitale II, Ecuador 573 (a genebank accession from CIMMYT) and Kitale composite A populations. Results suggested that estimates of genetic variance in the populations did not change, and that rapid rates of improvement using these methods could be expected to continue.

B14A, B73, Mo17: A classic example of successful introgression of exotic germplasm into temperate, B14A was derived from Cuzco x B14 germplasm and selected for rust resistance (Russell et al., 1971), and was an important contributor to US germplasm. B73 was developed from cycle five of Iowa Stiff Stalk Synthetic (BSSS) (Russell, 1972); its upright stature represented a major change in plant architecture and it has served as a major founder of the Stiff Stalk heterotic pattern commonly used in temperate programmes. B73 had superior heterosis in combination with Mo17, a Non Stiff Stalk heterotic pattern line (Zuber, 1973); this hybrid combination dominated the US mid-maturity temperate production area in the 1970s and the early 1980s.

Maize Amargo ('Bitter Maize'): Penny and Dicke (1956) reported that progeny of crosses with Maize Amargo had reduced feeding by European corn borer (ECB), and released lines which contributed to commercial hybrids. Marston (1931) described the introduction of Maize Amargo, a flint variety from Argentina in 1921, where it was reported to be resistant to locust attack, and its subsequent use in Michigan breeding programmes, contributing ECB resistance. Amargo may have originated in Hungary prior to introduction in Argentina; if so, it is yet another example of the positive impact of access to international germplasm.

CIMMYT: A comprehensive history of the origins of the CIMMYT maize bank was authored by S. Taba (2005). The collection started with landrace accessions collected by the joint Rockefeller Foundation-Government of Mexico programme, Office of Special Studies (OEE) in 1943 to improve productivity of basic food crops in Mexico. By 1947, the Mexican maize collection was about 2 000 accessions. (*Today, more than 28 000 accessions are held* (Costich, pers. comm., 2016).) CIMMYT's predecessor institutions (OEE and the Inter-American maize programme) extensively collected maize landraces in Mexico, Central America and the Caribbean, and also introduced maize germplasm accessions from other countries, through germplasm exchanges, until the early 1960s. From 1967 through the 1980s, the CIMMYT maize bank regenerated and incorporated a large part of the additional landrace accessions received from the US National Seed Storage Lab, Fort Collins, Colorado. These additional accessions originated from a project established in the 1950s by the National Academy of Sciences (NAS) and National Research Council (NAS-NRC), USA, which collected and studied Latin American and Caribbean maize landraces and preserved them for future use, within the NPGS and other institutions such as the Missouri Botanical Garden. The project was directed by a committee of geneticists, maize breeders, botanists and administrators, and produced a series of bulletins on races of maize (Goodman et al., 1988). The catalogues, 'Collections of Original Strains of Corn I and II' (NAS-NRC 1954 and 1955), listed 10 922 samples, some of which (7 629 samples from regional banks in Brazil and Colombia) were transferred to CIMMYT in 1967. CIMMYT compiled passport information for the accessions from the catalogues and also from the original collection sheets that were available in the bank archives (Taba, 2005). In order to help ensure that CIMMYT's genetic resource collections be regenerated, during 1982–1984 Pioneer Hi-Bred International Inc. regenerated 300 CIMMYT accessions each year at Homestead, Florida (Salhuana, 1984, 1988).

LAMP: Dr. William Brown recognized the importance of genetic diversity from Latin America, as well as the challenges associated with identifying and mobilizing useful genetic diversity from un-adapted germplasm and Pioneer Hi-Bred International contributed funding to execute the Latin American Maize Project, or LAMP, for a six-year period, 1987–1993. The primary objective of LAMP was to evaluate 1400 accessions found in Latin American and US germplasm banks, and determine their potential for use in breeding programmes. It became the first coordinated international effort to undertake the evaluation of the genetic resources of a major world crop. Based on the cooperative efforts of 12 countries, LAMP results identified landrace accessions that showed good yield potential and were suitable for enhancement, or pre-breeding, prior to their incorporation into breeding programmes (Pollak and Salhuana, 2001; Salhuana et al., 1991). Data from this comprehensive, multinational project can be accessed *via* the US NPGS. Without LAMP, there could be no Genetic Enhancement of Maize (GEM) Project, which we describe in the next case study.

GEM: Based on the results of the LAMP Project, many US breeders believed that it is important to continue pre-breeding efforts. The GEM Project was initiated in 1993 with the support from the American Seed Trade Association (ASTA) to develop germplasm from exotic, usually non-temperate adapted, sources and to release introgressed, temperate-adapted germplasm that could be readily used in a commercial programme. Advised by a Technical Steering Group comprising members from industry, public plant breeders and USDA-ARS, the project's mission is to increase the genetic diversity in US commercial maize breeding materials. In 1994 several public and private maize breeders sent a proposal to all ASTA members with maize breeding programmes, soliciting their interest in becoming GEM collaborators, and to public sector maize breeders as an opportunity to become collaborators (without the requirement of private sector collaborators to make breeding crosses). A subcommittee worked with the US Congress to secure permanent base funding for the USDA-ARS to support GEM, in addition to the in-kind support from the private sector (estimated at approximately $450 000 per year), plus their contributions of proprietary germplasm valued in the billions of dollars, advice and service.

The Raleigh, North Carolina, breeding programme focuses on identification of superior 50% exotic germplasm development, while the Ames, Iowa, programme focuses on 25% exotic germplasm (Pollak and Salhuana, 2001). Over the past 23 years, the Raleigh, Ames and public collaborators have released 300 lines derived from exotic germplasm resources via traditional breeding methods, and about 200 doubled haploid (DH) lines derived from more than 60 landrace accessions (http://www.public.iastate.edu/~usda-gem/). While topcross yield is a major criterion for release of the traditionally derived GEM lines, the DH lines are not evaluated for agronomic performance prior to release as their purpose is to provide a rich resource of maize genes and traits from exotic landraces in a temperate-adapted format, unfiltered by agronomic performance criteria.

GEM participants provide breeding support activities and resources including proprietary germplasm; evaluations for yield, agronomic, disease and insect pest resistance; and compositional traits. Participants have exclusive access to traditionally released lines for two years; following this period, GEM lines are made publicly available via the NPGS maize collection. The DH lines, designated as BGEM lines, are jointly released with Iowa State University, and are publicly available upon release. On average, about 1300 GEM germplasm samples are provided annually to 70–80 requestors. The number of participants has decreased in recent years as a result of private sector consolidation, while participation activity has increased. Today, there are 29 US private sector, nine international private sector, 22 US public and government sector, and five international public sector participants contributing to programme objectives.

A 2015 survey of users' 20 years of experience with GEM provided some feedback: private sector participants report some success with second or third cycle lines derived from GEM germplasm, with grain quality, silage quality and disease resistance being traits of primary interest (http://www.public.iastate.edu/~usda-gem/Publications/GEM_Presentations.htm). Public collaborators have elucidated novel starch properties, genes and pathways (Jiang et al., 2015; Wu et al., 2015; Butardo et al., 2012); developed improved lines and synthetics for silage quality (Gustafson et al., 2010; Frey et al., 2004);

identified native sources of resistance to corn rootworm (Hibbard, 2016); and developed early, diverse germplasm for the extreme northern portions of the US Corn Belt (Carena et al., 2009), and more.

North Carolina State University: North Carolina State University's maize breeding programme has developed and released maize inbreds for yield, disease resistance and other traits of agronomic importance, with great emphasis on use of exotic germplasm. The programme was originally led by Donald Thompson, and since 1983 by Major Goodman. Between 1980 and 1990, 150 NCSU-developed inbreds were released, with 86 having at least 50% exotic parentage and 40 of those having all-tropical parentage, useful resources for maize improvement and diversity in the United States and abroad (Nelson, 1990).

Focused on three heterotic groups, Stiff Stalks, non-Stiff Stalks and Tropical Exotics, and located in the Southern US environment where 50% exotic germplasm can be more readily adapted, breeding strategies focus on improving yield and disease resistance with topcross yield trial testing at the S3 or S4 stage of inbreeding. The Goodman Lab's website, http://www.cropsci.ncsu.edu/maize/germplasm.html, indicates that all NCSU maize inbreds are publicly available and can be accessed either through the NPGS maize collection or directly from the NCSU programme (some may have associated intellectual property restrictions). The website's publication list reflects the breadth and depth of NCSU's research on diversity, maize domestication, evolution and genetic structure, breeding methodology, molecular characterization of germplasm, disease resistance and the innovative use of gametic incompatibility factors to facilitate controlled parentage in production environments. Dr. Goodman oversaw the GEM programme in Raleigh in the past, and now works in partnership with USDA-ARS scientist, Dr. Matt Krakowsky.

University of Hawaii-Manoa: Dr. Jim Brewbaker has provided a series of introgression lines bridging tropical and subtropical Asian and Latin American germplasm with temperate germplasm, open-pollinated populations, and synthetics selected for resistance to specific diseases. This programme has been uniquely positioned to utilize tropical germplasm and to screen for tropical maize diseases, and focused on selection for adaptation, disease and insect resistance, and unique genetics. These dent, flint and sweet corn lines have proved to be of unique value for maize chlorotic mottle virus (MCMV) and sugarcane mosaic virus (SCMV) resistance research, and for classical genetic studies, and have been important for the efforts to identify germplasm resistant to MLN in Africa. Many of the Brewbaker lines have been provided to the US NPGS and to other global genebanks, and are available via University of Hawaii-Manoa's Foundation Seeds (http://www.ctahr.hawaii.edu/hfs/stocks.asp). Releases and research are described in Brewbaker (1979, 1981, 1997, 2003, 2009), Brewbaker et al. (1989), Ming et al. (1997) and Brewbaker and Josue (2007).

6.2 Examples of current germplasm use to combat emerging threats

*Maize lethal necrosis (MLN)***:** CIMMYT has a long history of OPV improvement, inbred line development, and now is a world leader in the use of molecular technology to understand and utilize exotic maize germplasm to address production challenges in the developing world. Most recently, massive collaborative efforts have been organized to understand MLN in Africa, a devastating disease first reported there in 2011 and caused by synergistic co-infection by two viruses, MCMV and a virus from the Potyviridae family, especially wheat streak mosaic virus or SCMV (formerly known as MDMV-B) (Wangai et al., 2012; Mahuku et al., 2015). MCMV and SCMV cause the most widespread disease of maize worldwide. SCMV is found worldwide, while MCMV is distributed widely in the Western Hemisphere and in Europe. In 2012, yield losses of up to 90% resulted in an estimated loss of 126 000 metric tons in Kenya alone, and MLN has spread to Rwanda and Democratic Republic of the Congo, with symptoms reported in Tanzania, Uganda, South Sudan and Ethiopia. Mahuku et al. (2015) reported on MCMV transmission by thrips, soil and seed, and the use of next-generation sequencing to compare the sequences of MCMV and SCMV. They provide an excellent summary of the chronology of reports of occurrence of MCMV in the world, and ranked lines which showed promise for genetic resistance in disease screening trials. Those lines have something in common – tropical germplasm in their lineage from areas of the world (such as Thailand and Brazil) where viruses are an omnipresent threat. Included in these are lines from Kasetsart University in Thailand, the Ohio State virus research programme and the North Carolina State Program.

A presentation on the history of global occurrences of MCMV/MLN (Dolezal, 2014) can be found at http://www.public.iastate.edu/~usda-gem/Publications/GEM_Presentations.htm. In the United States, seasonal outbreaks are followed by colder months, and good field phytosanitation aids in reducing the severity of viral diseases. Control is much more challenging in Africa where overlapping growing seasons, continual presence of insect vectors, high initial viral titres per plant and the continuous presence of maize stover make reduction or elimination of inoculum virtually impossible. In February 2014, a severe outbreak of MCMV occurred on sweet corn in Taiwan for the first time (Deng et al., 2014), further evidence of the global movement of plant pathogens and the threats presented to food production and food security.

Others: The parasitic weed *Striga* remains a serious threat to production (Section 5.1 addresses sources of resistance). Changes in disease occurrence or pests that challenge current US maize production include a new race of Southern corn rust, increased incidence of Northern corn leaf blight due to *Exserohilum turcicum* races overcoming Ht gene resistance, wheat mosaic virus (high plains disease), Goss' wilt, bacterial stalk rot, Diplodia leaf blight and an array of stalk rots. Diseases of high concern that are not yet problematic in the United States but challenge production in other areas of the world include black bundle disease, banded leaf sheath disease, rice black streak dwarf virus, late wilt, maize Río Cuarto virus, maize redness disease (phytoplasma), *Phaeosphaeria* leaf spot, downy mildews and more (Dolezal, 2014). Previously known in South Africa, bacterial leaf streak disease of maize, caused by *Xanthomonas vasicola* is newly reported in the Midwestern US (Korus et al., 2016; https://www.ksre.k-state.edu/news/news-stories/2016-news-releases/september/corn-leaf-streak.html). A relatively comprehensive list of maize diseases (last updated in 1993) can be found at http://www.apsnet.org/publications/commonnames/Pages/CornorMaize.aspx.

7 *In situ* diversity and conservation of maize genetic diversity

7.1 Important global regions of maize genetic diversity

The teosintes (all taxa in the genus *Zea*, other than *Z. mays* subsp. *mays*) represent the primary gene pool (Harlan and de Wet, 1971) for maize. Maize comprises approximately 300 landraces (Goodman et al., 1988) that were initially described and published in the 'Races of Maize' series and recently re-published in full on searchable CDs (USDA-ARS, 2005) and are available at https://www.ars.usda.gov/midwest-area/ames/plant-introduction-research/docs/races-of-maize/. The terms 'race' and 'landrace' have distinct meanings for maize. 'A landrace is a crop population with a distinct identity. Landraces can be classified into a racial category; some races include both landraces and commercial cultivars' (Perales and Golicher, 2014).

Races have been described according to morphology, geographical distribution and molecular data, with the intention to reflect phylogeny and potential genetic similarity (Perales and Golicher, 2014). Surprisingly few races of maize, just 13 or 4.3% of all currently described races, form the basis of global commercial maize agriculture today (Goodman and Taba, 2007).

The FAO SOW (1997) described areas of primary importance for maize diversity as Central American and Mexico, including the centre of domestication, and the areas of secondary importance as South America, the Caribbean, North America, East Asia (China–Japan), Southeast Asia (Indochina and Indonesia), South Asia (the Hindustani area), Central Asia, West Asia, near Eastern Asia area, the South and East Mediterranean, East Africa (Abyssinia), West Africa and Southern Africa. As biotic and abiotic challenges differ in these regions, uniquely diverse germplasm of value has resulted from all areas.

7.2 The dynamic nature and potential vulnerability of *Zea* diversity *in situ*

An understanding of the imperative need to conserve crop genetic resources *ex situ* arose from observations in the United States, Europe and Asia during the twentieth century of the whole-scale replacement of traditional farmer landraces by varieties bred by professional plant breeders. Change in variety usage on farms occurred very rapidly as farmers focused entirely on agronomic management of newly bred varieties. In so doing they abandoned the varietal conservation and adaptation roles that had previously been undertaken during at least 300 previous generations of farmers. Conscious actions by others to conserve landraces became immediate imperatives, otherwise those varieties would have been forever lost. It was considered inevitable during the 1960s that such trends of replacement would eventually and possibly rapidly occur globally, including in centres of origin or landrace diversity due to the critical need to increase agricultural productivity and to improve farm income.

Despite actions that were taken to conserve landraces *ex situ*, changes in cropping systems, increasing use of modern varieties and urbanization still caused loss of maize genetic resource diversity in Mexico, Costa Rica, Chile, Malaysia, the Philippines and Thailand. Only 20% of local varieties reported in 1930 were still known in Mexico. Farmers in Iringa, the United Republic of Tanzania, who used modern varieties no longer maintained local varieties in a pure form. It was estimated that 91% of US landrace field maize varieties had become extinct (Veteläinen et al., 2009). Most challenges for the continued cultivation of maize for any given site in Mexico came from the opportunity costs of not growing higher value crops such as tomatoes and onions, or lower cost and risk crops such as sorghum. The major challenge to continued, on-farm cultivation of maize landraces at one site was urban flight due to opportunities for higher incomes (Perales et al., 1988). By 2010, many maize landrace cultivars were reported lost in Albania as farmers replaced them with modern cultivars, indicating a loss of genetic diversity, but a cautionary note was added that 'loss of cultivars and loss of diversity are not synonymous and that genes in a lost cultivar may still exist in other currently available cultivars' (Veteläinen et al., 2009).

During the past three decades there has been considerable variability in adoption patterns of modern varieties within regions. Farmer-saved maize seed was grown by 60–100% of farmers in most Central American countries (with the exception of El Salvador) and by more than 50% of farmers in Bolivia, Colombia, Paraguay and Peru. Hybrid seed was more widely used in Argentina, Brazil, Ecuador, Uruguay and Venezuela. Similar patterns were found in Eastern and Southern Africa, where the adoption of hybrid maize ranged from 91% in Zimbabwe to 3% in Mozambique (Smale et al., 2013). Complex populations of maize landraces are cultivated by many Mexican farmers to meet a variety of needs including coping with environmental heterogeneity, combating pests and diseases, cultural and ritual needs, and to satisfy dietary and food preferences (Bellon, 2004).

In areas where farmers continue to use landraces, it is critical to better understand the rationale underlying choice in varietal selection and the farmer's role in varietal development (Perales et al., 1998). This information is needed to determine 1) whether varietal choice results from a failure of plant breeding support to provide varieties that would better meet their economic and household needs (Brush, 1989; Frankel et al., 1995; Byerlee, 1996; Perales et al., 1998) and 2) whether there is potential vulnerability to loss or erosion of genetic diversity as a result of farmers making different choices in varietal management.

Varietal choices by farmers are based on complex analyses including commercial factors, taste, soil type, comparative levels of performance, degree of crop management required, risk management and market availability. Factors affecting choice can vary in relative magnitude in place and in time (Perales et al., 1998). Farmers are most likely to continue cultivating local landraces when 1) other introduced varieties provide no competitive advantage to the farmer or 2) the competitive advantage of the landrace can be increased, for example through designation and recognition of a specialty product which is reflected in higher demand and prices.

There is growing recognition of the dynamic change in landrace usage and genetic constitution as a result of on-farm management. Landrace cultivation ensures the continued evolution of germplasm (Ford-Lloyd and Jackson, 1986). Varieties are lost *in situ* to be replaced by others (Almekinders et al., 1994). For example, farmers take further advantage of breeding conducted in the formal sector by crossing local landraces with introduced hybrids, thereby creating creolized varieties (Bellon and Brush, 1994; Lamola and Bertram, 1994; Wood and Lenne, 1997; Perales et al., 1998; Bellon and Risopoulos, 2001; van Heerwaarden et al., 2009; Westengen et al., 2014; Bitocchi et al., 2015; Fenzi et al., 2015). Creolized varieties can have significantly positive results on improved farm management and production, thereby lifting farm incomes (Bellon and Risopoulos, 2001; Perales et al., 1998; Zannou et al., 2004; Becerril and Abdulai, 2010; Coomes et al., 2015; Fenzi et al., 2015). According to van Heerwaarden et al. (2009) 'true landraces were rare' in sites they studied in Mexico, with creolized varieties providing most of the seed. And perhaps counter-intuitively, van Heerwaarden et al. (2009) discovered a loss of genetic diversity for Mexican maize as a result of the widespread cultivation of creolized varieties, a loss that was 'moderated by the distinct features offered by modern varieties' (van Heerwaarden et al., 2009).

Proponents of '*in situ* hybridization' argue that continued evolution is necessary to generate new genetic diversity as opposed to placing genes in cold storage, or '*ex situ*'. However, the process of creolization further highlights the dynamic nature of maize diversity even in regions where farmers are primarily dependent upon seed supplied through the informal seed sector. As a result of creolization, exotic alleles and loci are introduced so that the genetic constitution of a landrace grown ten or more years ago might be very different, and possibly even extinct, compared to the situation today (Dyer and Taylor, 2008). Continued evolution is necessarily disruptive of the current state. It is inevitable that in a dynamic context, while the level of genetic diversity may be maintained quantitatively, it will also be subject to qualitative changes in terms of genic and allele frequencies. Consequently, the dynamic nature of on-farm variety management leads us to conclude that the term '*in situ*' is neither applicable nor appropriate in the context of conservation of domesticated plant genetic resources. Indeed, the *in situ* approach has been generally understood in the context of protecting wild species of animals and plants, including crop wild relatives, rather than for domesticated species (Smale et al., 1998a). Furthermore, it is unreasonable to expect farmers to place more emphasis on the need to conserve diversity, especially including for future potential use by others, over the need to utilize all available genetic resources to increase their livelihoods. We propose periodic sampling of germplasm used *in situ* for conservation *ex situ*. The dynamic nature of plant breeding using conserved germplasm then renders the argument that genebanks are cold stores acting to freeze evolution as totally false and redundant.

8 *Ex situ* conservation of maize diversity in genebanks

8.1 Conservation of *Zea* and *Tripsacum* diversity *ex situ* in genebanks: historical and current status

The number of genebanks and the size of *ex situ* collections accelerated during the 1970s and 1980s in response to farmers' adoption of newly developed varieties in preference to continuing the cultivation of landraces. However, after the initial enthusiasm to build genebanks, allocation of resources to provide for the less glamorous, but essential, resources necessary to maintain collections waned (GAO, 1997; Imperial College, 2002) with 'genes and traits being lost as a result' (Cary Fowler quoted in Upfront, 2006). The global situation had become dire, 'fewer than half the maize seeds in store around the world were able to germinate. Many had not been dried properly before being put into storage. Others were lost when refrigeration units failed during power outages.' (Upfront, 2006).

During the late 1990s, Cary Fowler (who had previously coordinated the first FAO SOW survey on PGRFA) and Geoff Hawtin (Director of the International Plant Genetic Resources Institute, a CGIAR centre now known as Bioversity International) had become so alarmed about the fragile status of plant genetic resource conservation that they established a long-term funding mechanism (an endowment) to provide financial resources to support long-term global conservation and use of PGRFA. Thus came into being the Global Crop Diversity Trust (GCDT) (https://www.croptrust.org/). In 2006, the Trust entered into a relationship agreement with the Governing Body of the International Treaty on PGRFA (ITPGRFA). The agreement recognizes the Trust as an 'essential element' of the Treaty's funding strategy in regard to the *ex situ* conservation and availability of PGRFA.

Two major achievements of the Trust include the establishment of a global seed vault and the CGIAR Research Program (CRP) for Genebanks, including CIMMYT and IITA, the two CGIAR germplasm banks that house maize collections. The Svalbard Global Seed Vault is located on the Norwegian island of Spitsbergen in the remote Arctic Svalbard archipelago. The facility was established to provide security backup for a wide variety of germplasm from global genebank locations in a permafrost environment, providing insurance against the loss of seeds in genebanks or in the event of regional or global crises, such as the strife in Syria that has led to the relocation of ICARDA personnel and germplasm. Svalbard has been important for security of and access to ICARDA holdings. The seed vault is managed under terms spelled out in a tripartite agreement between the Norwegian government, the Trust and the Nordic Gene Bank (a cooperative effort of the Nordic countries under the Nordic Council of Ministers, http://www.nordgen.org/sgsv/) and available maize collections from a number of genebanks, including CIMMYT and the NPGS, are deposited there. The Trust now either directly or partially supports 14 (70%) of the 20 activity areas that comprised the original FAO Global Plan of Action developed in 1996, and the Second Plan of Action (FAO, 2011) including the maize CRP (http://www.cimmyt.org/maize-crp/), 1 of 15 CRPs (http://www.cgiar.org/resources/crp-documents/). Notably, the CRPs include many elements, and CRP proposals for 2017–2021 are currently being implemented.

There was no global perspective on the state of the world's genetic resources, including the status of *ex situ* collections, until the FAO sponsored two surveys (FAO SOW, 1997, 2010). These surveys provided baseline data which form the basis for determining prioritized strategies to assure the global conservation and use of maize germplasm.

Table 3 Biological status of genebank maize accessions (Genesys, 2014 data)

Status	No of accessions	Total (%)
Traditional cultivar/landrace	62,182	13.7
Inbred line	17,612	48.4
Breeding/research material	14,223	11.1
Advanced/improved cultivar	6,773	5.3
Segregating population	1,294	1.0
Synthetic population	1,144	0.9
Wild	677	0.5
Founder stock/base population	410	0.3
Other	395	0.3
Genetic stock	358	0.3
Weedy	90	0.1
Hybrid	60	0.1
Breeder's line	43	0.0
Natural	8	0.0
Semi-natural/sown	1	0.0
Not specified	23,196	18.1
Total	**128,466**	**100**

Source 2014 data: Plant genetic resources accession level data provided by: Genesys. All intellectual property rights (including copyright) in the Data are owned and retained by the said institution(s). Data accessed through GENESYS Global Portal on Plant Genetic Resources, http://www.genesys-pgr.org, 1952-09-08.

Recorded numbers of maize accessions stored *ex situ* were 261 584 (FAO, 1997) and 327 932 (FAO, 2010). Accessions were stored across a large number of genebanks. In 1997, 20 genebanks were reported to hold 1% or more of global accessions, by 2010 this number had increased to 24. Genebanks holding 1% or more of maize accessions are presented in Table 2. Of the major holders of maize germplasm, CIMMYT, China, India and the United States provided long-term storage conditions. The Russian Federation only provides medium-term storage. The 1997 survey showed that additional genebanks individually stored less than 1% of global accessions but collectively held 39% of global accessions; this percentage had risen to 45% of global accessions by 2010. The greater proportion of maize accessions being held by national genebanks contrasts with the ex situ conservation of other important crops including wheat, rice, potato, cassava, banana/plantain, sorghum, yam and sweet potato, where the largest *ex situ* collections are held by the International Agricultural Research Centers.

These figures do not include holdings of genebanks that have not ported their accession information to the Genesys portal. Currently, missing data include several important genebanks in Latin America. Of particular note is the very low number of wild maize relatives held in genebanks; teosinte species represent less than 2% of global maize collections. Table 3 illustrates the biological status of genebank holdings reported in Genesys, while Table 4 illustrates the plight of holdings of maize wild relatives. Teosinte is challenging to maintain, particularly in temperate areas, genebank supplies are extremely low, and few are distributable. Most teosintes in Mexico that still occur naturally survive by being protected in some way, perhaps in an abandoned fenced-in area or enclosure,

Table 4 *Zea* species representation in global genebanks

Species representation*	No of accessions	Holdings (%)
Zea mays	111,529	86.82
Zea mays	14,233	11.08
Zea mays mays	1,693	1.32
Zea mais	507	0.39
Zea mexicana	229	0.18
Zea sp.	105	0.08
Zea zea mays	92	0.07
Zea luxurians	33	0.03
Zea diploperennis	22	0.02
Zea perennis	14	0.01
Zea hybr.	6	0.00
Zea mays	2	0.00
Zea nicaraguensis	1	0.00
Total	**128,466**	**100**

*Taxonomy extracted from datasource has not been recently curated.
Source 2014 data: Plant genetic resources accession level data provided by: Genesys. All intellectual property rights (including copyright) in the Data are owned and retained by the said institution(s). Data accessed through GENESYS Global Portal on Plant Genetic Resources, http://www.genesys-pgr.org, 1952-09-08.

or as maize-mimetic weeds with genetic mechanisms to resist geneflow from maize (Doebley et al., 1987). It is imperative that appropriate *in situ* preserves or other methods be developed to ensure the *in situ* survival of these valuable genetic resources (Kermicle and Evans, 2010).

Of the 128 466 accessions cited, less than 40% are indicated as available for distribution, and less than 3% are safety duplicated in the Svalbard Global Seed Vault (data not shown). The percentage of wild relatives available is much lower, reflecting the urgent need to conserve them and make them available. Of these holdings, diverse materials from many areas of origin are well-represented. If these accessions are well-maintained, viable and available for distribution, a wealth of genetic material from a wide range of geographies, ecologies and habitats can be used for research and development.

We describe the global status of conserved maize germplasm by 2010 as akin to a patient who has escaped serious illness and who is not yet on life support. Most maize landraces had been sampled and were adequately stored, at least in the major global genebanks, including with backup collections. However, there remained significant gaps. Only 15% of wild species were in *ex situ* storage and their germplasm pools were under threat of loss due to climate and habitat change. Virtually all teosintes were considered under threat of extinction *in situ* due to current ranching and farming practices (Goodman and Taba, 2007). Further collection of landraces was required for the Caribbean and Central America. The entire collection of Dominica had been lost, as had much of the material collected by the International Board for Plant Genetic Resources in the 1970s. Unfortunately, on a global basis, many accessions (65%) comprised mixed categories or unknown material. For Africa, there was a complete gap in knowledge of available germplasm.

Many national genebanks lacked long-term storage and even capabilities to prepare seeds correctly for medium-term storage. A major genebank in the Russian Federation did not have the capability to provide long-term storage. Almost 10% of accessions were in imminent need of either regeneration or recollection. Most collections in Old World genebanks were not backed-up, there was a lack of capacity for regeneration, and the content and status of collections were largely unknown to the international community. The assembling of core collections had been initiated by several countries. However, documentation and information management systems were unable to provide data that were universally available and comprehensible on an accession basis. Detailed information from which these summaries are drawn can be found in the two FAO State of the World reports (FAOSOW, 1997; 2010).

For *ex situ* collections to become more than a defensive last stand against loss of genetic diversity, additional planning, resource allocation and cooperation are required. Networks involving national and international genebanks, and both publicly and privately funded researchers and maize breeders, are necessary to identify and incorporate additional genetic resources into global maize agriculture during the next decades and beyond into the unforeseeable future.

8.2 A strategic plan for *ex situ* conservation and use of a broader base of maize germplasm

The GCDT supported the development of a strategic plan for the conservation and use of maize genetic resource diversity through surveys and an expert meeting (Goodman and Taba, 2007). In addition to the shortcomings noted in the first FAO SOW report, the GCDT also noted that regeneration of some eco-specific accessions were challenging and that over-regulated phytosanitary restrictions caused major bottlenecks in distributing genetic diversity. Major elements of the plan of action described by the GCDT report (Goodman and Taba, 2007), which we endorse, are the following:

- Achieving excellence in global accession management through development of a comprehensive, easily usable meta database with standardized accession-level information, including passport data, photographs, seed availability, core membership, safety backup status and relative maturity.
- Education of a new generation of maize germplasm specialists.
- Establish global and regional cooperative networks to use the database, together with expertise available in national genebanks to determine which accessions require regeneration and which eco-geographic regions are under- or unrepresented followed by immediate remedial action, otherwise diversity that is only currently available *in situ* will be lost.
- Send core subsets for backup storage in the Svalbard genebank.
- Integrate conservation and management of maize germplasm with programmes to evaluate and facilitate the use of *ex situ* diversity in maize breeding.
- Rational level of regulation of phytosanitary evaluation requirements.
- Provide safety backup of *ex situ* teosinte collections, and create national/international reserves to protect remaining natural stands of teosinte; monitor *in situ* diversity of *Tripsacum* and provide backup in conservation gardens and in *ex situ* long-term storage.

Goodman and Taba (2007) concluded that 'Developing a worldwide strategy for preservation, documentation, distribution, and utilization of maize genetic resources will require attention to these and other, generally less difficult problems'. This is a positive assessment, provided of course, there is the political will and financial way to get the job done. They also express the frustration of many in the genetic resources world, noting that 'The time for holding additional maize germplasm meetings and for writing extensive papers on the woeful status of the maize germplasm system has long past.'

In the decade that passed since the 2006 meeting that generated the GCDT report (Goodman and Taba, 2007), some progress was made on a few of these points, but the goal of establishing a global strategy for maize genetic resources was not fully attained. The GCDT sponsored another meeting in 2016, convening a group of 39 maize genetic resource experts from around the world. Advances during the preceding decade were noted, particularly with respect to the development of the accession-level global database, Genesys, by the Crop Trust and the development by the USDA-ARS of the GRIN-Global germplasm bank data management system and its implementation in a growing number of international and national banks, including CIMMYT, INIA (Chile), INIAF (Bolivia), Australia, the Czech Republic, Portugal, Tunisia, and others. A new set of priorities was defined by the group, ranked by the number of votes received: by those in attendance:

- Political and financial constraints to access and conservation
- Pre-breeding for direct use of farmers and breeders to address the challenges of climate change
- Enhanced Global Collaboration on Maize Germplasm Conservation

- Conservation of Wild Relatives of Maize: *Teosinte* and *Tripsacum*
- Enhanced Global Accessions-Level Information Management
- Capacity development in maize genetic resources
- Enhanced germplasm characterization and evaluation need to be integrated to improve agricultural productivity
- Increased security and availability of conserved accessions globally
- Old World Collections: New Focus on Africa and Asia needed
- Filling gaps in collections from Americas
- Assess the status of collections held by the Vavilov Institute

As one might expect, many of the same priorities that were identified in 2006 are still considered to be priorities ten years later. However, some advances were made and new ideas have emerged. The Svalbard Global Seed Vault was just breaking ground in 2006, officially opened in 2008 and has received maize seed from multiple genebanks, including the two largest collections, CIMMYT and USDA. As mentioned above, databases, both within banks and on a global scale, are steadily improving. Pre-breeding has been established as a key activity that is critical in order to put diverse germplasm into use. An area of concern is the continued neglect of recruitment and training of the next generation of genetic resources experts. Interest in germplasm collection outside of the centres of diversity and origin, in regions of the world that have adopted maize as a staple food for humans and livestock, particularly Africa and Asia, has increased. Overshadowing all are the constraints on access and conservation of maize due to politics and finances.

9 Critical issues facing maize genebank curators today and in the immediate future

9.1 Status of *in situ* and *ex situ* diversity in domesticated and wild species of *Zea* and *Tripsacum*

Technological advances in DNA sequencing and recent advances in our understanding of the genetic control of important traits during the domestication process have provided new insights into the evolution of maize and provided a more comprehensive perspective of the organization of genetic diversity. As a result, it has been possible to identify gaps in existing *ex situ* collections and to highlight the need to conserve diversity present in the immediate ancestors of maize, the teosintes and also in the more distantly related species of *Tripsacum*. The cultivation of local landraces can be expected to be under continued pressure due to 1) climate change, 2) the introduction of more productive diversity, affecting farmer choice on variety usage, 3) farmer-directed introgression of germplasm introduced via formal breeding and 4) further abandonment of farms and increased urbanization (FAO SOW, 2010).

As a result, there will be greater calls upon maize curators to conserve not only OPVs of maize but also the wild relatives of maize, in order to retain as much of the full range of variation in the Panzea genome as possible (FAO SOW, 2010). It is therefore necessary to review the current global status and potential vulnerability of genetic diversity in the genus *Zea* and *Tripsacum*, both *in situ* and *ex situ*.

9.2 Increased need to meaningfully characterize and monitor trends in genetic diversity on farms and in breeding programmes

It is essential to meaningfully characterize genetic resources in order to understand their intrinsic properties and target their use for specific objectives. However, characterization of germplasm raises both conceptual and technical challenges (Meng et al., 1998; Smale et al., 1998b). A more comprehensive and multidimensional view of agro-biodiversity is required to enable 'a greater understanding of how diversity is created, maintained, and renewed' (Montenegro de Wit, 2016). Meng et al. (1998) described a framework for considering diversity of crop genetic resources. There are two important aspects or reasons to consider diversity in either spatial or temporal terms. Diversity can be considered as observable in the field (apparent) or not (latent), and thus relates to methods that can be used to characterize, measure and compare diversity. Germplasm can be characterized according to 1) morphology, including characteristics of agronomic interest to farmers and others that are used purely to differentiate one variety from another, but which are subject to a loss of clarity due to GxE interactions; 2) molecular marker or sequence data that allow for comparisons among varieties that are purely DNA based and thus robust to environmental effects; 3) variety names, which can be informative of agronomic performance (Montenegro de Wit, 2016); and 4) comparative pedigrees, but these may be incomplete or in error. In addition, comparisons of agro-ecological data may be predictive of relative levels of diversity, including adaptation to climate, soil type or resistances to pests and diseases or abiotic stresses (Corral et al., 2008; Perales and Golicher, 2014). Molecular marker or sequence data cannot provide information related to agronomic performance without considerable additional research, and may be irrelevant for most traits of interest to farmers and breeders, that is, those that are affected by multigenic factors. Equally important in those countries where people consume maize as a staple are evaluations based on nutrition and culinary uses, as well as cultural considerations.

Recent technological developments make possible the more comprehensive collection of sequence data, and data collection costs may be expected to decrease further. Measures of molecular markers, including DNA sequence data, have proven to be useful means to measure genetic diversity across a broad spectrum of varieties or taxa, with applications that include 1) the identification, both quantitatively and qualitatively, of genetic diversity that did not transfer due to bottlenecks imposed by selection or genetic drift; 2) the monitoring of change of genetic diversity over time; and 3) providing a basis to help elucidate genetic factors associated with traits of agronomic importance. DNA data provide a unifying means to characterize and compare genetic resources both within and across species and genera. And while morphological appearance and agronomic performance result from interactions of genotype x environment (including crop management), there is great utility in being able to compare germplasm using data that directly reflect the genome. Nonetheless, new discoveries, such as the role of epigenetic factors in controlling gene expression, will always deserve consideration in terms of characterizing genetic resources.

The need for a consensus on appropriate methodological approaches to monitor trends in genetic diversity in evolving landraces and varieties improved in formal breeding programmes remains a glaring need (Montenegro de Wit, 2016). This need is evidenced by 1) spirited debates in the scientific literature where wholly different conclusions as to whether genetic diversity is being maintained or declining are reached (van Heerwarden et al., 2009; Dyer et al., 2014; Brush et al., 2015), resulting in an impasse in defining strategies for optimum generation, conservation and use of diversity (Montenegro de Wit, 2016) and 2) the lack of published data showing trends in genetic diversity as revealed by molecular marker or DNA sequence data, either for landrace or commercial hybrid usage. While abundant pedigree data are available, for example to monitor trends for parental inbred lines in the United States (Mikel, 2011), they are insufficient to allow the monitoring of genetic diversity as it actually exists on farms in the form of hybrids. Calculation of the appropriate level of prioritization and funding of germplasm conservation and use programmes requires a firmer foundation, that unequivocally shows the need to increase the availability of genetic diversity to public and private breeding programmes, including to farmers. Further opportunities can be available as a result of 1) coupling genetic diversity with improved breeding capabilities and genomic sequence data that have already been developed as a result of public and private investments and 2) developing more synergistic relationships among farmers, farmer-breeders and breeders who collectively constitute the so-called 'informal' and 'formal' seed sectors. In this way we take full advantage of both *in situ* and *ex situ* systems of crop diversity management, improvement and use.

9.3 The interface between breeding and germplasm will bring increased demands upon genebanks for more relevant information on all accessions.

Flint-Garcia (2013), citing Tenaillon et al. (2001), notes that 'two randomly chosen inbred lines (of maize) are more diverse than the difference between the human and chimps.' CBD germplasm developed in the United States has also proven to be a significant contributor towards raising yields in Europe, Canada, South America, India and China (Hans Bhardwaj, pers. comm.; Smith et al., 2006; Glover et al., 2005; Li et al., 2005; Zhang et al., 2016). This race is thus a proven exotic source of germplasm, adding useful new diversity to maize breeding programmes around the globe. However, the inverse has proven to be more challenging, in that less than 2% of the United States widely used hybrid germplasm containing exotic germplasm (Goodman, 2005).

Fortunately, new opportunities are emerging with a promise to successfully introduce more beneficial exotic germplasm into breeding programmes and farms. These opportunities arise from the development of improved abilities to identify and select for positive genetic signals associated with important agronomic traits, including those under multigenic control (Campos et al., 2004; Prada, 2009; Yang et al., 2011; Mir et al., 2013; Cooper et al., 2014a,b; Sood et al., 2014; Gaffney et al., 2015; Gorjanc et al., 2016; Wallace et al., 2016). For example, Gorjanc et al. (2016) evaluated new breeding strategies to harness polygenic variation from landraces using GS coupled with stochastic simulation to improve the transfer and evaluation of exotic germplasm into already well-adapted germplasm. Their results suggest that sustained improvement initiated through the direct use of landrace germplasm is the most feasible approach with regard to agronomically useful traits under polygenic control.

As maize genetic resources are genotyped and researchers avail themselves of this phenotypic and genomic information, it is anticipated that new allelic diversity for traits will be identified and incorporated from these temperate-adapted resources. Romay et al. (2013) provided a thorough analysis of the genomic diversity of the available inbreds in the US maize collection and GEM releases. Available at Panzea, these SNP data are being utilized by many researchers for a variety of analytical objectives (http://www.panzea.org/about_us). Supported by the US National Science Foundation and other sources including the USDA and the home institutions of the investigators, Panzea investigators seek to understand the biology of rare alleles of maize and its wild relatives, and the genetic architecture of these species. In the process, they have pioneered public sector methods for genetic mapping, evolution and population genetics that have been applied to many species.

While large multinational breeding companies may have advanced their bioinformatics infrastructure to utilize genomic and phenotypic information for GS and genome-wide association studies, public researchers face many challenges in doing so. Panzea and other genomic resources are rapidly being adopted by public sector breeders and researchers to understand allelic diversity related to specific traits in adapted germplasm. A few examples would include the work of Peiffer et al. (2014), to understand the genetic architecture of plant height; of Brenner et al. (2012), to identify exotic versus temperate donor genome contributions with respect to cell wall digestibility; and of Owens et al. (2014), examining provitamin A biofortification of maize using genome-wide association and genomic prediction models of carotenoid levels. Use of GS to advance selection of materials resistant to MLN is but one example of potentially large positive impacts of adoption of this method (Semagn et al., 2015; Gowda et al., 2015; Gorjanc et al. 2016; Bian and Holland, 2017).

There is a limit to the degree or type of characterization a genebank can or reasonably should accomplish. Consequently, greater cooperation and communication among genebank curators, researchers and breeders will likely need to occur. Collation of published data on germplasm that resides in genebanks could be facilitated using search engines. It is therefore necessary to review progress towards a rational global information system for all crop genetic resources and to review the future roles and interactions of genebank curators with researchers and breeders to facilitate access to conserved germplasm. Van Treuren and van Hintum (2013) predicted that genebanks may be called upon to provide 'access to allele-level information, directed acquisition of novel variation, and to collaborate in high-throughput phenotyping'. Sackville-Hamilton (2013) foresaw a 'seamless linkage from genebank through genomic research and pre-breeding to breeding'. These activities cannot be successful without a massive investment in developing information management systems that support global inquiry, analyses and meaningful summarization of genetic, phenotypic, climatic, pedigree, phenomic and other relevant data (see DivSeek white paper at ftp.fao.org, www.divseek.org and divseek@croptrust.org).

Three primary information systems that inform on genebank holdings and associated information include Genesys, the Global Gateway to Genetic Resources, https://www.genesys-pgr.org/; GRIN-Global, a genebank information management system, http://www.grin-global.org/; and the ECPGR, European Cooperative Program for Plant Genetic Resources, http://www.ecpgr.cgiar.org/. The GRIN-Global System resulted from a collaboration between the USDA-ARS, the GCDT and Bioversity International. GRIN-Global is now implemented in eight countries and others are in various stages of implementation. GRIN-Global is database-neutral, accommodates multiple languages, is flexible to support further evolution due to a three-tiered architecture and facilitates interoperability with other information systems, including genomic databases. Genesys serves as a portal for any genebank to provide accession-level information, if they choose. In this way it is possible to identify specific accessions maintained within each participating genebank.

The mission of GOBii, the Genomic Open-source Breeding Informatics Initiative, is to develop open-source computational infrastructure and analysis capabilities, enabling implementation of genomic and marker-assisted selection in routine breeding programmes for staple crops in the developing world (http://cbsugobii05.tc.cornell.edu/wordpress/). Scientists of the CGIAR Centers, Cornell University and other collaborators are devoted to developing genomic systems that will include databases, imputation systems and breeder decision tools, thus facilitating use of un-adapted genetic resources.

Coordination and cooperation are essential to leverage the advances of these various initiatives in order to maximize the discovery and utilization of useful genetic diversity. An example of such an effort is the Seeds of Discovery (SeeD), a public project initiated by CIMMYT and the Mexican government (MasAgro, SAGARPA) to characterize maize landrace accessions in order to facilitate gene discovery and to establish distinct haplotypes, with the goal of making the genebank resources more attractive to breeders, and to establish foundations for harnessing favourable variation from landraces. The SeeD project explored 40 000 accessions originally, and established a breeder's core of 4000+ maize landrace accessions. The core has been genotyped using a large number of molecular markers and extensively phenotyped, including testcross performance (Anon, http://seedsofdiscovery.org, 2016). Bridging germplasm has been created, and many traits are being explored for novel variation, including biotic and abiotic stress resistance traits and grain composition traits.

FIGS is an approach designed to utilize characterization, provenance and passport information to target genetic resources for specific climate change-adaptive traits. Khazaei et al. (2013) showed that machine learning and development of algorithms applied to faba bean leaf morphology and plant water use datasets identified traits useful for identifying drought tolerance, and demonstrated that FIGS is an effective tool to enhance discovery of new genes for adaptation to abiotic stress. When applied to well-characterized maize of known provenance, the FIGS approach may offer promise for use in targeting specific maize germplasm resources for adaptive traits, especially in countries experiencing novel climates.

10 Conclusions

Ideally, every farmer should have the opportunity to select from a range of varieties that will collectively allow him or her to maximize the potential of their land and management skills in their agro-ecological and socio-economic environments. The idea that gene editing technologies may obviate the need to conserve germplasm for future use is fundamentally flawed and naïve in the extreme. This is because 1) the identification of genetic loci associated with the control or regulation of agronomic traits is first dependent upon the evaluation of naturally occurring diversity, 2) most traits of agronomic interest are under multigenetic control where individual genes and contributing loci remain anonymous and 3) although currently speculative, it would seem a reasonable hypothesis that a lack of genic collinearity is not restricted to just two maize inbred lines that are placed taxonomically and phylogenetically into a single racial complex, the CBD. Consequently, it is reasonable to assume that genic regions not present in CBD germplasm will be found among the greater proportion of maize germplasm that is arrayed in other races. The availability of a comprehensive array of germplasm diversity is a resource that will further enable advances as a result of plant breeding for traits regardless of their mechanisms of genetic control. Thus both genetic resource diversity and the tools to investigate and utilize those resources are vital and complementary means to advance maize agriculture worldwide.

Progress towards this ideal will require genebanks to undertake new responsibilities and roles, thereby facilitating access to farmers, both directly and indirectly, via seed fairs, informal and formal seed supply systems, and to plant breeders of all types including publicly or commercially funded, a global portfolio of useful and potentially useful germplasm.

Much knowledge has been gained regarding potential vulnerabilities that threaten stability of cultivated maize germplasm production since the initial impetus of widespread regional loss of diversity (Jarvis et al., 2007; Fu, 2015). These concerns provided the original raison d'être for construction of the *ex situ* genebanks. Today this knowledge provides the basis for describing a more comprehensive and important role for genebanks, now and in the future.

First, it is now understood that cultivation of landraces can contribute to resiliency, if new useful diversity can be accessed and integrated successfully into new and improved varieties. The constitution of landrace germplasm is of variable dynamics and can include significant flows from hybrids, depending on grower practices relative to seed saving and exchange. Consequently, our recommended term for what we regard has been too loosely described as '*in situ* conservation' is rather '*in situ* crop management'. Two needs to improve germplasm conservation and further use result from an understanding of the dynamic nature of landraces are 1) the need to periodically conserve *ex situ* the array of germplasm that represents a landrace and 2) the need to improve access to germplasm that could better meet the needs of farmers who rely upon landrace or creolized varieties. There is great potential to facilitate greater global use of inbred lines and hybrids developed in the formal breeding sector, including those that are, or have been, protected as intellectual property. The USDA already places into the global public domain maize inbred lines that were protected under PVP and germplasm that is protected under patents can be available following expiration of patents (Kurtz et al., 2016).

Second, it is now known that there is a lack of synteny within the domesticated maize genome and even between inbred lines developed from a single race of maize. Consequently, it is a reasonable hypothesis that additional genes and allelic diversity reside in the genus *Zea*, diversity that remains to be discovered.

Third, it is now known that genetic diversity residing in crop wild relatives of maize has potential use in the future. Consequently, it is imperative to assure the future availability of those resources using both *in situ* and *ex situ* conditions as appropriate and taking fully into account the vulnerability of natural habitats to climate change and changing use patterns.

Fourth, how genetic diversity is characterized and measured is complex and challenging. Initial concepts of measuring diversity were solely in the context of how to sample allelic diversity in farmers' fields to populate *ex situ* collections and more recently to compare genetic diversity between varieties, species or genera. Progress must be made towards a better understanding of how to usefully measure and characterize germplasm, including in terms of on-farm use.

The future role of the genebank must continue to evolve from the initial concept as an emergency resource and long-term repository of genetic diversity which otherwise would be forever lost. Genebanks should occupy a pivotal point in the global flow of germplasm, facilitating both conservation and providing access for use with the goal to improve productivity and farmer livelihoods, via the engagement of breeders, including farmer-breeders. Genebanks can provide additional knowledge to facilitate further use including agro-ecological, essential characterization data, and potentially additional data, including gene sequences, according to funding and prioritization of genebank activities.

Ultimately, the greatest challenges to achieving more complete access and use of improved germplasm on farms will not be technically or scientifically based. Major challenges will be to achieve mutually beneficial cooperation between stakeholders of varying perspectives and experiences. There can be few objectives more important to achieve than the improved nutritional, economic, political and ecological sustainability of life on earth. Positive progress on resourcing the fundamental components of this objective by policy makers and governments will only be possible following at least a minimal degree of consensus among a wide range of stakeholders. An essential task will be to markedly raise awareness, especially among the public and policy makers in the industrially developed world, of their ultimate dependence upon the sustained achievement of excellence in the stewardship and use of plant genetic resources in breeding programmes and on farms.

11 References

Alene, A. D., Menkir, A., Ajala, S. O., Badu-Apraku, B., Olanrewaju, A. S., Manyong, V. M. and Ndiaye, A. (2009) The economic and poverty impacts of maize research in West and Central Africa. *Agr. Econ.* 40:535–50.

Almekinders, C. J., Louwaars, N. P. and De Bruijn, G. H. (1994) Local seed systems and their importance for an improved seed supply in developing countries. *Euphytica*, 78(3):207–16.

Anon. (2016) Maize Phase II, CGIAR-Research Progam on Maize. Maize Agri-Food systems Proposal 2017–2022. CGIAR: Montpellier, France. https://library.cgiar.org/bitstream/handle/10947/4403/2.%20MAIZE%20AFS-CRP%20and%20FP%20Narratives%20Proposal%202017-2022%20(July%2030,%202016).pdf?sequence=1.

Anon. SeeDs of Discovery. Unlocking the genetic potential of maize and wheat. MasAgro project. Jointly sponsored by Secretaria de Agricultura Ganaderia Desarrollo Rural Pesca y Alimentacion (SAGARPA) and the International Wheat and Maize Improvement Center (CIMMYT) http://seedsofdiscovery.org/maize/ (accessed 16 April 2016). http://seedsofdiscovery.org/, Accessed November 2016.

Arteaga, M. C., Moreno-Letelier, A., Mastretta-Yanes, A., Vázquez-Lobo, A., Moreno-Estrada, A., Equiarte, L. E. and Piñero, D. (2016) Genomic variation in recently collected maize landraces from Mexico. *Genomics Data* 7:38–45.

Becerril, J. and Abdulai, A. (2010) The impact of improved maize varieties on poverty in Mexico: A propensity score-matching approach. *World Dev.* 38:1024–35.

Beissinger, T., Beissinger, M., Hirsch, C. N., Vaillancourt, B., Deshpande, S., Barry, K., Buell, C. R., Kaeppler, S. M., Gianola, D. and de Leon, N. (2014) A genome-wide scan for evidence of selection in a maize population under long-term artificial selection for ear number. *Genetics* 196(3):829–40, 1 March 2014. doi: 10.1534/genetics.113.160655.

Bellon, M. R. (2004) Conceptualizing interventions to support on-farm genetic resource conservation. *World Dev.* 32(1):159–72.

Bellon, M. R. and Brush, S. B. (1994) Keepers of maize in Chiapas, Mexico. *Econ. Bot.* 48:196–209.

Bellon, M. R. and Risopoulos, J. (2001) Small-scale farmers expand the benefits of improved maize germplasm: A case study from Chiapas, Mexico. *World Dev.* 29:799–811.

Ben-Ari, T. and Makowski, D. (2014) Decomposing global crop yield variability. *Environ. Res. Lett.* 9(11):10, 114011. doi:10.1088/1748-9326/9/11/114011.

Bennett, M. K. (1955) The food economy of the New England Indians, 1605–75. *J. Polit. Econ.* 63:369–97.

Bennetzen, J. L. and Freeling, M. (1997) The unified grass genome: synergy in synteny. *Genome Res.* 7(4):301–6.

Bennetzen, J., Buckler, E., Chandler, V., Doebley, J., Dorweiler, J., Gaut, B., Freeling, M., Hake, S., Kellogg, E., Poethig, R. S. and Walbot, V. (2001) Genetic evidence and the origin of maize. *La. Am. Antiq.* 12(1):84–6.

Bergquist, R. R. (1981) Transfer from *Tripsacum dactyloides* to corn of a major gene locus conditioning resistance to *Puccini sorghi. Phytopath.* 71:518–20.

Beverly, R. (1705) History of and present state of Virginia. http://docsouth.unc.edu/southlit/beverley/beverley.html

Bian, Y. and Holland, J. B. (2017) Enhancing genomic prediction with genome-wide association studies in multiparental maize populations. *Heredity* doi:10.1038/hdy.2017.4.

Biemont, C. and Vierira, C. (2006) Junk DNA as an evolutionary force. *Nature.* 443:521–4.

Bitocchi, E., Bellucci, E., Rau, D., Albertini, E., Rodriguez, M., Veronesi, F., Attene, G. and Nanni, L. (2015) European flint landraces grown *in situ* reveal adaptive introgression from modern maize. *PLos One* 10(4):e0121381, p. 20.

Blanco, M. and Gardner, C. (2014) *Germplasm Enhancement of Maize – Strategies and Synergy with Maize Curation.* USDA-ARS: Ames, Iowa, USA.

Blakey, C., Costich, D., Sokolov, V. and Islam-Faridi, M. (2007) Tripsacum genetics: from observations along a river to molecular genomics. *Maydica* 52, 81–99.

Bonzani, R. M. and Oyuela-Caycedo. A. (2006) The gift of the variation and dispersion of maize: Social and technological context in amerindian societies. In Staller, J. E., Tykot, R. H. and Benz, B. F. (Eds), *Histories of Maize.* Academic Press: Burlington, MA, USA, pp. 344–56.

Bracco, M., Cascales, J., Hernández, J. C., Poggio, L., Gottlieb, A. M. and Lia, V. V. (2016) Dissecting maize diversity in lowland South America: genetic structure and geographic distribution models. *BMC Plant Biol.* 16:186. doi:10.1186/s12870-016-0874-5.

Brenner, E. A., Blanco, M., Gardner, C. and Lübberstedt, T. (2012) Genotypic and phenotypic characterization of isogenic doubled haploid exotic introgression lines in maize. *Mol. Breed.* 30(2):1001–16.

Brewbaker, J. L. (2009) Registration of Nine Maize Populations Resistant to Tropical Diseases. *J. Plant. Reg.* 3:10–13. doi:10.3198/jpr2008.07.0396crc.

Brewbaker, J. L. (1979) Diseases of maize in the wet lowland tropics and the collapse of the Classic Maya civilization. *Econ. Bot.* 33:101–18.

Brewbaker, J. L. (1981) Resistance to maize mosaic virus. In Gordon, D. T., Knoke, J. K. and Scott, G. E. (Eds), *Virus and Virus-like Diseases of Maize in the United States.* Southern Coop. Ser. Bull. 247. Ohio Agric. Res. and Develop. Center: Wooster, OH, pp. 145–51.

Brewbaker, J. L. (1997) Registration of 13 tropically adapted parental inbred lines of maize resistant to maize mosaic virus. *Crop Sci.* 37:631–2.

Brewbaker, J. L. (2003) *Corn Production in the Tropics: The Hawaii Experience.* College of Tropical Agriculture and Human Resources, Univ. of Hawaii at Manoa.

Brewbaker, J. L. and Josue, A. D. (2007) Registration of 27 maize parental inbred lines resistant to maize mosaic virus. *Crop Sci.* 47:459–61.

Brewbaker, J. L., Logroño, M. L. and Kim, S. K. (1989) The MIR (Maize Inbred Resistance) trials: Performance of tropical-adapted maize inbreds. Hawaii Inst. Trop. Agric. Human Resources Research Ser. 62.

Brown, W. L. (1984) The coming debate over ownership of plant germplasm. In: J Proc. 39th Ann. Corn & Sorghum Res. Conf. Chicago. Dec. 1984. American Seed Trade Association: Alexandria, VA, pp. 44–51.

Brown, W. L. and Anderson, E. (1947) The Northern Flint corns. *Ann. Missouri Bot. Gard.* 34:1–22.

Brown, W. L. and Anderson, E. (1948) The Southern dent corns. *Ann. Missouri Bot. Gard.* 35:255–76.

Brown, W. L. and Goodman, M. M. (1977) Races of corn. In Sprague, G. F. (Ed.) *Corn and Corn Improvement.* Chapter 2, pp. 49–88. Amer. Soc. Agron.: Madison, Wisconsin, USA, p. 774.

Brunner, S., Fengler, K., Morgante, M., Tingey, S., Rafalski, A. (2005) Evolution of DNA sequence nonhomologies among maize inbreds. *Plant Cell* 17:343–60.

Brush, S. N. (1989) Rethinking crop genetic resource conservation. *Conserv. Biol.* 3:19–29.

Brush, S. B. (2005) Protecting traditional agricultural knowledge. *Wash. Univ. J. Law Policy* 17:59–109.

Brush, S. B., Bellon, M. R., Hijmans, R. J., Ramirez, Q. O., Perales, H. R. and van Etten, J. (2015) Assessing maize genetic erosion. *Proc. Natl. Acad. Sci.* 112(1):E1. doi:10.1073/pnas.1422010112.

Buckler, E. S. and Stevens, N. M. (2005) Maize origins, domestication and selection. In Motley, T. J., Zerega, N. and Cross, H. B (Eds), *Darwin's Harvest. New Approaches to the Origins, Evolution, and Conservation of Crops*. Columbia University Press: NY, pp. 67–90.

Burke, M. B., Lobell, D. B. and Guarino, L. (2009) Shifts in African crop climates by 2050, and the implications for crop improvement and genetic resources conservation. *Glob. Environ. Change* 19(3):317–25. http://dx.doi.org/10.1016/j.gloenvcha.2009.04.003.

Butardo, Jr., V. M., Daygon, V. D., Colgrave, M. L., Campbell, P. M., Resurreccion, A., Cuevas, R. P., Jobling, S. A., Tetlow, I., Rahman, S., Morell, M. and Fitzgerald, M. (2012) Biomolecular analyses of starch and starch granule proteins in the high-amylose rice mutant Goami 2. *J. Agr. Food Chem.* 60(46):11576–85. doi:10.1021/jf303205p.

Byerlee, D. (1996) Modern varieties, productivity, and sustainability: recent experience and emerging challenges. *World Dev.* 24:697–718.

Campos, H., Cooper, M., Habben, J. E., Edmeades, G. O and Schussler, J. R. (2004) Improving drought tolerance in maize: A view from industry. *Field Crops Res.*, 90:19–34.

Carena, M. J., Pollak, L., Salhuana, W. and Denuc, M. (2009) Development of unique and novel lines for early-maturing hybrids: moving GEM germplasm northward and westward. *Euphytica* 170(1):87–97. doi:10.1007/s10681-009-9985-5.

Cassaday, K., Smale, M., Fowler, C. and Heisey, P. W. (2001) Beneifts from giving and receiving genetic resources: the case of wheat. *Plant Genet. Resour. Newsl.* 127:1–10.

Cassman, K. G., Dobermann, A., Walters, D. T. and Yang, H. S. (2003) Meeting cereal demand while protecting natural resources and improving environmental quality. *Annu. Rev. Environ. Resour.* 28:315–58. doi:10.1146/annurev.energy.28.040202.122858

CIMMYT (1994) The CIMMYT Maize Germplasm Bank: Genetic Resource Preservation, Regeneration, Maintenance, and Use. Maize Program Special Report. Mexico, D. F.

Cook, S. F. and Borah, W. (1979) *Essays in Population History: Mexico and California*. Univ. California Press: Berkeley, California, USA, p. 333.

Coomes, O. T., McGuire, S. J., Garine, E., Caillon, S., McKey, D., Demeulenaere, E., Jarvis, D., Aistara, G., Barnaud, A., Clouvel, P. and Emperaire, L. (2015) Farmer seed networks make a limited contribution to agriculture? Four common misconceptions. *Food Policy* 56:41–50.

Chia, J.-M., Song, C., Bradbury, P., Costich, D., de Leon, N., Doebley, J., Elshire, R. J., Gaut, B., Geller, L., Glaubitz, J. C., Gore, M., Guill, K. E., Holland, J., Hufford, M. B., Lai, J., Li, M., Liu, X., Lu, Y., McCombie, R., Nelson, R., Poland, J., Prasanna, B. M., Pyhäjärvi, T., Rong, T., Sekhon, R. S., Sun, Q., Tenaillon, M. I., Tian, F., Wang, J., Xu, X., Zhang, Z., Kaeppler, S. M., Ross-Ibarra, J., McMullen, M. D., Buckler, E. S., Zhang, G., Xu, Y. and Ware. D. (2012) Maize HapMap2 identifies extant variation from a genome in flux. *Nature Genet.* 44:803–7. doi:10.1038/ng.2313.

Cooper, M., Gho, C., Leafgren, R., Tang, T. and Messina, C. (2014) Breeding drought-tolerant maize hybrids for the US corn-belt: discovery to product. *J. Exp. Bot.* 65(21):6191–204. doi: 10.1093/jxb/eru064.

Cooper, M., Messinna, C. D., Podlich, D., Totir, L. R., Baumgarten, A., Hausmann, N. J., Wright, D. and Graham, G. (2014) Predicting the future of plant breeding: complementing empirical evaluation with genetic prediction. *Crop Pasture Sci.* 65:311–36. http://dx.doi.org/10.1071/CP14007.

Corral, J. A. R., Puga, N. D., Sanchez-Gonzales, J. de J., Parra, J. R., Eguiarte, D. R. G., Holland, J. B. and Garcia, G. M. (2008) Climatic adaptation and ecological descriptors of 42 Mexican maize races. *Crop Sci.*, 48:1502–12.

Costich, D. (2016) Personal communication.

da Fonseca, R. R., Smith, B. D., Wales, N., Cappellini, E., Skoglund, P., Fumagalli, M., Samaniego, J. A., Carøe, C., Ávila-Arcos, M. C., Hufnagel, D. E., Korneliussen, T. S., Vieira, F. G., Jakobsson, M., Arriaza, B., Willerslev, E., Nielsen, R., Hufford, M. B., Albrechtsen, A., Ross-Ibarra, J., and Gilbert, M. T. P. (2015), The origin and evolution of maize in the Southwestern United States. *Nature Plants*. doi: 10.1038/NPLANTS2014.3.

Dalrymple, D. G. (1978) *Development and Spread of High-yielding Varieties of Wheat and Rice in the Less Developed Nations.* Sixth Ed., U.S. Department of Agriculture Office of International Cooperation and Development in cooperation with the U.S. Agency for International Development: Washington D.C., p. 134.

Darrah, L. L. personal communication. (2016)

Darrah, L. L., Eberhart, S. A. and Penny, L. H. (1972) A maize breeding methods study in Kenya. *Crop Sci.* 12:605–8.

De Lange, E., Balmer, D., Mauch-Mani, B. and Turlings, T. C. J. (2014) Insect and pathogen attack and resistance in maize and its wild ancestors, the teosintes. *New Phytol.* 204:329–41.

Doebley, J. F. (1983) The taxonomy and evolution of *Tripsacum* and teosinte, the closest relatives of maize. In Gordon, D. T., Knoke, J. K. and Naul, L. R. (Eds), *Proc. Intl. Maize Virus Disease Colloquium and Workshop held at the Ohio State University, Ohio Agricultural Research and Development Center, Wooster, Ohio. August 1982.* Ohio State University, pp. 15–28.

Deng, T. C., Chou, C. M., Chen, C. T., Tsai, C. H. and Lin, F. C. (2014) First Report of Maize chlorotic mottle virus on Sweet Corn in Taiwan. *Plant Dis.* 98:1748.

Doebley, J., Goodman, M. M. and Stuber, C. W. (1987) Patterns of isozyme variation between maize and Mexican annual teosinte. *Econ. Bot.* 41: 234–46.

Doebley, J. and Iltis, H. H. (1980) Taxonomy of *Zea* (Gramineae). I. A subgeneric classification with key to taxa. *Amer. J. Bot.* 67:982–93.

Doebley, J. F., Wendel, J. F., Smith, J. S. C., Stuber, C. W. and Goodman, M. M. (1988). The origin of cornbelt maize: the isozyme evidence. *Econ. Bot.* 42, 120–31.

Dolezal, W. (2014) Gem, Global maize disease status – threats and opportunities. Power point presentation to the 2014 GEM Cooperator meeting, American Seed Trade Association Meeting, December, 2014. Chicago, IL. http://www.public.iastate.edu/~usda-gem/Publications/GEM_Presentations.htm.

Dubreuil, P., Warburton, M. L., Chastanet, M., Hoisington D. and Charcosset, A. (2006) More on the introduction of temperate maize into Europe: large-scale bulk SSR genotyping and new historical elements. *Maydica.* 51:281–91.

Duvick, D. N. and Cassman, K. G. (1999) Post – Green Revolution Trends in Yield Potential of Temperate Maize in the North-Central United States. *Crop Sci.* 39:1622–30.

Dyer, G. A., Lopez-Feldman, A., Yunez-Naude, A. and Taylor, J. E. (2014) Genetic erosion on maize's center of origin. *Proc. Natl. Acad. Sci. U. S. A.* 111:14094–9. doi:10.1073/pnas.1407033111.

Dyer, G. A. and Taylor, J. E. (2008) A crop population perspective in maize seed systems in Mexico. *Proc. Natl. Acad. Sci. U. S. A.* 105:470–5.

Eberhart, S. A., Harrison, M. N. and Ogada, F. (1967) A comprehensive breeding system. *TAG Theor. Appl. Genet.* 37(4):169–74.

Eichten, S. R., Briskine, R., Song, J., Li, Q., Swanson-Wagner, R., Hermanson, P. J., Waters, A. J., Starr, E., West, P. T., Tiffin, P., Myers, C. L., Vaughn, M. W. and Springer, N. M. (2013) *Plant Cell* 25(8):2783–97. doi: 10.1105/tpc.113.114793.

Eyre-Walker, A., Gaut, R. L., Hilton, H., Feldman, D. L. and Gaut, B. (1998) Investigation of the bottleneck leading to the domestication of maize *Proc. Natl. Acad. Sci. U. S. A.* 95:4441–6.

Falcon, W. P. and Fowler, C. (2002) Carving up the commons-emergence of a new international regime for germplasm development and transfer. *Food Policy* 27:197–222. doi:10.1016/S0306-9192(02)00013-1.

FAO (2011) Second Global Plan of Action for Plant Geneti Resources for Food and Agriculture. FAO: Rome, Italy. http://www.fao.org/docrep/015/i2624e/i2624e00.htm.

FAO SOW (1997) The state of the world's plant genetic resources for food and agriculture. Food and Agriculture Organization of the United Nations: Rome, p. 511.

FAO (2009a) Global agriculture towards 2050. Food and Agriculture Organization of the United Nations: Rome.

FAO (2009b) How to Feed the World in 2050. Food and Agriculture Organization of the United Nations. http://www.fao.org/fileadmin/templates/wsfs/docs/expert_paper/How_to_Feed_the_World_in_2050.pdf.

FAO SOW (2010) The second report on The State of the World's Plant Genetic Resources for Food and Agriculture. Commission on Genetic Resources for Food and Agriculture: Rome, p. 370.

FAOSTAT (2013) Statistics Database. Food and Agriculture Organization of the United Nations: Rome, Italy. http://faostat.fao.org/beta/en/#data/QC

FAOSTAT (2014) Statistics Database. Food and Agriculture Organization of the United Nations: Rome, Italy. http://faostat.fao.org/beta/en/#data/QC

FAOSTAT (2017) *Statistics Database. Food and Agriculture Organization of the United Nations: Rome, Italy.* http://faostat.fao.org/beta/en/#data/QC

Farnham, D. E., Benson, G. O. and Pearce, R. B. (2003) Corn perspective and culture. In White, P. J. and Johnson, L. A. (Eds), *Corn: Chemistry and Technology*, 2nd Edition. Amer. Assoc. Cereal Chem. Inc.: St. Paul, Minnesota, USA, pp. 1–33.

Fenzi, M., Jarvis, D. L., Reyes, L. M. A., Moreno, L. L. and Tuxill, J. (2015) Longitudinal analysis of maize diversity in Yucatan, Mexico: influence of agro-ecological factors on landraces and modern variety introduction. *Plant Genet. Resour.* doi:10-1017/S1479262115000374.

Fischer, R. A., Byerlee, D. and Edmeades, G. (2014) *Crop Yields and Global Food Security*. ACIAR: Canberra, ACT.

Flint-Garcia, S. A. (2013) Genetics and consequences of crop domestication. *J. Agr. Food Chem.* 61:8267–76.

Flint-Garcia, S. A., Bodnar, A. L. and Scott, M. P. (2009) Wide variability in kernel composition, seed characteristics, and zein profiles among diverse maize inbreds, landraces, and teosinte. *Theor. Appl. Genet.* 119:1129–42. doi:10.1007/s00122-009-1115-1.

Foley, J. A., Ramankutty, N., Brauman, K. A., Cassidy, E. S., Gerber, J. S. and et al. (2011) Solutions for a cultivated planet. *Nature* 478:337–42.

Ford-Lloyd, B. V. and Jackson, M. T. (1986) *Plant Genetic Resources: An Introduction to their Conservation and Use*. Edward Arnold: London, U.K.

Fowler, C. 2006. *The New York Times*: Upfront.

Fowler, C., Smale, M., and Gaiji, S. (2000) Germplasm flows between developing countries and the CGIAR: an initial assessment. Paper prepared for the Global Forum on Agricultural Research. FAO and IPGRI: Rome, Italy.

Fowler, C., Smale, M. and Gaiji, S. (2001) Unequal exchange? Recent transfers of agricultural resources and their implications for developing countries. *Dev. Policy Rev.* 19:181–204. doi:10.1111/1467-7679.00130.

Frankel, O. H., Brown, A. H. D. and Burdon, J. J. (1995) *The Conservation of Plant Biodiversity*. Cambridge Univ. Press: Cambridge, UK.

Frey, T. J., Coors, J. G., Shaver, R. D., Lauer, J. G., Eilert, D. T. and Flannery, P. J. (2004) Selection for silage quality in the Wisconsin quality synthetic and related maize populations. *Crop Sci.* 44(4):1200–8.

Frison, C. and Halewood, M. (2006) Annotated bibliography addressing the international pedigrees and flows of plant genetic resources for food and agriculture, System-wide Genetic Resources Program. Bioversity International Maccarese: Rome. https://www.bioversityinternational.org/fileadmin/_migrated/uploads/tx_news/Annotated_bibliography_addressing_the_international_pedigrees_and_flows_of_plant_genetic_resources_for_food_and_agriculture_1087.pdf (accessed 5 February 2016).

Fu, H. and Dooner, H. K. (2002) Intraspecific violation of genetic colinearity and its implications in maize. *Proc. Natl. Acad. Sci. U. S. A.* 99:9573–8.

Fujisaka, S., Williams, D. and Halewood, M., Eds. (2011) The impact of climate change on countries' interdependence on genetic resources for food and agriculture. Background Paper No. 48. Food and Agriculture Organization of the United Nations (FAO): Rome, Italy. http://hdl.handle.net/10568/34971.

Gaffney, J., Schussler, J., Löffler, C., Cai, W., Paszkiewicz, S., Messina, C., Groeteke, J., Keaschall, J. and Cooper, M. (2015) Industry scale evaluation of maize hybrids selected for increased yield in drought stress conditions of the U.S. corn belt. *Crop Sci.* 55:1608–18. doi:10.2135/cropsci2014.09.0654.

GAO. (1997) United States General Accounting Office (GAO) (1997) *U. S. Department of Agriculture: Information on the Condition of the National Plant Germplasm System*. Report to Congressional Committees. DIANE publishing, p. 89.

García-Ruiz, A. (2016) Changes in genetic selection differentials and generation intervals in US Holstein dairy cattle as a result of genomic selection, *PNAS* 113(28):7690–2.

GCDT (2015) Genebank saved from Syrian civil war. Global Crop Diversity Trust, Bonn, Germany. https://www.croptrust.org/in-the-news/wired-genebank-saved-syrian-civil-war/ (accessed 2 October 2015).

Gepts, P. (2004) Who owns biodiversity, and how should the owners be compensated? *Plant Physiol.* 134:1295–307. doi:10.1104/pp.103.038885.

Gibbon, D., Dixon, J. and Flores, D. (2007) Beyond drought tolerant maize: study of additional priorities in maize. Report to Generation Challenge Program. Impacts, Targeting and Assessment Unit. Centro Internacional de Mejoramiento de Maíz y Trigo: Mexico D.F., p. 42.

Global Crop Diversity Trust. (2016) Excerpt from mission statement, https://www.croptrust.org/our-mission/crop-diversity-why-it-matters/.

Glover, M. A., Willmot, D. B., Darrah, L. L., Hibbard, B. E. and Zhu, X. (2005) Diallel analyses of agronomic traits using Chinese and US maize germplasm. *Crop Sci.* 45(3):1096–102.

GOBii, Genomic Open-source Breeding Informatics Initiative. Accessed July 2016. http://cbsugobii05.tc.cornell.edu/wordpress/.

Godfray, H. C. J., Beddington J. R., Crute I. R., Haddad, L., Lawrence D. and et al. (2010) Food security: The challenge of feeding 9 billion people. *Science* 327: 812–18.

Goodman, M. M. (2005) Broadening the US maize germplasm base. *Maydica* 50(3/4):203.

Goodman, M. M. (1978) A brief survey of the races of maize and current attempts to infer their racial relationships. In Walden, D. B. (Ed.) *Maize Breeding and Genetics*. Chapter 10, pp. 143–58. John Wiley & Sons: New York, USA, p. 794.

Goodman, M. M. (1999) Broadening the genetic diversity in maize breeding by use of exotic germplasm. In Coors, J. G. and Pandey, S. (Eds.). *The Genetics and Exploitation of Heterosis in Crops*, pp. 139–48. ASA-CSSA-SSSA: Madison, Wisconsin, USA, p. 524.

Goodman, M. M., Brown, W. L., Sprague, G. F. and Dudley, J. W. (1988) Races of maize. In *Corn and Corn Improvement. Third edition*. ASA, CSSA, SSSA: Madison, Wisconsin, USA, pp. 33–79.

Goodman, M. M., Moreno, J., Xastillo, F., Holley, R. N. and Carson, M. L. (2000) Using tropical maize germplasm for temperate breeding. *Maydica* 45:221–34.

Goodman, M. M. and Taba, S. (2007) *Global Strategy for the ex situ Conservation and Utilization of Maize Germplasm*. [http://repository.cimmyt.org/xmlui/handle/10883/17829].

Gore, M., Chia, J-M., Elshire, R. J., Sun, Q., Ersoz, E. S., Hurwitz, B. L., Peiffer, J. A., McMullen, M. D., Grills, G. S., Jeffrey Ross-Ibarra, J., Ware, D. H. and Buckler, E. S. (2009) A First-Generation Haplotype Map of Maize. *Science* 326:1115–17.

Gorjanc, G., Jenko, J., Hearne, S. J. and Hickey, J. M. (2016) Initiating maize pre-breeding programs using genomic selection to harness polygenic variation from landrace populations. *BMC Genomics* doi:10.1186/s12864-015-2345-g.

Gowda, M., Das, B., Makumbi, D., Babu, R., Semagn, K., Mahuku, G., Olsen, M. S., Bright, J. M., Beyene, Y. and Prasanna, B. M. (2015) Genome-wide association and genomic prediction of resistance to maize lethal necrosis disease in tropical maize germplasm. *Theor. Appl. Genet.* 128:1957–68.

Grassini, P., Eskridge, K. M. and Cassman, K. G. (2013) Distinguishing between yield advances and yield plateaus in historical crop production trends. *Nat. Commun.* doi:10.1038/ncomms3918.

Gurney, A. L., Grimanelli, D., Kanampiu, F., Hoisington, D., Scholes, J. D. and Press, M. C. (2003) Novel sources of resistance to *Striga hermonthica* in *Tripsacum dactyloides*, a wild relative of maize. *New Phytol.* 160:557–68.

Gustafson, T., Coors, J. and de Leon, N. (2010) Evaluation of S2-topcross selection for maize (Zea mays L.) silage yield and quality in the Wisconsin Quality Synthetic population. *Crop Sci.* 50:1795–804.

Hallauer, A. R. and Carena, M. J. (2014) Adaptation of tropical maize germplasm to temperate environments. *Euphytica* 196:1–11.

Hansey, C. N., Vaillancourt, B., Sekhon, R. S., de Leon, N., Kaeppler, S. M. and Buell, C. R. (2012) Maize (*Zea mays* L.) Genome Diversity as Revealed by RNA-Sequencing. *PLoS One*. 7(3):e33071. doi:10.1371/journal.pone.0033071.

Hard, R. J. and Roney, R. (1998) A massive terraced village complex in Chihuahua, Mexico, 3000 years before present. *Science* 279:1661–4.

Harlan, J. R., and de Wet, J. M. J. (1971) Towards a rational classification of cultivated plants. *Taxon* 20:509–17.

Hart, J. P. and Means, B. K. (2002) Maize and Villages: A summary and critical assessment of current northeast early late prehistoric evidence. In Hart, J. P. and Reith, C. B. (Eds), Chapter 18. *Northeast Subsistence-settlement Change: AD 700–1300*. University of the State of New York, The State Education Dept.: Albany, New York, USA, pp. 345–58.

Harrison, M. N. (1969) Major problems affecting productivity of maize. In Conference on Agricultural Research Priorities for Economic Development in Africa: Contributed papers: soil and water management: crop production and protection (Vol. 2, p. 99). National Academies.

Harrison, M. N. (1970) Maize improvement in East Africa. In Leakey, C. L. A. (Ed.) *Crop Improvement in East Africa*. Commonweath Agr. Bur.: Farnham Royal, Burks, U.K, pp. 21–59.

Hibbard, B. (2016) Personal communication.

Hilton, H. and Gaut, B. S. (1998) Speciation and domestication in maize and its wild relatives: Evidence from the *Globulin-1* gene. *Genetics* 150:863–72.

Hooker, A. L. (1981) Resistance to *Helminthosporium turcicum* from *Tripsacum floridanum* incorporated into corn. *Maize Genet. Coop. Newsl.* 55:87–8.

Horner, E. S. (1990) Registration of maize germplasms FS8A(S), FS8A(T), FS8B (S), and FS8B(T). *Crop Sci.* 30: 4: 964. doi:10.2135/cropsci1990.0011183X003000040057x.

Hufford, M., Bilinski, P., Pyhajarvi, T. and Ross-Ibarra, J. (2012) Teosinte as a model system for population and ecological genomics. *Trends Genet.* 28(12):606–15.

Hung, H.-Y., Shannon, L. M., Tian, F., Bradbury, P. J., Chen, C., Flint-Garcia, S. A., McMullen, M. D., Ware, D., Buckler, E. S., Doebley, J. F. and Holland, J. B. (2012) *ZmCCT* and the genetic basis of day-length adaptation underlying the post-domestication spread of maize. *Proc. Natl. Acad. Sci. U. S. A.* doi:10.1073/onas.1203189109.

Iltis, H. H. and Doebley, J. F. (1980) Taxonomy of *Zea* (Gramineae). II. Subspecific categories in the *Zea mays* complex and a generic synopsis. *Amer. J. Bot.* 67:994–1004.

Imperial College Crop Diversity at Risk (2002) *The Case for Sustaining Crop Collections*. Department of Agricultural Science, Imperial College: Wye, Ashford, Kent, England, p. 32.

Jamann, T. M., Sood, S., Wisser, R. J. and Holland, J. B. 2017. High-throughput resequencing of maize landraces at genomic regions associated with flowering time. *PloS One* 12(1):e0168910.

Jarvis, D. I., Padoch, C. and Cooper, H. D. (2007) *Managing Biodiversity in Agricultural Ecosystems*. Columbia University Press, p. 512. http://www.jstor.org/stable/10.7312/jarv13648.

Jiang, H., Campbell, M., Wu, Y., Du, S., Srichuwong, S. and Jane, J. (2015) Dosage effect of high-amylose modifier gene(s) on the starch structure of maize amylose-extender mutant. *J. Agr. Food Chem.* 63(2):433–9. doi: 10.1021/jf504099x.

Kermicle, J. L. and Evans, M. M. (2010) The Zea mays sexual compatibility gene ga2: naturally occurring alleles, their distribution, and role in reproductive isolation. *J. Heredity*. 101:737–49.

Khazaei, H., Street, K., Bari, A., Mackay, M. and Stoddard, F. L. (2013) The FIGS (Focused Identification of Germplasm Strategy) approach identifies traits related to drought adaptation in Vicia faba genetic resources. *PLoS One*, 8(5):e63107. https://doi.org/10.1371/journal.pone.0063107

Khoury, C. K., Achicanoy, H. A., Bjorkman, A. D., Navarro-Racines, C., Guarino, L., Flores-Palacios, X., Engels, J. M. M., Wiersema, J. H., Dempewolf, H., Sotelo, S., Ramírez-Villegas, J., Castañeda-Álvarez, N. P., Fowler, C., Jarvis, A., Rieseberg, L. H. and Struik. P. C. (2016) Origins of food crops connect countries worldwide. *Proc. Royal Soc. B.* London, UK. 283:-9. doi: 10.1098/rspb.2016.0792

Kim, S. K., Efron, Y., Khadr, F., Fajemisin, J. and Lee, M. H. (1987) Registration of 16 maize-streak resistant tropical maize parental inbred lines. *Crop Sci.* 27, 824–5.

Kloppenburg, J. and Kleinman, D. L. (1988) Preface: Plant genetic resources: The common bowl. In Kloppenburg, J. R. (Ed.), *Seeds and Sovereignty: The Use and Control of Plant Genetic Resources*. Duke Univ. Press: Durham, NC.

Korus, K., Lang, J. M., Adesemoye, A. O., Block, C. C., Pal, N., Leach, J. E. and Jackson-Ziems, T. A. (2017) First Report of *Xanthomonas vasicola* Causing Bacterial Leaf Streak on Corn in the United States. *Plant Dis.* doi: 10.1094/PDIS-10-16-1426-PDN

Kristensen, T. N., Hoffmann, A. A., Pertoldi, C. and Stronen, A. V. (2015) What can livestock breeders learn from conservation genetics and *vice versa? Front. Genet.* 6(38):1–12. http://doi.org/10.3389/fgene.2015.00038.

Kroeber, A. L. (1947) *Cultural and Natural Areas of Native North America*. Volume 38. Univ. California Press: Berkeley, California, USA, p. 242.

Kurtz, B., Gardner, C. A. C., Millard, M. J., Nickson, T. and Smith, J. S. C. (2016) Global access to maize germplasm provided by the US National Plant Germplasm System and by US plant breeders. *Crop Sci.* 56:931–41.

Labate, J., Lamkey, K. R., Mitchell, S. E., Kresovich, S., Sullivan, H. and Smith, J. S. (2003) Molecular and historical aspects of corn belt dent diversity. *Crop Sci.*, 43(1):80–91.

Lamola, L. M. and Bertram, R. B. (1994) Experts Gather in Mexico to Seek New Strategies in Preserving Agrobiodiversity. *Diversity* 10:15–17.

Lennon, J. R., Krakowsky, M., Goodman, M., Flint-Garcia, S. and Balint-Kurti, P. J. (2016) Identification of alleles conferring resistance to Gray Leaf Spot in maize derived from its wild progenitor species teosinte. *Crop Sci.* 56:209–18.

Li, Y., Ma, X., Wang, T., Li, Y., Liu, C., Liu, Z., Sun, B., Shi, Y., Song, Y., Carlone, M., Bubeck, D., Bhardwaj, H., Whitaker, D., Wilson, W., Jones, E., Wright, K., Sun, S., Niebur, W. and Smith, S. (2011) Increasing maize productivity in China by planting hybrids with germplasm that responds favorably to higher planting densities. *Crop Sci.* 51:2391–400.

Lim, S. M. and Hooker, A. L. (1971) Southern Corn Leaf Blight: Genetic control of pathogenicity and toxin production in Race T and Race O of *Cochliobolus heterostrophus. Genetics* 69(1):115–17.

Little, E. A. (2002) Kautantouwit's legacy: Calibrated dates on prehistoric maize in New England. *Am. Antiq.* 67:109–18.

Lobell, D. B., Cassman, K. G. and Field, C. B. (2009) Crop yield gaps: Their importance, magnitude and causes. *Annu. Rev. Environ. Resour.* 34:179–204. doi:10.1146/annurev.environ.041008.093740.

Lu, Y., Shah, T., Hao, Z., Taba, S., Zhang, S. and et al. (2011) Comparative SNP and haplotype analysis reveals a higher genetic diversity and rapider LD decay in tropical than temperate germplasm in maize. *PLoS One* 6(9): e24861. doi:10.1371/journal.pone.0024861.

Mahuku, G., Lockhart, B. E., Wanjala, B., Jones, M. W., Kimunye, J. N., Stewart, L. R., Cassone, B. J., Sevgan, S., Nyasani, J. O., Kusia, E. and Kumar, P. L. (2015) Maize lethal necrosis (MLN), an emerging threat to maize-based food security in sub-Saharan Africa. *Phytopathology*, 105(7):956–65.

Makarevitch I., Waters, A. J., West, P. T., Stitzer, M. and Hirsch, C. N. (2015) Transposable elements contribute to activation of maize genes in response to abiotic stress. *PLoS Genet.* 11(1):e1004915. doi:10.1371/journal.pgen.1004915.

Mangelsdorf, P. C. and Reeves, R. G. (1931) Hybridization of maize, Tripsacum, and Euchlaena. *J. Heredity*, 22(11):329–43.

Mano, Y. and Omori, F. (2013) Relationship between constitutive root aerenchyma formation and flooding tolerance in *Zea nicaraguensis*. *Plant Soil* 370: 447–60. doi:10.1007/s11104-013-1641-0.

Mano, Y., Omori, F., Tamaki, H., Mitsuhashi, S. and Takahashi, W. (2016) DNA marker-assisted selection approach for developing flooding-tolerant maize. *JARQ* 50:175–82.

Mano, Y., Omori, F., Takamizo, T., Kindiger, B., Bird, R. McK., Loaisiga, C. H. and Takahashi, H. (2007) QTL mapping of root aerenchyma formation in seedlings of a maize x rare teosinte 'Zea nicaraguensis' cross. *Plant Soil* 295:103–13. doi:10.1007/x/11104-007-0266-9.

Marston, A. H. (1931) Breeding European corn borer resistant corn. *J. Am. Soc. Agron.* 23:960–4.

Matsuoka, Y., Vigourouz, Y., Goodman, M. M., Sanchez, J. S., Buckler, E. and Doebley. J. (2002) Single domestication for maize shown by multilocus microsatellite genotyping. *Proc. Natl. Acad. Sci. U. S. A.* 99:6080–4.

McCann, J. (2005) *Maize and Grace*. Harvard University Press.

McClintock, B. (1956) Controlling Elements and the Gene. *Cold Spring Harb. Symp. Quant. Biol.* 21:197–216.

McClintock, B. (1984) The significance of responses of the genome to challenge. *Science* 226:792–801.

Meng, E. C. H., Taylor, J. E. and Brush, S. B. (1998) Implications for the conservation of wheat landraces in Turkey from a household model of varietal choice. In Smale, M. (Ed.), *Farmers Gene Banks and Crop Breeding: Economic an, Berlin, Germany*, pp. 127–42. Kluwer Academic Publishers, p. 270.

Mikel, M. A. (2011) Genetic Composition of Contemporary U.S. Commercial Dent Corn Germplasm. *Crop Sci.* 51:592–9.

Ming, R., Brewbaker, J. L., Pratt, R. C., Musket, T. A. and McMullen, M. D. (1997) Molecular mapping of a major gene conferring resistance to maize mosaic virus. *TAG Theor. Appl. Genet.* 95(1):271–5.

Mir, C., Zerjal, T., Combes, V., Dumas, F., Madur, D., Bedoya, C., Dreisigacker, S., Franco, J., Grudloyma, P., Hao, P. X. and Hearne, S. (2013) Out of America: tracing the genetic footprints of the global diffusion of maize. *Theor. Appl. Genet.* 126:2671–82.

Montenegro de Wit, M. (2016) Are we losing diversity? Navigating ecological, political, and epistemic dimensions of agrobiodiversity conservation. *Agr. Hum. Values* 33:625–40.

Morgante M., De Paoli E. and Radovic S. (2007) Transposable elements and the plant pan-genomes. *Curr. Opin. Plant Biol.* 10:149–55.

Nelson, P. T. (1990) Genetic and phenotypic characterization of maize germplasm resources: Ex-PVP inbreds, NCSU inbreds, and elite exotic inbreds. Ph.D. Dissertation. North Carolina State University: Raleigh, NC.

OECD/FAO (2012) OECD-FAO Agricultural Outlook 2012–2021, OECD Publishing and FAO. Available: http://dx.doi.org/10.1787/agr_outlook-2012-en.

Owens B. F., Lipka, A. E., Magallanes-Lundback, M., Tiede, T., Diepenbrock, C. H., Kandianis, C. B., Kim, E., Cepela, J., Mateos-Hernandez, M., Buell, C. R., Buckler, E. S., DellaPenna, D., Gore, M. A. and Rocheford, T. (2014) A foundation for provitamin A biofortification of maize: genome-wide association and genomic prediction models of carotenoid levels. *Genetics* 198: 1699–716.

Palacious, X. F. (1998) Contribution to the estimation of countries' interdependence in the area of plant genetic resources. Background Study Paper 7, Rev. 1. FAO Commission on genetic resources for food and agriculture. FAO: Rome, Italy.

Paliwal, R. L. (2000) Origin, evolution, and spread of maize. In Paliwal, R. L., Granados, G., Lafitte, H. R., Vlollc, A. D. (Eds), *Tropical Maaize: Production and Improvement*, pp. 5–11. FAO: Rome, p. 376.

Peiffer, J. A., Romay, M. C., Gore, M. A., Flint-Garcia, S. A., Zhang, Z., Millard, M. J., Gardner, C. A., McMullen, M. D., Holland, J. B., Bradbury, P. J. and Buckler, E. S. (2014) The genetic architecture of maize height. *Genetics*, 196(4), 1337–56.

Penny, L. H. and Dicke, F. F. (1956) Inheritance of resistance in corn to leaf feeding of the European Corn Borer1. *Agron. J.* 48:200–3. doi:10.2134/agronj1956.00021962004800050002x.

Perales, H. and Golicher, D. (2014) Mapping the diversity of races in Mexcio. *PLos One*. doi:10/1371/journal.pone.0114657.

Perales, H., Brush, S. B. and Qualset, C. O. (1998) Agronomic and economic competitiveness of maize landraces and *in situ* conservation in Mexico. In Smale, M. (Ed.), *Farmers Gene Banks and Crop Breeding: Economic Analyses of Diversity in Wheat, Maize and Rice*, pp. 127–42. Kluwer Academic Publishers: Berlin, Germany, p. 270.

Pingali, P. (2006) Westernization of Asian diets and the transformation of food systems: Implications for research and policy. *Food Policy* 32: 281–98.

Piperno, D. R., Ranere, A. J., Holst, I., Iriarted, J. and Dickau, R. (2009) Starch grain and phytolith evidence for early ninth millennium B. P. maize from the Central Balsas River Valley, Mexico. *Proc. Natl. Acad. Sci. U. S. A.* 106:5019–24.

Pollak, L. M. and Salhuana, W. (2001) The germplasm enhancement of maize (GEM) project: Private and public sector collaboration. In Cooper H. (Ed), *Broadening the Genetic Base of Crop Production*. CABI: Wallingford (United Kingdom), pp. 319–29.
Ponsart, C., Le Bourhis, D., Knijn, H., Fritz, S., Guyader-Joly, C., Otter, T., Lacaze, S., Charreaux, F., Schibler, L., Dupassieux, D. and Mullaart, E. (2014) Reproductive technologies and genomic selection in dairy cattle. *Reprod. Fertil. Dev.*, 26(1):12–21.
Prada, D. (2009) Molecular population genetics and agronomic alleles in seed banks: searching for a needle in a haystack? *J. Exp. Biol.* 60:2541–52.
Prischmann, D. A., Dashiell, K. E., Schneider, D. J. and Eubanks, M. W. (2009) Evaluating Tripsacum-introgressed maize germplasm after infestation with western corn rootworms (Coleoptera: Chrysomelidae). *J. Appl. Entomol.* 133:10–20. doi: 10.1111/j.1439-0418.2008.01311.x.
Ramirez, M., Ortiz, R., Taba, S., Sebastian, L., Peralta, E., Williams, D. E., Ebert, A. W. and Vezina, A. (2013) Demonstrating interdependence on plant genetic resources for food and agriculture. In Halewood, M., Noriega, I. L. and Louafi, S. (Eds), *Crop Genetic Resources as a Global Commons: Challenges in International Law* and *Governance*. Bioversity International: Routledge Abingdon, Oxfordshire, UK, pp. 39–61.
Ray, D. K., Gerber, J. S., MacDonald, G. K. and West, P. C. (2014) Climate variation explains a third of global crop yield variability. *Nat. Commun.* 6. doi:10.1038.ncomms 6989.
Ray, D. K., Mueller, N. D., West, P. C. and Foley, J. A. (2013) Yield trends are insufficient to double crop production by 2050. *PLoS One* 8:e66428.
Ray, D. K., Ramankutty, N., Mueller, N. D., West, P. C. and Foley. J. A. (2012) Recent patterns of crop yield growth and stagnation. *Nat. Commun.* 3:7. doi: 100.1038/comms 2296.
Rebourg, C., Chastanet, M., Gousenard, B., Welcker, C., Dubreuil, P. and Charcosset, A. (2003) Maize introduction into Europe: the history reviewed in the light of molecular data. *Theor. Appl. Genet.* 106:895–903.
Rich, P. J. and Ejeta, G. (2008) Towards effective resistance to *Striga* in African maize. *Plant Signal. Behav.* 3:618–21.
Romay, M. C., Millard, M. J., Glaubitz, J. C., Peiffer, J. A., Swarts, K. L., Casstevens, T. M., Elshire, R. J., Acharya, C. B., Mitchell, S. E., Flint-Garcia, S. A., McMullen, M. D., Holland, J. B., Buckler, E. S., Gardner, C. A. (2013) Comprehensive genotyping of the USA national maize inbred seed bank. *Genome Biol.* :R55. doi: 10.1186/gb-2013-14-6-r55.
Romero Navarro, J. A., Willcox, M., Burgueno, J., Romay, C., Swarts, K., Trachsel, S., Preciado, E., Terron, A., Delgado, H. V., Vidal, V., Ortega, A., Banda, A. E., Montiel, N. O. G., Ortiz-Monasterio, I., Vicente, F. S., Espinoza, A. G., Atlin, G., Wenzl, P., Hearne, S., Buckler, E. S.(2017) A study of allelic diversity underlying flowering-time adaptation in maize landraces. *Nat. Genet.* 49(3):476–80. http://dx.doi.org/10.1038/ng.3784.
Russell, W. A., Penny, L. H., Sprague, G. F., Guthrie, W. D. and Dicke, F. F. (1971) Registration of maize parental lines (Reg. Nos. PL 1 to 13). *Crop Sci.* 11(1):143.
Russell, W. A. (1972) Registration of B70 and B73 parental lines of maize (Reg. Nos PL16 and PL17). *Crop Sci.* 12(5):721. doi:10.2135/cropsci1972.0011183X001200050085x.
Sackville-Hamilton, N. R. (2013) Genebanks and food security. Abstracts Pre-breeding-fishing in the gene pool. European Plant Genetic Resources Conf. NordGen, SLI, Alnarp, Sweden.
Salhuana, W. (1984) *Information of the CIMMYT Accessions Grown by Pioneer Hi-Bred International, Inc. at Homestead, Florida*. Internal document, Pioneer International Hi-Bred Inc.: USA.
Salhuana, W. (1988) Seed increase and germplasm evaluation. In: Recent Advances in the conservation and Utilization of Genetic Resources. Proc. Global maize Germplasm Workshop. CIMMYT: Mexico, D. F., pp. 29–38.
Salhuana, W., Jones, Q. and Sevilla, R. (1991) The Latin American Maize Project. Model for rescue and use of irreplaceable germplasm. *Diversity* 7(182):40–2.
Semagn, K., Beyene, Y., Babu, R., Nair, S., Gowda, M., Das, B., Tarekegne, A., Mugo, S., Mahuku, G., Worku, M., Warburton, M. L., Olsen, M. and Prasanna, B. M. (2015) Quantitative Trait Loci Mapping and Molecular Breeding for Developing Stress Resilient Maize for Sub-Saharan Africa. *Crop Sci.* 55(4): 1449–59. doi:10.2135/cropsci2014.09.0646.

Smale, M., Bellon, M. R. and Pingali, P. L. (1998a) *Farmers Gene Banks and Crop Breeding: Economic Analyses of Diversity in Wheat, Maize and Rice*, Kluwer Academic Publishers: Berlin, Germany, p. 270.

Smale, M., Byerlee, D. and Jayne, T. (2013) Maize revolutions in sub-Saharan Africa. In Otsuka K. and Larson D.F. (Eds), *An African Green Revolution*. Springer: Netherlands, pp. 165–95.

Smale, M., Solieri, D., Cleveland, D. A., Louette, D., Rice, E. B. and Aguirre, A. (1998b) Collaborative plant breeding as an incentive for on-farm conservation of genetic resources: Economic issues from studies in Mexico. In Smale, M. (Ed.), *Farmers Gene Banks and Crop Breeding: Economic Analyses of Diversity in Wheat Maize and Rice*. Kluwer Academic Publishers, p. 239.

Smith, J. S. C., Desbons, P., Gogerty, J. and Niebur, W. S. (2006) Changes in parentage and genetic diversity of widely used maize hybrids grown in the northern United States and France from 1930 to the present. *Maydica* 51, 57–77.

Smith, J. S. C., Cooper, M., Gogerty, J., Loeffler, C., Borcherding, D. and Wright, K. (2014) Yield gains contributed by plant breeding in US maize 1930–2011. In Smith, J. S. C., Specht, J., Diers, B. and Carver, B. (Eds), *Yield Gains in Major US Field Crops*. Crop Science Society of America: Madison, WI, pp. 125–71.

Sood, S., Flint-Garcia, S., Willcox, M. C. and Holland, J. N. (2014) Mining natural variation for maize improvement: Selection on phenotypes and genes. Chapter 22. In Tuberosa, R., Graner, A. and Frison, E. (Eds), *Genomics of Plant Genetic Resources*. Springer Science and Business Media: Dordrecht, The Netherlands, pp 615–49. doi:10.1007/978-94-7572-5_25.

South American Maize market – Growth and Trends (2016) Mordor Intelligence.com. https://www.mordorintelligence.com/industry-reports/south-american-maize-market.

Springer, N. M., Ying, K., Fu, Y., Ji, T., Yeh, C. T., Jia, Y., Wu, W., Richmond, T., Kitzman, J., Rosenbaum, H. and Iniguez, A. L. (2009) Maize inbreds exhibit high levels of copy number variation (CNV) and presence/absence variation (PAV) in genome content. *PLoS Genet.*, 5(11):e1000734.

Strigens, A., Schipprack, W., Reif, J. C. and Melchinger, A. E. (2013) Unlocking the genetic diversity of maize landraces with doubled haploids opens new avenues for breeding. *PLoS One* 8:e57234. doi:10.1371/journal.pone.0057234.

Swanson-Wagner, R., Briskine, R., Schaefer, R., Hufford, M. B., Ross-Ibarra, J., Myers, C. L., Tiffin, P. and Springer, N. M. (2012) Reshaping of the maize transcriptome by domestication. *Proc. Natl. Acad. Sci. U. S. A.* 109: 11878–83.

Taba, S. (Ed.) (2005) Latin American Maize Germplasm Conservation: Regeneration, In situ Conservation, Core Subsets, and Prebreeding; Proceedings of a Workshop held at CIMMYT, 7–10 April 2003. CIMMYT: Mexico, D.F.

Tenaillon, M. I., Sawkins, M. C., Long, A. D., Gaut, R. L., Doebley, J. F. and Gaut. B. S. (2001) Patterns of DNA sequence polymorphisms along chromosome 1 of maize (*Zea mays ssp mays* L.). *Proc. Natl. Acad. Sci. U. S. A.* 98:9161–6.

Teixeira, J. E. C., Weldekidan, T., de Leon, N., Flint-Garcia, S., Holland, J. B., Lauter, N., Murray, S. C., Xu, W., Hessel, D. A., Kleintop, A. E. and Hawk, J. A. (2015) Hallauer's Tusón: a decade of selection for tropical-to-temperate phenological adaptation in maize. *Heredity*, 114(2), 229–40.

Tilman, D., Balzer, C., Hill, J. and Befort, B. L. (2011) Global food demand and the sustainable intensification of agriculture. *Proc. Natl. Acad. Sci. U. S. A.* 108: 20260–4.

Troyer, A. F. (1999) Background of US hybrid corn. *Crop Sci.* 39(3):601–26.

USDA-ARS (2005) Races of maize publications (collection). https://www.ars.usda.gov/midwest-area/ames/plant-introduction-research/docs/races-of-maize/

USDA-FAS (2017) World Agricultural Production Circular Series, Table 04, Corn Area, Yield, and Production. USDA Foreign Agricultural Service, Office of Global Analysis, April 2017. 28pp. https://apps.fas.usda.gov/psdonline/circulars/production.pdf.

Van Treuren, R. and van Hintum, T. (2013) Genebanking in the sequence era. Abstracts Pre-breeding-fishing in the gene pool. European Plant Genetic Resources Conf. NordGen, SLI, Alnarp, Sweden.

Van Heewaarden, J., Hellin, J., Visser, R. F. and van Eeuwijk, F. A. (2009) Estimating maize genetic erosion in modernized smallholder agriculture. *Theor. Appl. Genet.* 119:875–88.

van Heerwaarden, J., Doebley, J., Briggs, W. H., Glaubitz, J. C., Goodman, M. M., de Jesus Sanchez Gonzalez, J. and Ross-Ibarra, J. (2011) Genetic signals of origin, spread, and introgression in a large sample of maize landraces. *Proc. Natl. Acad. Sci. U. S. A.* 108:1088–92.

Veteläinen, M., Negri, V. and Maxted, N. (2009) European Landraces On-farm Conservation, Management and Use. Bioversity Technical Bulletin No. 15. Bioversity International: Rome, p. 344.

Wallace, H. A. and Brown, W. L., 1956. Corn and its Early Fathers. Iowa State University Press, Ames, IA, p. 141.

Wallace, J., Zhang, X., Beyene, Y., Semagn, K., Olsen, M., Prasanna, B. M. and Buckler, E. S. (2016) Genome-wide association for plant height and flowering time across 15 tropical maize populations under managed drought stress and well-watered conditions in Sub-Saharan Africa. *Crop Sci.* 56:2365–78. doi: 10.2135/cropsci2015.10.0632.

Wang, F. (2013) Flows of germplasm resources into and out of China. In Halewood, M., Noriega, I. L. and Louafi, S. (Eds), *Crop Genetic Resources as a Global Commons: Challenges in International Law and Governance*. Chapter 3. Bioversity International: Routledge Abingdon, Oxfordshire, pp. 62–77.

Wang, L., Yang, A., He, C., Qu, M. and Zhang, J. (2008) Creation of new maize germplasm using alien introgression from Zea mays ssp. Mexicana. *Euphytica* 164:798–801.

Wangai, A. W., Redinbaugh, M. G., Kinyua, Z. M., Miano, D. W., Leley, P. K., Kasina, M., Mahuku, G., Scheets, K. and Jeffers, D. (2012) First report of maize chlorotic mottle virus and Maize Lethal Necrosis in Kenya. *Plant Dis.* 96(10):1582.

Warburton, M. L., Reif, J. C. Frisch, M., Bohn, M., Bedoya, C., Xia, X. C., Crossa, J., Franco, J., Hoisington, D., Pixley, K., Taba, S. and Melchinger, A. E. (2008) Genetic diversity in CIMMYT nontemperate maize germplasm: Landraces, open pollinated varieties, and inbred lines. *Crop Sci.*, 48:617–24.

Wen, W., Guo, T., Tovar, V. H. C., Li, H., Yan J. and Taba, S. (2102) The strategy and potential utilization of temperate germplasm for tropical germplasm improvement: a case study of maize (Zea mays L.). *Mol. Breed.* 29(4):951–62. doi:10.1007/s11032-011-9696-1.

Westengen, O. T., Ring, K. H., Berg, P. R. and Brysing, A. K. (2014) Modern maize varieties going local in the semi-arid zone in Tanzania. *Evol. Biol.* 14:1. doi:10.1186/1471-2148-14-1.

Westengen, O. T., Berg, P. R., Kent, M. P. and Brysting, A. K. (2012) Spatial structure and climatic adaptation in African maize revealed by surveying SNP diversity in relation to global breeding and landrace panels. *PLos One* 7(10):e47832. doi:10.1371/journal.pone.0047832.

White, S. E. and Doebley, J. F. (1999) The molecular evolution of terminal ear1, a regulatory gene in the genus Zea. *Genetics*, 153(3), 1455–62.

Wood, D. and Lenne, J. M. (1997) The conservation of agrobiodiversity on-farm: questioning the emerging paradigm. *Biodiver. Conserv.* 1:109–29.

World of Corn (2016) National Corn Growers Association, http://www.worldofcorn.com/pdf/_WOC_2017.pdf

World of Corn (2017) National Corn Growers Association, http://www.worldofcorn.com/pdf/_WOC_2017.pdf

Wu, Y., Huang, R., Auger, D., Campbell, M. and Yen, Y. (2015) A High-amylose modifier locus revealed by analysis of a homozygous mapping population. *Crop Sci.* 55(3):1163–9.

Yang, X., Gao, S., Xu, S., Zhang, Z., Prasanna, B. M., Li, L., Li, J. and Yan, J. (2011) Characterization of a global germplasm collection and its potential utilization for analysis of complex quantitative traits in maize. *Mol. Breed.* 28:511–26.

Yu, X., Li, X., Guo, T., Zhu, C., Wu, Y., Mitchell, S. E., Roozeboom, K. L., Wang, D., Wang, M. L., Pederson, G. A. and Tesso, T. T., Schnable, P. S., Bernardo, R. and Yu, J. (2016) Genomic prediction contributing to a promising global strategy to turbocharge gene banks. *Nature Plants*, 2:16150. doi:10.1038/nplants.2016.150. http://www.nature.com/articles/nplants2016150#supplementary-information ER -.

Zannou, A., Ahanchede, A., Struik, P., Richards, P., Zoundjihekpon, J., Tossou, R. and Vodouhe, S. (2004) Yam and cowpea diversity management by farmers in the Guinea-Sudan transition zone of Benin. *Neth. J. Agric. Sci.* 52:393–420.

Zhang, X., Zhang, H., Li, L., Lan, H., Ren, Z., Liu, D., Wu, L., Liu, H., Jaqueth, J., Li, B. and Pan, G. (2016) Characterizing the population structure and genetic diversity of maize breeding germplasm in Southwest China using genome-wide SNP markers. *BMC genomics*, 17(1):697. doi:10.1186/s12864-016-3041-3.

Zuber, M. S. (1973) Registration of 20 maize parental lines (Reg. Numbers PL 18 to 37). *Crop Sci.* 13(6):779–80. doi:10.2135/cropsci1973.0011183X001300060085x.

Chapter 2

Key challenges in maize breeding in sub-Saharan Africa

Baffour Badu-Apraku, IITA, Nigeria; M.A.B. Fakorede and R.O. Akinwale, Obafemi Awolowo University, Ile-Ife, Nigeria

1 Introduction

Maize breeding in sub-Saharan Africa (SSA) has improved tremendously in the last two or three decades in terms of development, release and use of improved varieties. Through the efforts of international centres such as the International Institute of Tropical Agriculture (IITA) and International Center for Maize and Wheat Improvement (CIMMYT), and the strong partnership with and among the scientists working in National Agricultural Research Systems (NARS), farmers have access to improved maize varieties which are being used to address critical needs for food security, human nutrition and poverty alleviation. Despite the impressive growth and progress made in maize research, particularly in the NARS, many breeding programmes are still facing myriads of challenges which limit efficiency in their operations. These include lack of research facilities such as well-equipped laboratories and efficient irrigation systems, lack of well-trained personnel and funds to execute high-throughput techniques such as marker-assisted selection (MAS) technology.

In this chapter, we present some of the major challenges to maize breeding activities in SSA, with emphasis on the Western, Central, Eastern and Southern Africa sub-regions. The areas considered include:

http://dx.doi.org/10.19103/AS.2016.0001.03

- Research facilities;
- Human resources;
- Use of broad germplasm resources;
- Challenges in hybrid development;
- New plant breeding techniques;
- Participatory plant breeding research.

2 Research facilities

The prime objective of maize improvement is to identify, in the heterogeneous germplasm, individuals in which favourable alleles are present at the highest proportion of loci involved in the expression of key traits (Goodman et al., 1987). Plant breeders work to improve the performance and quality of existing crop varieties as well as creating new ones. In addition, breeders need to preserve the elite germplasm that has been developed, and advocate its fair and effective use by farmers, as well as by publicly and privately employed maize breeders. Breeders work with germplasms having beneficial traits such as grain yield, tolerance or resistance to biotic (such as diseases, insect pests, weeds, especially obnoxious and parasitic weeds) and abiotic (such as drought, low soil fertility – especially N and heat) stresses, maturity rating and several others. To positively respond to the needs of maize farmers and consumers, breeders need access to key facilities. Public sector breeders working in NARS must have adequate space for their offices, laboratories and greenhouses/screen houses at the research station. Facilities for cultivating, drying and processing seed are necessary. Extensive farmlands at the station for breeding nurseries, seed production in well-isolated pieces of land and plots for conducting evaluation trials/experiments are indispensable in maize breeding work. In addition, breeders need outlying research farms, managed by the breeders, and facilities to make frequent interaction with research collaborators and farmers who volunteer to carry out on-farm trials, such as the baby component of mother–baby trial (MBT), which is a technology transfer concept involving active interaction of researchers and farmers on evaluation of prospective technologies for transfer to the participating as well as neighbouring farmers. This concept is presented in greater detail later in this chapter. Pollination materials, including shoot and tassel bags, along with waterproof tags, indelible markers, paper clips and stapler with the staple pins are indispensable materials for maize pollination. The shoot and tassel bags are made in rather unique designs from some special waterproof paper to make them tolerant to the weather conditions in the field, especially rainfall. Maize breeders must order the bags from the United States; the bags are expensive, funds are often limiting and delivery takes a long time. Breeders in WCA, therefore, improvise with locally available materials such as nylon or plastic bags for covering shoots and brown paper tassel bags. In most cases, the improvised materials are inferior to the standard bags made in the United States and contamination may frequently occur during pollination, resulting in seed mixtures during harvesting, which is very much undesirable during seed production, especially breeder seeds.

An important facility that is missing in nearly all NARIs of WCA is small-scale seed processing unit, along with standard cold room for seed storage at the right environmental conditions. In some cases, seeds are stored under ambient room conditions, or in air-conditioned rooms, refrigerators, the freezer compartment of refrigerator, and deep

Table 1 Outputs from linear regression of maize seedling root length on duration of seed storage under five storage conditions (Ajayi and Fakorede, 2001)

Storage condition	Regression equation	Coefficient of determination (r^2)
Ambient room conditions	$\hat{Y}$ = 11.21–0.28x	61.82***
Air-conditioned room	$\hat{Y}$ = 12.37–0.29x	54.31**
Refrigerators	$\hat{Y}$ = 11.69–0.28x	31.50*
Freezer of refrigerator	$\hat{Y}$ = 11.68–0.30x	60.79***
Deep freezer	$\hat{Y}$ = 10.91–0.20x	36.37*

*, ** and *** r^2 significantly different from zero at 0.05, 0.01 and 0.001 levels of probability.

freezers, all of which have high relative humidity, unless dehumidifiers are installed in them. In a study conducted at the Seed Science Laboratory of Obafemi Awolowo University, Ile-Ife, Nigeria, maize was stored under the five conditions and germination and root development were monitored biweekly for six months, beginning at the time of storage (Ajayi and Fakorede, 2001). For all storage conditions, the value of most traits decreased as duration of storage increased. For example, root length of resulting seedlings decreased by 0.2 to 0.3 cm per day as storage duration increased under the five storage conditions (Table 1). However, the rate of decrease was lower for seeds stored in the deep freezer than the other storage environments. In addition, Ajayi and Fakorede (2001) found that planting seeds immediately after they were removed from cold storage (deep freezer and refrigerator in this case) reduced germination and increased abnormal seedlings. Equilibrating the seeds with ambient conditions (23–28°C) up to 14 days before planting maximized germination and minimized the occurrence of abnormal seedlings (Fig. 1). The researchers concluded that the effect of the differences between storage and ambient conditions on seeds should be minimized before the seeds are planted to ensure near-perfect emergence in the field. Although Ajayi and Fakorede's study did not include storage in a cold room, which would have served as the standard for purposes of comparison, their results suggest that seeds should be removed from cold storage at least a few days before planting in the field.

Recurrent drought, particularly at the beginning of the growing season, unpredictable dry spells at any growth stage of maize during the growing season and extended terminal drought have come to stay in WCA. Apart from recurrent drought since the 1950s, climatic change has manifested in WA in several other ways, including a consistent decrease in total annual rainfall over time, delayed onset and early recession of the rains resulting in reduced duration of the cropping season, false start of the rainy season, increased drought probability during the cropping season and increased annual maximum temperature (Jagtap 1995; Fakorede and Akinyemiju, 2003).

In the rainforest agro-ecology of Nigeria, for example, Fakorede and Akinyemiju (2003) found that total rainfall for March, April and May at Ile-Ife consistently decreased from about 500 mm in 1981 to <100 mm in 2000 at the rate of about 22 mm per year, with a coefficient of determination of 76% (Fig. 2).

In another study, Fakorede et al. (2004) quantified drought per day (termed drought day, DD) and used the value to compute drought probability of each day for the 1975–2000 climatic data of the T&R Farm at Ile-Ife, a typical rainforest location. It was hypothesized

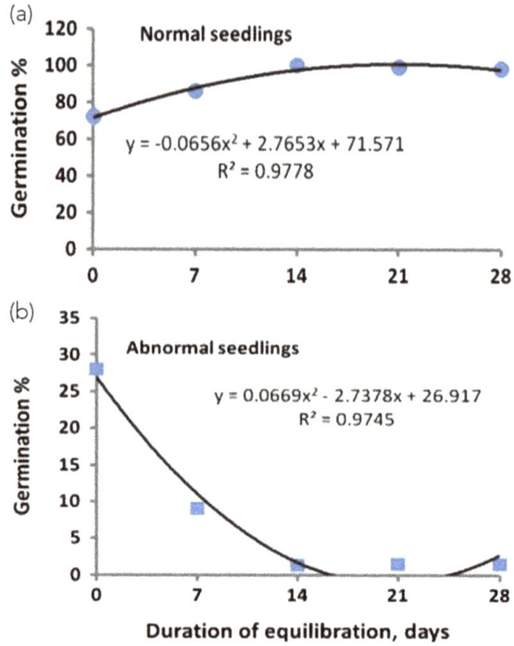

Figure 1 Germination of maize seeds as affected by duration of equilibration of the seeds with ambient field conditions before planting.

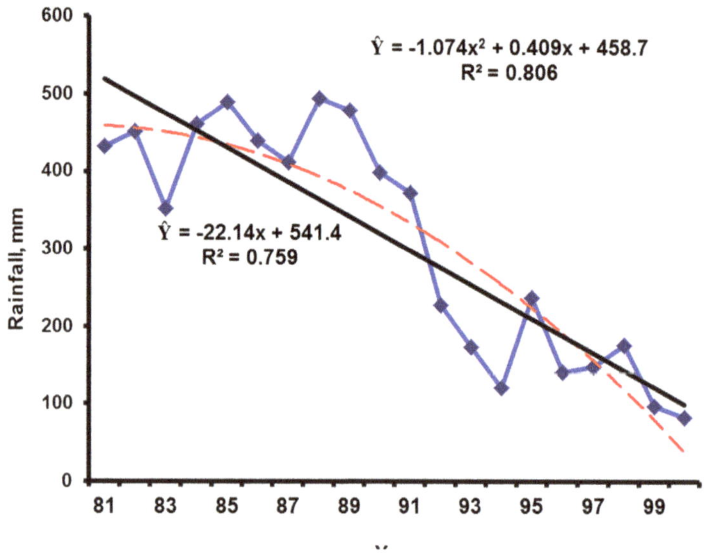

Figure 2 Total annual rainfall for March, April and May at the Teaching and Research Farm of Obafemi Awolowo University, Ile-Ife, Nigeria, 1981–2000.

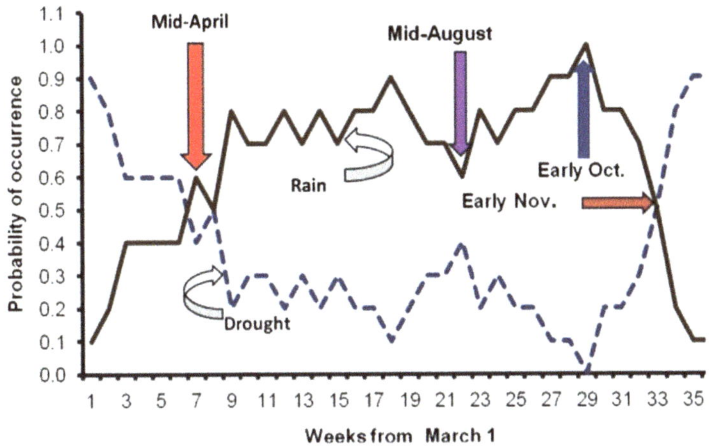

Figure 3 Probability of drought and rain from the first week in March until the end of November at the Teaching and Research Farm of Obafemi Awolowo University, Ile-Ife, Nigeria, 1975–2000.

that, for a 24-hr period, a DD has occurred if potential evaporation (PE) is greater than the total rainfall. Drought probability, *P*, for a period, *t*, was then calculated as:

$$P = (\sum DD)/t$$

Rainfall data were obtained with a manual rain gauge and PE data from a shielded Piché evaporimeter in a Stephenson's screen of standard height. The results showed that *P* was very high in March (from wk 1), the conventional beginning of the cropping season; decreased to about 0.2–0.3 during the first week in May (wk 9); and remained at about that level until the beginning of October (wk 29), except for a slight upsurge during the second week in August (wk 22), which corresponds to the peak of the August dry spell (Fig. 3).

From the viewpoint of moisture availability, therefore, the effective maize planting season starts about the middle of April and the late season crop must complete flowering about the middle of October to avoid the chances of terminal drought at this location. In effect, the duration of the season has now reduced to 6–7 months from the long-established 8–9 months, a definite negative impact of the climate change. This is the advantage of developing early/extra-early and or drought-tolerant (DT) varieties that can escape or withstand the unpredictable terminal drought characteristic of all agro-climatic zones of WA, in addition to sporadic dry spells that may occur at any growth stage of the crop. The amount of water required at different growth stages of maize has been investigated for early maturing maize in a preliminary study (Fakorede, 2015, unpublished; Fig. 4). Total moisture required for the crop from planting to maturity was 530 mm. The greatest demand for water was from about 5 days before tasseling, reaching a climax at tassel emergence and remaining high during anthesis, silking and the early part of grain filling (the lag phase) – a total of about 25 days. Nearly 40% of total moisture required for the growth cycle of maize must be available at this time. Moisture stress at this period has been found to cause greater reduction in grain yield than at any other growth stage. For example, results of studies conducted in the US Corn Belt (Shaw, 1977) and in eastern

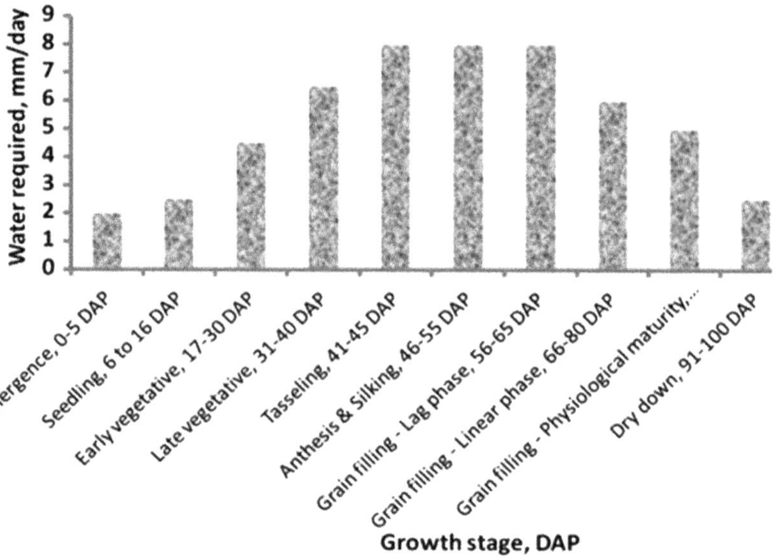

Figure 4 Moisture requirements of an early maturing maize variety at different growth stages.

Africa led to the conclusion that stress from about two weeks before to two weeks after flowering most drastically reduced maize yield. Moisture stress during the grain-filling period, to a lesser extent, also reduced grain yield. Effect of moisture stress at the seedling and early vegetative growth stages was relatively small.

In a 12-year study conducted at the Teaching and Research Farm of Obafemi Awolowo University, Ile-Ife, Nigeria (Fakorede and Akinyemiju, 2003), total rainfall, number of rain days and temperature at different stages of maize growth and development had no significant correlation with grain yield (Table 2). Pre-season weather did not affect the performance of maize during the early season. Also, total rainfall, number of rain days and maximum temperature at the different stages and the whole growth cycle did not influence maize yield that season. However, the higher the number of rain days during the early vegetative growth stage, the lower the grain yield. Conversely, the higher the probability of drought during this growth stage, the higher the yield. Similarly, *r*-values between drought probability and yield were negative for the late vegetative to flowering and grain-filling stages as well as the whole growth cycle. Thus, some level of drought during the early vegetative stage was favourable to maize yield whereas drought during the flowering and grain-filling stages was detrimental. The *r*-values between grain yield and drought probability at flowering, grain filling and for the whole cycle during the late (second) season were similar in sign but larger in magnitude than the early season. In other words, the impact of drought was more severe in the late than the early season. Similarly, maximum temperature effects on yield were higher.

Results of these studies are consistent with those reported by early works on this topic (Shaw, 1977; Edmeades et al., 1995 *inter alia*). The implication of the results is that, in breeding for drought tolerance, moisture stress should be imposed during the flowering and grain-filling stages for optimum impact. Maize breeders have, therefore, been screening for drought tolerance during the dry season under carefully monitored irrigation and imposed water

Table 2 Correlation coefficients between climatic variables at different growth stages with grain yield of maize during the early and late (second) seasons at the Teaching and Research Farm of Obafemi Awolowo University, Ile-Ife, Nigeria

			Early vegetative	Late veg to flowering	Grain filling	Whole cycle
			----------days after planting (DAP)----------			
Climatic variable	Pre-season	Planting to emergence	1–30	31–60	61–100	1–100
Early season						
TR mm	0.06	− 0.21	− 0.42	0.21	0.03	0.02
No. rain days	0.05	0.00	− 0.63*	− 0.32	− 0.22	− 0.37
Drought prob.	0.10	0.62*	0.86**	− 0.88**	− 0.55*	− 0.59*
Max temp. °C	0.38	0.15	− 0.17	− 0.10	0.16	− 0.09
Late season						
TR mm	–	0.06	0.31	0.66*	0.67*	0.58*
No. rain days	–	0.12	0.07	− 0.21	− 0.15	0.43
Drought prob.	–	0.24	0.24	− 0.95**	− 0.76**	− 0.66*
Max temp. °C	–	0.31	0.45	− 0.56*	− 0.63*	− 0.43

*, ** Significantly different from zero at 0.05 and 0.01 level of probability, respectively.

stress as from 28 days after planting until harvest maturity. But this effort has been primarily at the international research centres in SSA. Most NARS research stations do not have the necessary weather stations and irrigation systems to conduct fine research in this area that would have been used as the basis for taking advantage of the genotype × environment interaction to develop high-yielding varieties and boost maize production in the region.

The few studies conducted on the relationship between weather and crop growth and productivity clearly point to an urgent need to conduct more extensive studies on this topic. The results were collected from studies conducted with outmoded facilities. At the OAU Teaching and Research Farm, for example, the weather data were collected from instruments such as thermometers for maximum and minimum temperatures, and Piché evaporimeter for evaporation, placed in Stephenson's screen erected since the early 1960s without replacement with more recent models. Rainfall was determined with a manual rain gauge, the US Class A open pan evaporimeter was also used to determine total evaporation and the sunshine hour was determined with Stokes sunshine recorder. More modern and precision instruments such as automatic weather stations (AWS) are now readily available in international markets, but most national maize improvement programmes in SSA have no access to them because of funding limitations. The AWS is an automated version of the traditional weather station, either to save human labour or to enable measurements from remote areas. The system records data as close as possible to real time and saves the data for later recovery. Most AWS have thermometers for measuring temperature, anemometer for measuring wind speed, hygrometer for measuring humidity and barometer for measuring pressure. Some other types of AWS may have in addition ceilometer for measuring cloud height, rain gauge for measuring rainfall, pyranometer or other types of sensors for monitoring incident solar radiation, present weather sensor and/or visibility sensor. Initially, AWS were often placed where electricity and communication lines were available. Nowadays, solar panel, wind turbine and mobile phone technology have made it possible to have wireless AWS that are not connected to the electrical grid or telecommunications network. This makes it possible to install AWS anywhere, including remote areas. Availability of AWS will greatly facilitate a quick and better understanding of crop physiology, an indispensable discipline to maize improvement, but is presently poorly understood.

Perhaps worse than all other parts of the world, power outage is very common and unpredictable in WCA countries, especially in Nigeria. Power outage often causes rapid loss of seed viability. This has led to repeated loss of germplasm, thereby forcing the breeders to restart their research – a very frustrating experience. The solution to the problem is far from the reach of research scientists, so policymakers must be involved. However, if funds are not limiting, power generators could be purchased to alleviate the problem to a reasonable extent.

3 Human resources

There is an acute shortage of well-trained scientists and technicians to effectively carry out research activities in WCA. A study conducted in 2006 by the Forum for Agricultural Research in Africa (FARA) on the assessment of the requirements for efficient, effective and productive NARS in Africa showed that 62% of the national agricultural research institutes (NARIs) had less than 100 scientists. At that time, only Kenya, South Africa and Nigeria had NARIs with more than 500 scientists. The situation is worse when it comes to well-trained breeders and technicians in the various maize breeding programmes in SSA. To date (ten years after the FARA study), there are still some African countries that have no maize breeder at all.

Two different human resources are needed for managing the technical aspects of plant breeding: research scientists and research technicians. Capacity development of the human resources can also be at several levels, including formal education with specific fields of specialization, on-the-job learning or skill acquisition, usually involving superior officers leading the junior ones and giving them opportunities for hands-on experience, personalized learning nowadays involving online virtual classes using social media and group courses of short (a few days) or medium (several months) duration.

3.1 Research technicians

These are indispensable support personnel in all aspects of agricultural research. The academic qualification for these individuals in WCA is usually a high school certificate with only a few additional years of education at college or polytechnic post high (secondary) school, such as a two-year programme leading to the award of National Diploma Certificate. This level of education is general, with little or no specialization in specific fields or aspects of agricultural research. Such technicians are exposed only to the basic sciences and must necessarily learn on the job to be effective in handling all aspects of maize research, from seed preparation for planting, through management of field trials, pollination, data collection, harvesting, post-harvest operations and data analysis.

Research technicians in maize breeding have benefited a great deal from on-the-job skill acquisition programmes and short-term courses. In many cases, IITA and, to some extent CIMMYT, along with international donor agencies have sponsored such training courses. For example, one objective of the West and Central Africa Collaborative Maize Research Network (WECAMAN) was to enhance the research capacity and capability of the participating NARS in the sub-region. One of the ways WECAMAN used to pursue this objective was through training courses. One of the training courses was for research technicians in maize breeding. With financial support from the US Agency for International Development and later, the United Nations Development Program and International Fund for Agricultural Development, the network has conducted training activities for more than 20 years in 11 member countries. The Technician Training Course was initiated in 1988 at Ouagadougou, Burkina Faso, to 'mitigate the problem of lack of skilled research support personnel in the sub-region' (Badu-Apraku et al., 2004). The course was designed to upgrade the skills of national maize programme technicians in the management of field experiments. The Technician Training Course was 5 months long, usually from July to November (the cropping season), and it allowed participants to observe and experience the latest developments in maize production research throughout a full growing season. The Technician Training Course was practical-oriented but was fully backed up with theory to improve the trainees' understanding of maize breeding and production strategies and techniques. The broad topics covered in the course included identification, layout and implementation of the various experimental designs; data collection and analysis; seed production; and report writing. Apart from the lectures and practical classes, each trainee was assigned a project to manage from planting to harvesting. The trainee collected and analysed the data, summarized the results, wrote a report and submitted bound copies of the report to the network coordinator, who was the principal resource person. Initially, we expected each member country of the network to send one trainee to the Technician Training Course each time the course was offered, but for one reason or another that did

Table 3 Year, number of participants and countries represented at the WECAMAN Technician Training Course, 1988–2000

Year	No. of participants	Countries represented
1988	6	Central African Republic, Guinea–Conakry, Chad, Benin, Burkina Faso, Mali
1989	3	Ghana, Guinea–Bissau, Chad
1990	6	Benin, Burkina Faso, Cameroon, Ghana, Mali, Nigeria
1994	8	Ghana, Nigeria, Cameroon, Benin, Togo, Mali, Côte d'Ivoire, Burkina Faso
1996	9	Burkina Faso, Nigeria, Mali, Togo, Ghana, Côte d'Ivoire, Cameroon, Benin, Kenya
1998	7	Ghana, Togo, Cameroon, Mali, Chad, Senegal, Guinea
1999	7	Benin, Nigeria, Senegal, Chad, Burkina Faso, Guinea, Côte d'Ivoire
2000	4	Ghana, Nigeria, Burkina Faso, Côte d'Ivoire
Total	**50**	

Table 4 Number of participants at the DTMA training course on variety testing and seed production of open-pollinated and hybrid varieties, 2009–2015

Year	2009	2010	2011	2012	2013	2014	2015	Total
Number of participants	30	26	33	27	24	64	17	221

not happen. The impact of the training was obvious from the improved quality of data from regional trials and the capacity of technicians to maintain research activities effectively during the absence of national scientists. Fifty national technicians had participated in the Technician Training Course (Table 3) before the termination of WECAMAN in 2006 (Badu-Apraku et al., 2012).

After the termination of WECAMAN in 2006, training of maize breeding research and seed production technicians has continued with financial support from some other donors. The Drought Tolerant Maize for Africa (DTMA) project, with financial support from the Bill & Melinda Gates Foundation (BMGF), has been conducting 1 to 2 week training courses for maize breeding and seed production in anglophone (Ghana and Nigeria) and francophone (Benin and Mali) countries since 2006. Annually since 2009, the DTMA project has conducted a training course for seed company staff members on variety testing and seed production of open-pollinated and hybrid varieties and, as of 2015, a total of 221 seed technicians, producers and marketers have been trained (Table 4).

In addition, a 3-week training course for plant breeding technicians from Ghana and Nigeria was sponsored by the Alliance for Green Revolution in Africa, and was conducted in late July to mid-August 2015 at IITA. The second phase of the training was held in late July to mid-August 2016 at ICRISAT-Bamako, Mali, primarily for Burkina Faso, Mali and Niger, but with participants also from Cameroon, Chad and Senegal. In each case, the training covered maize and six other crops, and several new training innovations were incorporated in the course.

3.2 Research scientists

Agricultural scientists now readily design complex experiments and collect data on many traits, in the hope that data analysis can be handled relatively easily since user-friendly computer software packages are now readily available. In addition, new areas of research and improved research methods are continuously being developed by researchers, thereby creating a dynamic field of study for agricultural scientists and their technicians to master. Mastery of these innovations will make it possible for African scientists to produce meaningful, thought-provoking research, using the most appropriate and accurate methodology.

Agricultural research and development in Africa has continued to be dynamic and innovative, especially in recent times. Seed companies are on the increase and this has necessitated the establishment of seed laws within and across countries and within and across sub-regions to facilitate free movement of improved germplasm across borders. Varietal release mechanisms are being legalized in the WA countries and this has necessitated the development of stringent rules for submission of candidate varieties for release. In Nigeria, for example, there is a National Varietal Release and Registration Committee (NAVRRC) served by a Technical sub-Committee (TSB) that evaluates all applications and candidate varieties for release. To prevent similar varieties from being proposed for release by different breeders or institutions, TSB has stipulated compliance with the rules of *value for cultivation and use* (VCU) and *distinctness, uniformity and stability* (DUS) for varieties to be released. The scientists need the faithful assistance of technicians to meet the target for the acceptance of their candidate varieties for release.

Training at individual (e.g. MSc and PhD fellowships) and group (e.g. short courses involving a fixed number of participants) levels has been an integral part of the training programmes of IITA and CIMMYT since their inception. Training at individual level usually includes undergraduate internships, MSc and PhD fellowships, while group training includes short courses involving a fixed number of participants. Most trainings, however, are based on individual crops or themes such as farming systems, or a specific stress factor in a crop, such as *Striga* control in maize. Trainings geared towards degree programmes are usually handled in collaboration with universities in the sub-region. Recently, the West African Center for Crop Improvement (WACCI) was established at the University of Ghana, Accra, with a similar programme at the Kwame Nkrumah University of Science and Technology (KNUST), Kumasi, Ghana, to serve as centres for training plant breeders and seed systems specialists at masters and PhD levels. The Pan Africa University (PAU) also established a centre at the University of Ibadan, Nigeria. The three programmes at WACCI, KNUST and PAU very actively involve the plant breeders in the sub-region in the handling of the necessary courses, supervision of thesis research and in conducting the thesis defence of the student. Similarly, a PhD programme has also been established at southern Africa to serve as a centre for training plant breeders in eastern and southern Africa. Another training intervention available to plant breeders in the sub-region is the University of California – Davis Plant Breeding Academy. The Seed Biotechnology Center at the University of California, Davis, organized a professional development course to teach the principles of plant breeding to seed industry personnel. This two-year course addresses the reduced numbers of plant breeders being trained in academic programmes. Participants meet for six sessions over two years. Reading and exercises continue between sessions via internet to allow participants to maintain their current positions while being involved in the course. This course develops the skills and abilities of current industry personnel to enable them to become independent breeders or more valuable contributors to larger breeding programmes. The course targets personnel currently involved in plant breeding programmes who lack the academic background in genetics theory and practice to advance as independent breeders. Current breeders who desire a refresher course or would like to broaden their expertise are also potential participants.

In advanced countries where plant breeders are indispensable in satisfying the need of farmers, there has been complaints from the seed industry that academic training of plant breeders no longer meets demand (van Deynze and Bradford, 2006). The situation is fast spreading to SSA now that seed companies are establishing and farmers' demand for improved technology is increasing. Breeders receive similar training from universities in and outside of the sub-region, but their on-the-job responsibilities may be much different, although they are interdependent. This really influences their outputs. In WCA to date, nearly all maize breeders are employed in the public sector which, in this case, consists of educational institutions, such as universities, and research institutes. Very few breeders are employed in the private sector which is primarily the seed industry. The expectations for breeders in public and private sectors are different. Breeders in the public sector must focus on teaching, securing funds to support their research and carrying out publishable research in addition to germplasm and cultivar improvement. Often public breeding programmes are partially supported by funding from related basic research activities.

At the research institutes, funding comes from the government and donor agencies, the researchers have no teaching responsibilities and until very recently, on-the-job promotion was not dependent on publishing research papers. In contrast, breeding in the private sector demands an active life style with global travel, teamwork and responsibilities for business planning and intellectual property (IP) rights in addition to breeding and cultivar development. Education based on public experiences and careers can be vastly different from the skills and knowledge needed for most plant breeders who work in the private sector. Although both public and private breeders require the same fundamental training, the priorities and ancillary skills required differ. There is an urgent need for additional or complementary approaches to support education and training for plant breeders to meet the demand of both public and private breeding programmes.

In Nigeria, agriculture students spend five years for the bachelor degree because the penultimate year (400 level) is set aside for internship programme in which the students are exposed to hands-on experience on the farm. Furthermore, the students are taken on field trips to research stations such as IITA and NARIs, they are exposed to professional conferences and must complete a final year research project which must be written up as a technical report. This approach gives the students a practical experience that exposes them to real-life situations under the mentorship of academic and, at times, industrial supervisors. The government of Nigeria offers three additional hands-on experiences for students before they can be employed, that is, the Student's Industrial Works Experience Scheme, the Industrial Attachment (IT) and the National Youth Service Corps. Although none of these is specific to plant breeding per se, the participants are exposed to real-life situations that they often see as quite different from university campus life. The challenge now is to make the hands-on experience specific to plant breeding. Perhaps, seed companies can take the initiative by employing fresh graduates as staff-in-training for a period before they are fully absorbed as full-time staff. Taking a cue from the US system, universities in the sub-region employ young degree holders as graduate assistants (GAs) on the academic staff. As a rule, GAs only assist in teaching courses, supervising undergraduate laboratory classes and in conducting research. Compulsorily, they must register for MS degree during which they are supervised by a senior academic staff who, compulsorily, they must understudy. Through this approach, universities are able to put in place a very successful succession plan for academic staff.

For nearly three decades, the human resources and facilities available in WCA have been successfully employed by WECAMAN to conduct extensive training courses, special-purpose seminars and workshops for scientists from its member countries. One advantage of using available human resources in the sub-region is reduced costs, especially by eliminating certain expenses such as airfares from other regions. A total of 225 national scientists participated in WECAMAN's training courses and workshops from 1995 to 2007 (Table 5).

The experiences accumulated in WCA through teaching of undergraduate and graduate students in the university, along with training courses handled over the years by IITA and NARIs, have made it possible to identify areas where plant breeders need further training for on-the-job effectiveness. One area that has been very effective is personalized training of researchers whereby the trainee and the trainer interact one-on-one for several months at the research base of the trainer, that is, IITA-Ibadan in our case. The visiting scientist is usually a master's or doctorate degree holder who has been on the job for a few years. WECAMAN used the visiting scientist scheme in 1998–2006 to achieve this objective and about 15 scientists actively benefited from it. The DTMA project has also continued on the scheme, and another 15 scientists from five West African countries (Benin and Niger, 1 each; Mali, 3; Ghana and Nigeria, 5 each) took part from 2008 to 2015.

Table 5 Capacity building activities of the WECAMAN partners, 1995–2007

Course title	Year	Duration	No. of participants
Seed production	1995	2 weeks	27
Striga control and technology transfer	1995	2 weeks	8
Preparation of extension material	1996	1 week	15
Advanced statistical computing	1996	2 weeks	17
Farmer participatory methods of on-farm testing	1998	2 weeks	11
Impact assessment workshop	1998	2 weeks	6
Workshop on maize quality, processing and utilization	1998	2 weeks	13
Breeding for stress tolerance in maize	1999	1 week	13
Striga resistance breeding and marker-assisted selection in cereals	1999	1 week	10
Advanced statistical computing for breeders and agronomists	2000	2 weeks	14
Impact assessment of maize stress management technologies	2000	2 weeks	9
Travel workshop	2001	1 week	15
Biotechnology workshop on maize	2002	1 week	11
QPM development and seed delivery workshop	2003	2 weeks	32
Training workshop on statistical computing for breeders, agronomists and social scientists	2007	1 week	24
Total 19 courses			225

During the last three phases of the DTMA project (2007–15) funded by BMGF and executed by CIMMYT with a subcontract to IITA for the West African component of the project, a major effort was devoted to the establishment of a well-functioning research for developing a pipeline to ensure the production and dissemination of DT maize germplasm and to build the capacity of African NARES and seed companies (STMA Project proposal, 2015). Significant progress has been made in upgrading key breeding sites in SSA, besides enhancing the scientific/technical capacity of the scientists and technicians of partner institutions. Regional planning meetings, on-site backstopping consultation visits, graduate student project supervision, breeding and dissemination team awards have been used as strategies to improve the research capacity and technology dissemination capabilities of partners in eastern, southern and western Africa. Technician training course, monitoring and consultation visits have been important elements in improving trial quality. Seed business management training courses have also been conducted to strengthen the capability of NARES and the local seed sector organizations. The STMA Project is presently intensifying the effort devoted to capacity building of partners, including the seed production enterprises to ensure greater sustainability and scalability of technologies. The STMA Project is continuing to focus on building the capacity of the NARES and seed companies to accelerate widespread dissemination of multiple stress-tolerant maize varieties and hybrids in partner countries.

Apart from the general courses such as genetics, statistics, crop physiology, agro-climatology, seed science, plant pathology and entomology in which plant breeders need regular training to update their knowledge and skill, they also need training in new areas, including molecular tools, IP rights and protection, data management, ethics of research, research management and report writing. Regardless of the area covered in the training, trainers are always concerned about how well the trainees absorb the material presented and will be able to apply the knowledge and skill gained from the training. Therefore, when carrying out a training course, many conduct a *pre-course test* before the presentations and a *post-course test* at the end of the course. The difference in scores of the two tests is an indication of how much the participants gained in the course. In an undergraduate crop science course taught with a training (i.e. hands-on) approach, the students, on average, scored 45.1% with a standard error of the mean (SEM) of 0.82 in the post-course test relative to 31.2 ± 1.03 scored in the pretest (Table 6). In general, performance in the post-test was better than that of the pretest (Table 6 and Fig. 5). Although performance in the pre-course test positively influenced the post-course test score (r = 0.97), Fig. 6 clearly indicated that those who had relatively high scores in the pre-course test had lower gains in scores of the post-course test, an indication that such students had little or no gains from the training. Indeed, some of the trainees had negative gains in the post-course test score (Fig. 6). Similar trends were observed in training course organized at IITA-Ibadan, 2015, and ICRISAT-Bamako, 2016, for plant breeding technicians from anglophone and francophone countries in WCA, respectively.

Technical report writing is one of the most inadequately developed of all the skills that scientists use in their research activity. Although most researchers know that writing papers is an integral part of their job (*publish or perish*), fewer than 5% of researchers had any formal instruction in scientific writing as part of their training. Guided by this urgent need for acquisition of technical writing skill by students, the Faculty of Agriculture, Obafemi Awolowo University, Ile-Ife, Nigeria, introduced a course into the 5-year bachelor's degree

Table 6 Statistics of the scores obtained in pre- and post-tests administered on undergraduate agriculture students in a crop science course at Obafemi Awolowo University, Ile-Ife, Nigeria, 2016

Statistic	Pretest score	Post-test score
Mean ± SEM, %	31.2 ± 1.03	45.1 ± 0.82
Range	2.5–70	18.8–76.9
Variance	170.94	107.69
Standard deviation	13.07	10.38
Mode	25.0	52.5
25%tile	22.5	38.1
50%tile = Median	30.0	44.4
75%tile	40.0	52.5
Proportion scoring ≤40%	77.6%	31.7%
Proportion scoring >50%	6.8%	31.7%

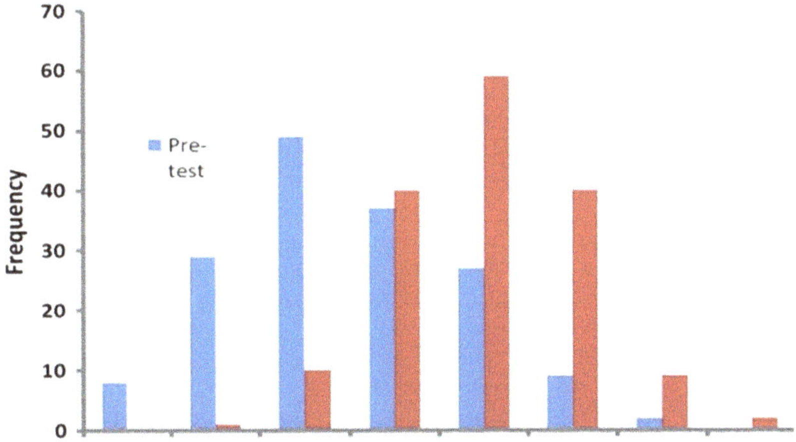

Figure 5 Frequency distribution of scores in pre- and post-course tests administered on undergraduate crop science students at Obafemi Awolowo University, Ile-Ife, Nigeria, 2016.

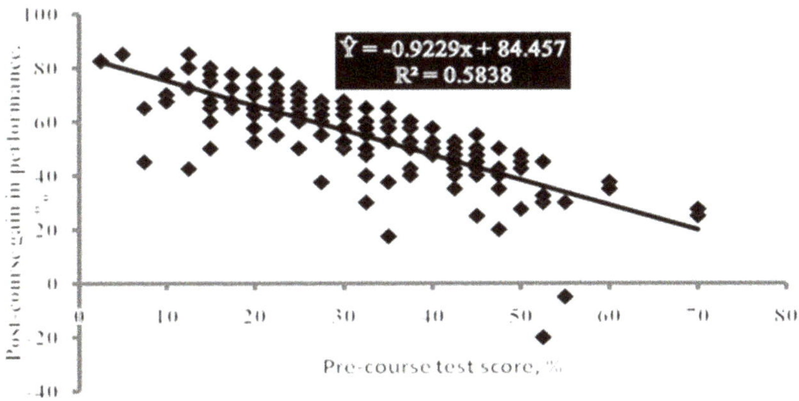

Fig. 6 Relationship between performance in pre-course evaluation and gain made in post-course evaluation by undergraduate crop science students at Obafemi Awolowo University, Ile-Ife, Nigeria, 2016.

curriculum in agriculture, entitled CPP 408 – *Report Writing in Crop Science*, a 400-level course, open to graduate students. A search through the literature showed little or no published papers addressing training or teaching methods in technical writing skills specific to agriculture. Rather, emphasis has been on experimental designs, data analysis and interpretation, including use and misuse of statistics such as multiple comparisons (e.g. Saville and Rowarth, 2008). Whenever offered, the course was taught to a fairly large class (usually about 300 students), using *Content Analysis*, a research tool used to determine the presence of certain words or concepts within texts or sets of texts. A study was conducted to evaluate the performance of the students in the course and its effectiveness on the students' performance in the compulsory final year research project (coded CPP 514) the students executed a year later (Fakorede et al., 2014). Students were assigned full-text research papers from journals and were guided into identifying and scoring on a standard scoring sheet the presence (1) or absence (0) of concepts expected in each section of technical papers. For example, the *Abstract* was expected to contain statements on five concepts: justification, objective(s), methodology, quantitative results and conclusion, all within the number of words specified by the *Guide to Authors* of the technical journal to be used as the outlet of the paper. At the end of the semester, each student was assigned 25 papers from standard journals to score as done in class, statistically analyse the scores and write up the research in a technical journal format. Performance of the students in CPP 408 ranged from 20 to 63% with a mean of 41.3 $\pm$ 1.04. Their performance in CPP 514 was better, ranging from 45 to 75% with mean = 66$\pm$1.15. Performance in the two courses was positively correlated (r = 0.48; r^2 = 0.23). This study revealed that undergraduate students could be effectively taught technical report writing, using content analysis. Further refinements of this approach will likely improve skill acquisition of students for improvement in writing technical reports. The possibility of using this approach to train researchers in skilful report writing needs to be investigated.

The outcomes of the training courses and workshops on the research capability and capacity of NARS included increased research output, as evident from the number of collaborative research projects executed at the national level; improvement of capabilities in data analysis, as shown by the quality and number of technical reports and papers presented during regional maize workshops; high-quality presentation of scientific papers at conferences; preparation of well-structured, focused and convincing research proposals; improvement of data recovery from collaborating NARS; and release of improved varieties and agronomic practices. Assessments of WECAMAN's activities indicated that the technicians and scientists had greatly improved their research skills through these courses, seminars and workshops. Furthermore, the Biennial West and Central Africa Regional Maize Workshops were successfully organized in 1995, 1997, 1999, 2001 and 2005. The regional workshops served as fora for the presentation of scientific papers by national and international scientists from WCA. The papers of the Regional Maize Workshops were compiled into six books of proceedings and were published.

Human resources development in plant breeding appears limitless in SSA. As new findings and research methods are discovered over time, the need to train more people while updating the skill of older ones will also have to increase. New methods of teaching students and training professionals will have to be put in place. Co-teaching of university courses may be necessary to reduce costs while not reducing the standard of the education. Online courses using social media are urgently needed in the education and professional training of plant breeding technicians and researchers. Fortunately, there are five international CGIAR crop research and development centres or institutes in WCA: IITA, CIMMYT, Africa Rice, ICRISAT and the AVRDC. These centres and several others are also present in the eastern and southern African sub-regions. The institutes can initiate the move to satisfy institutional needs that would lead to revitalizing higher education in SSA through partnership engagements. Similarly, universities can form networks to facilitate effective human resources development at minimal cost. An example is the Regional University Forum for Capacity Building in Agriculture (RUFORUM), a network of universities primarily focusing on all aspects of higher agriculture education in eastern and southern Africa but with some extension to western and central Africa. The network is run by Africans with African priorities, but fully integrated into the extensive international science and policy agenda. The forum awards scholarships for graduate studies as well as funds research in agriculture. In reality, RUFORUM is more than a network of member universities in Africa; rather, it is a vibrant community of eager students, academics and scholars, engaged farmers and their communities, policymakers, private sector actors and entrepreneurs. Now in existence for more than 12 years, RUFORUM holds biennial conferences to celebrate excellence, share collective achievements, and explore effective and efficient approaches to better serve society, review and discuss new ideas and initiatives for appropriate action. Similar networks specific to subject-matter areas such as plant breeding are urgently needed in Africa.

4 Use of tropical germplasm for maize improvement

Plant genetic resources are essential parts of agro-biodiversity and these include primitive forms of cultivated plant species, modern varieties, farmers' varieties (landraces), obsolete cultivars, breeding lines and genetic stocks, and wild and weedy species (IPGRI, 1993). They constitute invaluable assets necessary to combat food insecurity especially in the SSA. Tropical maize germplasm is an enormous reservoir of wide genetic resources, which hold great potentials for rapid development of the maize seed industry in the sub-region. Maize is one of the most diverse crops both genetically and phenotypically, especially in tropical regions such as SSA. This may be due to its allogamous mode of reproduction and because its centre of origin and diversity is in the tropics. Furthermore, the genetic diversity within this crop species could be attributed to long selection procedures practised by local farmers and breeders resulting in the development of landraces and improved varieties, respectively. Maize landraces have been cultivated for a long time under natural and human selection in different environments and under different cultural methods. This has resulted in the development of a large number of maize landraces adapted to the specific environmental conditions of their habitats and desired use by humans (Ruiz de Galarreta and Alvarez, 2001; Reif et al., 2006). However, the potential of maize germplasm is still largely underutilized and underexploited. Over-reliance on improved germplasm and modern cultivars and consequent neglect of the landraces, wild and weedy species narrow down the genetic basis of maize in the sub-region causing genetic erosion. Availability of such genetic resources is being threatened due to adoption of intensive monocropping of only high-yielding stress-resistant genetic materials which have narrow genetic bases, land degradation and climate change. Other major challenges to tropical germplasm include poor understanding of the genetic complexity in the germplasm and inadequate knowledge of the potentials inherent in the germplasm due to low research effort at studying the genetic materials. Most breeders in the sub-region use elite varieties as the base population in their improvement programme, abandoning landraces and wild relatives as a 'good for nothing' germplasm source. Little effort is put into the collection and conservation of these germplasm in recent times. If not for immediate use, landraces and wild relatives have a diverse genetic base and possess favourable alleles that could be explored for future improvement of the crop. Although breeders always maintain variability in their maize population while improving the population for desired traits, a peak is usually reached when variability in the population is exhausted and further improvement stops. At this point, landraces and wild relatives become the best bet to further advance the population. Studies have been carried out on assessment of genetic diversity in maize landraces (Gonul et al., 2012; Bracco et al., 2009; Hartings et al., 2008) and improved varieties (Badu-Apraku et al., 2006, 2007; Ajmone-Marsan et al., 1998). *Striga* tolerance in early and extra-early maize germplasm has been increased by introgression of favourable genes from teosinte (*Zea diploperennis*) a wild relative of maize (Badu-Apraku et al., 2016 in press). A few more studies have shown that maize landraces could be used to further improve elite varieties. Meseka et al. (2015) reported significant yield improvement in a study of twelve backcross populations formed by introgression of alleles from diverse maize landraces into an elite variety, under well-watered environments and drought stress. Olosunde (2015) also reported superiority of landraces over elite varieties for some seed quality traits such as conductivity and seedling vigour, which suggest that the landrace could be used as sources of improving quality of the elite varieties.

Another key challenge to conservation and use of tropical maize germplasm is rapid seed deterioration especially in the humid tropical climates which is characterized by high rainfall intensity, high relative humidity and high temperature, which encourage disease build up and accelerated respiration rate in stored seeds, all affecting regeneration cycle of accessions stored in gene banks. In addition, there are great losses in diversity in storage due to poor storage facilities, irregular power supply, equipment failures or damages and fire outbreaks, among other things as extraneous factors in most private and public gene banks in the sub-region due to financial constraints.

5 Heterosis, heterotic grouping and identification of testers in maize germplasm

A heterotic group is defined as a collection of germplasm that exhibits a higher degree of heterosis (on the average) when crossed with germplasm from an external group than when crossed with members of its own group (Lee, 1995). Melchinger and Gumber (1998) also defined heterotic group as a collection of related or unrelated genotypes from the same or different populations, which display similar combining ability and heterotic response when crossed with genotypes from other genetically distinct germplasm groups. Heterotic pattern refers to a specific pair of two heterotic groups, which express high heterosis and consequently high hybrid performance in their cross. Melchinger and Gumber (1998) recommended the following criteria for the choice of heterotic pattern in hybrid breeding: (i) high mean performance and large genetic variance in the hybrid population; (ii) high *per se* performance and good adaption of the parent population to the target regions; and (iii) low inbreeding depression, if hybrids are produced from inbred lines. Establishing heterotic pattern is of prime importance in the development of successful maize hybrid programme (Barata and Carena, 2006). Distinct heterotic patterns have been identified among temperate maize germplasm and potentials of the pattern explored maximally to improve productivity. Popular among such heterotic patterns are European flint × US Lancaster, which is commonly used in Europe and Reid Yellow Dent × Lancaster extensively exploited in the United States, China (Huang and Li, 2001; Fan et al., 2002) and many other parts of the world (Melani and Carena, 2005). Identifying distinct heterotic groups among tropical germplasm has been challenging. Researchers have invested considerable time in establishing heterotic groups among tropical germplasm. Paterniani (1990) noted that the lack of information on the subject matter in the tropics might be due to the existence of the large number of races and cultivars that are yet to be studied, thus making it difficult to have well-defined heterotic patterns in tropical maize. One of the earliest attempts at studying heterotic patterns among tropical maize germplasm identified a combination of Tuxpeno with Eto and other Caribbean flints as a promising heterotic pattern (Goodman, 1985). The studies also established Tuson × Tuxpeno, Cuban flint × Tuxpeno and Suwan 1 × Tuxpeno as the best combinations for hybrid production in the tropics. Another study on the heterotic patterns using late/intermediate maturing inbred lines from IITA could not establish clear heterotic patterns (Kim and Ajala 1996; Kim et al., 1999; Menkir et al., 2003), and the reason adduced was that the inbred lines were derived from source populations formed by mixing different germplasm without taking into consideration the need to maintain heterotic groups intact (Kim et al., 1999). Menkir et al. (2003) recommended a combination of divergent testers with molecular markers as a better alternative to classifying tropical maize inbred lines. Menkir et al. (2004) made an attempt to classify 38 tropical maize inbred lines into heterotic groups using two testers (TZMI 102 and TZMI 1407) and molecular markers. The testers successfully classified 23 of the inbreds into two heterotic groups. The line × testers method used was found to be more efficient in classifying the inbreds than the molecular markers. The results of the classification based on AFLP and simple sequence repeat (SSR) markers were largely consistent with each other but the molecular markers classified the inbreds into groups different from those classified by the testers. The researchers concluded that the molecular marker-based grouping might, at best, serve as a basis for designing and carrying out combining ability studies in the field. A similar study by Barata and Carena (2006) using diallel analysis and SSR markers corroborated the findings of Menkir et al. (2004) and recommended extensive field evaluation as being more appropriate in assigning unrelated maize inbred lines into heterotic groups. Another study by Menkir et al. (2005) examined genetic relationships among 41 *Striga*-resistant late and early inbreds derived from four diverse germplasm. The inbreds exhibited a broad range of genetic similarity values based on the results of the two molecular markers, and the markers clearly differentiated the inbreds according to their source populations and maturity classes. Wu et al. (2007) classified 27 maize inbreds into four distinct heterotic groups using North Carolina Design II (NCII). Agbaje et al. (2008) used a line × tester method to classify 35 early maturing yellow endosperm inbred lines into heterotic groups with two testers TZi 4001 and Ku1414 evaluated under *Striga*-infested and *Striga*-free conditions at Mokwa and Abuja and in *Striga*-free environment at Ile-Ife, Nigeria. None of the inbred lines could be classified into heterotic groups under any of the evaluation environment, evidently because the testers were not sufficiently effective to discriminate among the inbreds. Furthermore, Badu-Apraku et al. (2011) were not able to identify either definite heterotic groups or ideal testers in a diallel study among nine yellow-grained early maize inbreds using the genotype main effect plus genotype-by-environment interaction (GGE) biplot analysis. The reason they adduced for this was the over-dominating effect of specific combining ability (SCA) relative to the general combining ability (GCA). Nevertheless, distinct tester groups were identified. In their study of heterosis and genetic distance among 17 lowland white-grained tropical maize under drought stress and non-stress conditions using diallel and RFLP markers, Betran et al. (2003) found that the degree of inbreeding of the parental lines could affect their response to stress. They also observed that the environment significantly affected the correlations of genetic distance with lower values observed under more stressed conditions. They equally opined that optimal or non-stress environments where grain yield is maximal could be more appropriate to measure SCA effects and the predictive value of genetic distance.

Conventionally, SCA effects of hybrid yield have always been used to classify genetic materials into heterotic groups. This is based on the assumption that SCA of two lines from different heterotic groups is greater than those from the same group. However, the reliability of this method is dependent on the number of materials investigated and adequate sampling of the genetic background. Fan et al. (2008) used heterotic group's specific and general combining ability (HSGCA), a combination of SCA and GCA effects, to assign some tropical maize inbreds into heterotic groups instead of the traditional method involving SCA effects only. This proposition was on the basis that SCA effects were often greatly influenced by the interaction between two inbred lines and between hybrids and environments, which often leads to assigning the same inbred line into different heterotic groups in different studies (Fan et al., 2008). Results of the study showed that HSGCA method was more effective than the use of SCA and molecular marker methods in classifying tropical maize germplasm into distinct heterotic groups. These methods of heterotic grouping, that is, yield-SCA or HSGCA, are deficient in that they are based on only one trait, grain yield. However, grain yield is a complex trait controlled by polygenes and has low heritability especially under stress environments. Therefore, an alternative heterotic grouping method that incorporate yield and important adaptive traits under stress environments was proposed by Badu-Apraku et al. (2013) designated as heterotic grouping based on GCA of multiple traits (HGCAMT) and it has been reported to be comparable in efficiency to HSGCA, yield-SCA and SNP marker-based genetic distance under stress environment and more effective than other methods across multiple stress environments (Badu-Apraku et al., 2016 a,b). In addition, HGCAMT is more useful in classifying inbreds into heterotic groups based on NCII data, where HSGCA method is not practicable. Another important challenge in the identification of heterotic groups is that all earlier heterotic grouping methods classified a small percentage of the total genetic materials involved in such study. This renders many of the newly developed inbreds useless in hybrid development. However, to a large extent, the use of Ward's minimum variance cluster analysis for data computed from HSGCA and HGCAMT has helped overcome this challenge. These grouping methods are able to classify a larger percentage of inbred lines into heterotic groups.

The choice of proven testers is another crucial factor that determines progress in maize hybrid development. The seed industry in SSA is still at the infant stage and adoption and use of hybrids by farmers are greatly constrained due to the extent of research, number of available seed companies, poor financial capability of existing seed companies to maintain genetic materials and low financial status (purchasing power of the farmers). Tropical inbred lines exhibit poor vigour compared with temperate inbreds and production of single-cross hybrid seeds produced from these inbreds is considerably small. This in turn leads to high production cost of the hybrid seeds and most farmers cannot afford to buy them. Hybrids require the application of inputs such as fertilizer, herbicides and pesticides to express their high production potentials but most farmers in the sub-region are resource-poor and cannot afford them. Therefore, scientists have invested considerable research efforts and resources into the identification of single-cross testers to facilitate the development of three-way and double-cross hybrids and inbred testers for top-cross hybrid development rather than for the development of single-cross hybrids for release to farmers. Proven testers were not available in the tropical germplasm at IITA. At the initial stage of hybrid development, testers from temperate germplasm such as Mo17 and B73 were used. But the results of crosses of tropical lines with temperate testers were not producing desirable results. After a while, testers were identified among the late/intermediate maize germplasm which were adopted for early and extra-early maize germplasm. Agbaje et al. (2008) used TZi 4001 and Ku1414 in a line × tester study to identify heterotic groups but none of the testers could sufficiently cluster the lines into clear heterotic groups. Several hybrid groups produced by diallel or line × tester crosses (i.e. NCII) have been studied under stress and non-stress conditions during the last decade. For example, Badu-Apraku et al. (2016a) examined the combining ability of selected IITA and CIMMYT inbreds for grain yield and other traits and classified the lines into heterotic groups using heterotic SCA and GCA (HSGCA), single-nucleotide polymorphism-based genetic distance (SNP-GD) and GCA effects of multiple traits (HGCAMT) methods. The authors reported that the HGCAMT is an efficient heterotic grouping method and offers a great opportunity for grouping the numerous IITA inbreds that are yet to be grouped. Furthermore, the results of this study confirmed that molecular markers could be used in grouping IITA inbreds that are yet to be field tested. The efficiency of the heterotic grouping methods, SCA effects of grain yield, HSGCA, HGCAMT and SNP-based GD methods varied with the type of genetic material under study. The results of these studies have led to identification of several testers under contrasting stress conditions and more recently, some excellent testers have been identified across multiple stresses (Annor and Badu-Apraku 2016; Badu-Apraku et al., 2015, 2016). Through genetic studies conducted during the last decade, the HSGCA, HGCMAT and molecular markers such as SNP and SSR have been confirmed as invaluable for the grouping of the numerous tropical inbreds that have been developed in SSA (Badu-Apraku et al. 2015, 2016). The challenge now is to use these proven grouping methods to classify the numerous inbreds that are yet to be field tested to allow planned crosses to be made and to facilitate the development of superior hybrids. There is also a need to refine the heterotic groups that have been identified and to identify more efficient testers. Furthermore, research efforts need to be devoted to the development of proven testers for improving nutritional quality of maize in terms of provitamin A, quality protein content and micronutrients.

Another important aspect of hybrid development is the prediction of hybrid performance from *per se* performance of their inbred parents. This will help minimize time and resources expended on evaluating parent performance in hybrid combination. Many attempts have been made to predict the hybrid performance using many methods. The possibility of predicting hybrid performance is greatly influenced by the environment. Genetic and molecular approaches have been used to predict hybrid performance. Badu-Apraku et al. (2011b) reported higher correlation between mid-parent value of extra-early white inbreds and their F_1 hybrid performance under *Striga* infestation and drought stress than under optimal conditions. Akinwale (2012) reported highly significant correlation between mid-parent value under *Striga* infestation and F_1 grain yield under artificial *Striga* infestation indicating that inbred line *per se* performance is a good indicator of its performance in hybrid combination under *Striga* infestation.

On the use of molecular markers and their efficiency in predicting hybrid performance in tropical maize germplasm, available reports are contradictory. Several studies have reported that GD estimate had no significant correlation with hybrid performance for crosses between lines from different heterotic groups (Godshalk et al., 1990; Melchinger, 1993; Ajmone-Marsan et al., 1998). Dhliwayo et al. (2009) also reported that SSR-based GD were not significantly associated with hybrid performance, SCA effects or mid-parent heterosis for grain yield of maize, but they effectively grouped the lines according to known pedigree relationships and approximately according to heterotic patterns used by CIMMYT and IITA. Hence, they concluded that even though SSR-based genetic distances were not useful for predicting hybrid grain yield, they could be helpful for preliminary classification. George et al. (2010) observed that the utility of GD as a predictor of hybrid value is best up to a certain threshold, as correlations with GD became inconsistent when the inbred parents exhibit great divergence (i.e. GD > 0.77). Mohammadi et al. (2008) pointed out that genetic distance *per se* is not a reliable predictor of hybrid yield performance and suggested that the use of specific marker loci with tight linkage to the chromosomal segments that determine the expression of a trait of interest is important in order to improve prediction of hybrid performance based on parental molecular divergence. They also recommended the use of more sophisticated statistical methods such as the best linear unbiased prediction in combination with molecular marker data for choosing potentially promising inbred parents for hybrid production.

6 Application of new plant breeding techniques

Conventional plant breeding is based on increasing the chances of selecting individuals from populations generated from genetic mating designs. Selection has usually been carried out at the whole-plant level (i.e. the phenotype), which represents the net result of the interaction between genotype and environment. Phenotypic selection has resulted in tremendous genetic gains in crops but is severely limited when faced with traits that are largely modulated by the environment (Cooper et al., 2006). Furthermore, the nature of some traits can make the testing procedure itself complex, unreliable and/or expensive. Under such circumstances, indirect selection, based on genetic markers, becomes an efficient complementary breeding tool. The idea is that where individual genes influencing target traits can be identified and linked with one or more markers, the marker loci can be used as a surrogate for the trait, resulting in greatly enhanced breeding efficiency. MAS provides tremendous opportunities for enhancing genetic gains and breeding efficiency in maize (Prasanna et al., 2010; Xu et al., 2012). The MAS techniques are free of the political and regulatory issues that have dominated the application of transgenic/GM technologies, and progress is relatively faster with qualitative traits. For instance, microsatellite or SSR markers located within the opaque2 (o2) gene, that confers the nutritional advantage of quality protein maize (QPM), has provided opportunities for accelerating the pace of QPM conversion programmes through MAS (Prasanna et al., 2010). Another classical example is the successful employment of molecular markers for specific genes in the carotenoid biosynthetic pathway to significantly enhance the content of provitamin A in elite maize germplasm (Babu et al., 2012). With the discovery of useful allelic diversity for LycE and CrtRB1 and development of molecular markers (Harjes et al., 2008; Yan et al., 2010), source lines with >15 μg/g of provitamin A carotenoids are identified using MAS, and have now been routinely used as parents for new crosses at CIMMYT. This has led to the selection of lines with 40–250% higher provitamin A carotenoid concentrations than lines without the favourable allele (Babu et al., 2012).

Presently, the use of markers to track transgenes or pyramid favourable alleles determining a significant proportion of the phenotypic variance is possible for many crops, including maize. The number of reports asserting the successes of marker-assisted breeding (MAB) in dealing with polygenic traits has been significantly increasing. In addition, it is now generally widely accepted that the role of MAB goes beyond the manipulation of elite alleles at a few loci in bi-parental segregating populations. Presently, there is a need for validation of the genetic gain of favourable alleles in a suitable biological context and environment. Once this is achieved, allele-based markers can be easily developed and employed. This validation step remains a major hurdle in the establishment of a large set of markers appropriate for deployment in maize breeding programmes. However, considering the technological and methodological progress achieved in genomics in recent years, it is clear that the potential of MAB to complement phenotypic selection and improve maize productivity is set to increase significantly in the near future.

Despite the impressive progress made with MAS in maize breeding in the developed world, it is yet to be fully employed in maize research programmes in most developing countries, particularly the NARS programmes of SSA. This is due to the fact that for MAS to be a viable option for a plant breeding programme, adequately equipped laboratory facilities must be in place and appropriately trained scientist/personnel must be available.

Information is limited on the nutritional characteristics of the QPM and provitamin A varieties that have been released in WCA. This is due to lack of adequate facilities and equipment for nutritional analysis of the maize kernels in the various NARS breeding programmes as well as the international centres. Analysis of the amino acid (tryptophan and lysine) and carotenoid concentration (lutein, zeaxanthin, β-cryptoxanthin and β-carotene) profiles are routinely sourced from laboratories in Europe or the United States, making it expensive and time-consuming.

The *in vivo* haploid induction technique allows breeders to rapidly develop homozygous inbred lines from heterozygous source germplasm of maize. The efficiency of maize breeding programmes can be enhanced and varietal development accelerated through adoption of the doubled-haploid (DH) technique, assuming that resource allocation will be optimized and appropriate breeding procedures will be adopted (Seitz, 2005; Longin et al., 2006; Gordillo and Geiger, 2008). The DH technique is successfully used in many maize breeding programmes in temperate areas. However, the adoption of the technique has lagged behind in tropical regions where it is only recently gaining attention due to its advantages. DH lines are produced by *in vivo* induction of maternal haploids and are routinely used in maize breeding programmes. The *in vivo* haploid induction technique allows breeders to rapidly develop homozygous inbred lines from heterozygous source germplasm of maize. Another major challenge facing maize breeding programmes (both NARs and international centres) in WCA is lack of adoption of new technologies such as the DH technique to facilitate the production of homozygous lines for efficient and effective hybrid development. Undoubtedly, cost- and time-effective development of homozygous lines is an important component of every maize breeding programme. Traditionally, homozygous lines are obtained by self-pollination of heterozygous material for 6–7 generations, which is a time-consuming and expensive process (Prasanna et al., 2012). The induction and subsequent doubling of maternal haploids is an efficient alternative to generate homozygous lines in quick timeframe (two generations). The DH technology shortens the breeding cycle significantly by developing 100% homozygous lines within 2–3 seasons compared to conventional breeding that takes at least 7–8 seasons to develop inbred lines with 98–99% homozygosity. Other major advantages of DH lines compared to selfed lines include (i) maximum genetic variance between lines for *per se* and testcross performance from the first generation; (ii) perfect fulfilment of DUS criteria for variety protection; (iii) reduced expenses for selfing and maintenance breeding; (iv) simplified logistics; and (v) increased efficiency in MAS, gene introgression and stacking genes in lines. One cycle of DH line development with two stages of testcross evaluation takes only four years if off-season nurseries are available. Cycle length can be shortened to three years if the first three breeding steps (recombination, haploid induction and DH plant production) are completed in a single year. Furthermore, genome-wide MAS can effectively be incorporated into DH-line-based breeding schemes. Because of the genetic, methodological and logistic advantages, further progress in maize breeding is expected to increase considerably with the development of DH lines. Yet, the success of employing DH lines depends on a robust and efficient haploid induction technology as well as on breeding strategies that make optimum use of the breeder's genetic, technical and monetary resources (Gordillo and Geiger, 2008). For instance, CIMMYT has adapted the DH technology, in collaboration with the University of Hohenheim (Germany), to accelerate the development of homozygous lines in diverse, elite, highly adapted genetic backgrounds (Prasanna et al., 2012). The DH technology is a powerful means to accelerate the introgression of novel germplasm into elite maize breeding lines. It enhances 'forward breeding' and provides an opportunity to have an earlier look at the potential of new lines, greater knowledge about their environmental adaptability before they are fully tested and further used as parental lines for hybrid development and commercial cultivation. Use of DH technology can potentially enhance the efficiency of recurrent selection or genomic selection (GS)-based schemes for traits with low heritability, particularly for breeding programmes without access to off-season nurseries (Bouchez and Gallais, 2000).

Furthermore, the DH technology enables shifting of resources from the labour-intensive task of repeated inbreeding to generate inbred lines, spending more time on evaluation of the DH lines for yield and other adaptive traits, and using the identified lines for producing hybrids and synthetics. The *in vivo* haploid induction technology has provided an exciting avenue to increase the rate of progress in hybrid maize breeding. Theoretical and experimental research results encourage the breeder to take advantage of this new genetic tool. Considering the benefits of DH technology, its implementation in research programmes in SSA particularly, it will undoubtedly result in significant progress in developing maize for desired major traits in WCA. For instance, in collaboration with University of Hohenheim, CIMMYT has recently developed tropicalized haploid inducer lines (Prigge et al., 2012; Prasanna et al., 2012) to meet the needs of the breeding programmes especially in Africa, Latin America and Asia. These haploid inducers are being shared with public and private institutions worldwide through appropriate material transfer agreements. Furthermore, through the funding support of BMGF, a DH facility was established in Kenya under the DTMA project in 2012 for use by breeders in eastern, southern and West Africa. However, a major challenge militating against the adoption of the DH technology by the national maize programmes in West and central Africa is the outbreak of the maize lethal necrosis disease in eastern and southern Africa in 2012. This development resulted in a ban on maize seed imports from eastern Africa into most of the countries in West Africa. As a result, DH lines recently developed in CIMMYT–Kenya for IITA maize scientists cannot be imported to Nigeria. There is therefore the need to make arrangements for the inducer lines to be licensed by the University of Hohhenheim to IITA, so as to make it possible to develop the DH facility at IITA for the WCA sub-region. Thus, while commercial seed industries across the world have benefited significantly from the DH technology, the uptake among the institutions of the NARS and the small- and medium-scale enterprise seed companies in SSA is significantly low.

In summary, the major challenges militating against the adoption of the DH technology include lack of adequate facilities and equipment, the cost of establishing the DH facility and well-trained personnel to effectively use the technology in breeding programmes of SSA. Indeed, acquiring inducers and other important logistics for the DH technology are major hurdles limiting adoption. However, it is generally agreed that the efficiency of maize breeding programmes in SSA could be enhanced and varietal development accelerated through adoption of the DH technique assuming that resource allocation will be optimized and appropriate breeding procedures will be adopted (Seitz, 2005; Longin et al., 2006; Gordillo and Geiger, 2008).

7 Participatory plant breeding

Plant breeders develop crop varieties with the ultimate goal of satisfying the demand of the end users. For the maize crop, end users consist of farmers, marketers, processors, consumers and policymakers, all of who act and interact to ensure food security, human and livestock health and nutrition, and human employment. Farmers, the topmost in the chain of crop variety end users, although may lack western education, are rational people with significant knowledge of agricultural practices. The professional plant breeder needs all of these stakeholders in developing a clear vision in the breeding process, and their input is important. In the context of plant breeding in the developing world, participatory plant breeding (PPB) involves close farmer–researcher collaboration to bring about plant genetic improvement within a species. PPB helps in building farmer skills to facilitate farmer selection and seed production efforts, deliberately addresses the needs of a specific farmer or farmer group, and ultimately enhances the rapid adoption of improved varieties, which increases crop production and profitability of farmers.

For many years, there has been much controversy among breeders as to which of stressed and non-stressed (optimum) environments would be better for the development of varieties that would meet the farmer's needs. Varieties developed from breeding programmes under optimal conditions seem to be very effective in farming systems that are fairly homogeneous, but less effective when the reality of the farmer is more complex and stress-prone. Results of our studies involving breeding for stress tolerance have consistently shown that varieties developed under stress possess value addition when grown under optimal conditions, but the reverse is not so productive. Therefore, selecting under the farmer's conditions by the farmer in person is expected to result in adapted high-performing varieties.

As thoroughly discussed by Badu-Apraku et al. (2012), maize varietal development is normally done through on-station research and is researcher-managed, with the goal of increasing maize production at the farm level. The researcher evaluates potential new varieties in small plots (e.g. four 0.75 m rows, 5 m long = 15 m^2) with optimum levels of production inputs, such as fertilizer and herbicide application for effective weed control. The production environment in farmers' fields may be very different from the near-ideal situation in on-station research fields. Varieties that perform well on-station may not carry through in the farmers' fields. Farmers adopt new varieties that are economically superior to the existing one(s). They are interested in net benefits and in protecting themselves against risk. Before changing from one variety to another, farmers consider many factors such as agro-ecological requirements, the availability of additional production resources (farmland, credit, labour, skill, equipment, etc.) and the potential additional income resulting from the change. They also consider the compatibility of the new variety with their farming system. To make good recommendations, researchers must keep in mind the farmers' goals and evaluate new varieties under their production conditions and practices.

Theoretically, PPB is based on the hypothesis that the 'art' component of crop improvement involves selection, which may be with or without the knowledge or understanding of genetics and other science subject matters. Farmers existing during the primitive times and non-educated farmers saving their own seed for planting in the next season are examples of selection without the knowledge of genetics. Professional plant breeders carry out selection with understanding of genetics and other sciences. Two studies carried out in WCA clearly indicated the need to use PPB approach in maize improvement. Fakorede et al. (1997) reported the findings of a survey conducted across Nigeria in 1993 and 1994 involving 232 farm communities with an average of about 30 farmers per community. The responses to the 116-item questionnaire administered in the survey convinced the maize researchers that farmers knew exactly what traits would improve and stabilize maize productivity in their fields, and with training, they would successfully carry out selection. The second study involved a participatory research approach developed by the l'Equipe sur les Systemes de Production et la Gestion des Ressources Naturelles de Sikasso in Mali (Kamara et al., 1997). Both researchers and farmers worked together on station and on farm at several locations to test and assess varieties at the vegetative growth stage and after harvesting. The farmers made clear choices of varieties they preferred and the seeds of such varieties were to be produced for them as from the following season. As a matter of fact, that study marked the beginning of a maize revolution in Mali, now the second largest producer of maize in WCA, following Nigeria, but the country with the highest yield per unit land area.

Varietal evaluations carried out in the farmers' fields are referred to as on-farm trials, which may be researcher-managed, farmer-managed or jointly managed. The varieties are evaluated under real farmers' conditions and the trials create opportunities for communication and interaction among all stakeholders (farmers, breeders, extension staff and seed companies), as well as enabling farmers to participate in the evaluation of varieties under situations very similar to their production environments and practices.

There are several variants of maize on-farm trials. The oldest, more conventional and most commonly used involves two or three test varieties and the farmer's variety, planted in replicated plots of relatively large size if researcher-managed or without replication if farmer-managed. Apart from the seeds of test varieties which are supplied by the researcher, the volunteer farmer provides all the inputs and manages the trial from land preparation to harvest as he normally does in his farm. The researcher visits the trial often to take data, if necessary, as well as interacting and exchanging views with the farmer. The farmer has the benefit of keeping the harvest at the end of the season.

Another type of on-farm trial is called the mother–baby trial (MBT) (Banziger and de Meyer, 2000; Sieglinde, 2002). Although relatively new, it has gained popularity as an effective method of evaluating maize varieties under the farmers' conditions. An MBT involves evaluating a relatively large number of potential new varieties in a trial (mother trial) and subsets of the varieties (baby trials) in several satellite farmers' fields. Mother trials are researcher-managed; baby trials are farmer-managed. Mother trials may be conducted by the researcher under the farmer's typical management practices or with researcher-specified inputs. The concept of MBTs is designed to facilitate conversations among farmers, extensionists and researchers. This is important in order to develop, assess and test various agronomic practices in a manner that incorporates diverse farmer priorities, and can evaluate performance across a range of management practices and edaphic conditions in a quantifiable and repeatable manner. It has been used most widely in PPB, as a systematic approach to incorporate farmer assessment of crop varieties. Crop improvement efforts have used MBTs in several countries in SSA, including Malawi, and southern Africa (Kanyama-Phiri et al., 2000; Kamanga et al., 2001). This quantitative method to systematically integrate farmer assessment and stakeholder input into research programmes has led to development of agronomic recommendations and adoption of improved maize, rice, wheat and legume genotypes in SE Asia, Africa and the Americas. The MBT design approach has proven effectiveness to improve upon recommendations that often overlook smallholder farmers' distinct agro-ecological and socio-economic conditions, and therefore improve the relevance of agronomic research. It must be pointed out that the MBT design is often regarded as costly and complex due to system heterogeneity; yet the lack of incorporation of farmer priorities and circumstances result in technologies that are not applicable to the targeted population (Ramisch, 2012).

Participatory research such as PPB is another form of on-farm trial. This method involves farmers early in the process of varietal development. Progeny trials, selection and recombination to form new varieties not only are carried out in the farmers' fields but also involve both farmers and researchers working together.

Advantages of on-farm evaluation trials include the following (Badu-Apraku et al., 2012):

- Testing the acceptability and profitability of the newly developed varieties before they are promoted on a larger scale;
- Obtaining realistic input–output data for cost–benefit analysis;
- Providing important diagnostic information about farmers' problems;
- Enabling researchers to identify the impediments to a rapid and higher adoption of improved technologies;
- Providing an avenue for a better understanding of the processes involving the integration of farmers' indigenous knowledge into the scientific knowledge of researchers.

The MBT method makes it clearly a type of PPB. It enables farmers to participate in the evaluation of treatments under situations very similar to those of the farmers. New varieties are evaluated under farmer conditions and the farmer has an opportunity to compare the performance of the varieties under his conditions and researcher-managed conditions. In addition, an opportunity for communication and interaction among all stakeholders (farmers, breeders, extensionists and seed companies) is created at the farm level, a very effective approach to technology transfer. The expected outcome of this approach, therefore, is that the farmer is able to identify varieties he or she could use right away or as soon as possible after the breeder has finalized the improvement of the potential varieties. Obviously, MBT makes it possible for farmers to help breeders set breeding objectives; identify traits for improvement; determine genotype × environment interaction; understand farmers production constraints (economic, social, biotic, abiotic, etc.); and share decision-making, responsibilities and activities.

Unfortunately, this method has not gained much popularity in maize research in WCA because it is assumed to be more appropriate for the self-pollinated crops, such as rice and the legumes. Also, since seed production is gradually becoming a private sector affair, seed companies consider that it would be unprofitable to produce seeds of so many varieties, each one for only a relatively small number of farmers. The community seed production scheme may alleviate this constraint to a large extent.

8 Conclusion and future trends

During the last two decades, the international research centres, CIMMYT and IITA, working closely with the NARS of WCA, Eastern and Southern Africa and many donors have substantially expanded the scientific and training capacity of NARS in the region. The NARS have successfully produced and helped to develop a large number of maize varieties and disseminated new technologies to farmers. Wide adoption of available maize technologies has led to increased maize production in the region and significant returns to the investments in maize research have been made by donors and partner countries. Studies show that in 2005, of the 7 million ha planted with maize in nine WCA countries, over 4 million ha, representing about 60% of the area, had been planted with improved varieties. Adoption of new varieties increased from less than 5% of the maize area in the 1970s to about 60% in 2005, yielding an aggregate rate of return to research and development investment of 43%. The estimated number of people moved out of poverty through the adoption of new maize varieties rose gradually in the 1980s to nearly one million people per year from the mid-1990s. Over half of this impact has been attributed to maize research, particularly maize breeding. Notwithstanding these successes, researchers in the region still face challenges of increasing maize yields. Although total maize production has increased in the subregion, the increase has resulted from extensive rather than intensive cultivation. For example, FAO's statistics show that average maize yields in the WCA subregion have increased from the long-standing 0.8 t/ha to nearly 2 t/ha during the last three decades, although average yields in researcher-managed on-station and on-farm research were over 5 t/ha. Therefore, one of the present challenges for researchers is to reduce the yield gap between researchers' and farmers' production. This needs to be undertaken against challenges of maize breeding, climate change, increasing urban populations and environmental problems associated with land degradation. Prominent among the challenges of maize breeding in SSA are lack of adequate research facilities, human resources, broad-based germplasm resources, challenges in hybrid development, lack of use of new plant breeding techniques including the adoption of participatory plant breeding research techniques.

Looking to the future, maize still remains crucial for food security in SSA. The predominance of the crop in many farming systems and diets implies that yield gains have the potential to catalyse a Green Revolution. Although considerable strides have been made towards achieving this potential, food insecurity for millions of rural households persists even where progress in maize production has been achieved. With domestic maize production often not keeping up with the pace of expanding urban populations for food, there is increased consumption of other cereal crops such as rice and wheat, much of which is imported. At the same time, the demand for maize for industrial use and as a feed for livestock continues to grow rapidly. International and regional research needs to continue to support national programmes through regional alliances that build on the comparative advantages of the NARS in developing full scientific capabilities in SSA (Badu-Apraku et al. 2012).

There is a need for accelerated gains in breeding for improved grain yield of maize varieties and hybrids both under favourable and unfavourable environments in the CIMMYT and IITA maize breeding programmes in ESA and WCA to further drive the maize yield higher in SSA. This may be achieved in the breeding programmes in SSA by increasing the size of the breeding programmes to enable higher selection intensity, ensuring adequate genetic variation, increasing the accuracy of selection through higher heritability and accelerating breeding cycles. For example, molecular markers for resistance to turcicum leaf blight (TLB), grey leaf spot (GLS) and maize lethal necrosis (MLN) are currently at the validation stage in CIMMYT (Masuka et al. 2016). Four promising haplotypes for TLB are presently under testing (S. Nair, unpublished), while 16 promising genomic regions for resistance to GLS have been identified, representing 32 SNPs. Also, 24 SNPs of medium effect have been identified and validated for resistance to MLN using a combination of genome-wide association analysis (GWAS), GS and association mapping (Gowda et al. 2015). Application of these markers to screen for tolerant materials before expensive multi-location field testing is expected to allow resources for yield testing to be focused on materials with good levels of disease resistance, to improve the selection intensity for yield, and hopefully increase the rate of genetic gains from selection.

A DH facility recently installed by CIMMYT in Kenya allows completely homozygous lines to be developed from heterozygous parents in two seasons, thus enabling greater selection efficiency. This combined with higher between-entry variation compared with early generation testing methods improves overall realized heritability, resulting in higher genetic gain from selection (Cairns et al. 2013). Tropically adapted maize inducer lines with a haploid induction rate of up to 10% and a new marker system were recently developed in CIMMYT (Prigge et al. 2011; Chaikam et al. 2015) and a DH facility established in Kenya to allow the routine incorporation of DH into breeding programmes in ESA (CIMMYT, 2014). The current haploid induction rate is 5.5%, production capacity of over 50,000 haploid lines per year (S. Bumagat, 2016, unpublished). However, current costs (~30 USD/line) limit further utilisation of this technology. Efforts are presently being made to reduce DH production costs through a) more efficient haploid induction, b) more efficient chromosome doubling, c) optimization of protocols to more closely reach production targets and d) more efficient haploid discrimination systems. It is contemplated that a similar facility will soon be established in IITA to speed up the inbred and hybrid development programmes in WCA and increase the gains from selection in the breeding programmes of the sub-region.

Inbred development in a maize hybrid breeding routinely begins by crossing two inbreds. Progenies are then advanced by selfing or DH technology and evaluated in testcross trials to determine general and specific combining abilities. Testcross selection within a biparental cross is based on phenotypic information and requires a season of field trials (Hallauer, 1990). GS offers a faster way to predict GCA without phenotyping. Jacobsen et al. (2014) has developed a GCA model for GS within A/B crosses. The GCA model relies on information from previously phenotyped and genotyped crosses with inbreds A and B as one of the parents and can be used in advanced breeding programmes that use elite inbreds as the parents of new crosses. The ability to predict GCA has the potential to significantly reduce time and costs in inbred development and thus speed up breeding process. Efforts are presently being made in the CIMMYT and IITA maize breeding programmes to adopt GS to speed up the breeding process in the breeding programmes.

Improved data management and decision support tools are important in facilitating the meta-analysis of big data set. However, the importance of data management and decision support tools has been underutilized in breeding programmes in SSA. With advancing technology, the volume and diversity of data increase; hence, meta-analysis of over 20,000 CIMMYT regional trials in southern Africa showed a non-linear relationship between warming and maize yields (Lobell et al. 2011). For each degree day above 30°C, maize yields were reduced by 1% under optimal rain-fed conditions and by 1.7% under drought conditions. Based on these results breeding for heat stress has been incorporated into the CIMMYT ESA breeding programme (Cairns et al. 2013). Presently, the CIMMYT maize breeding programme has developed a fieldbook, a data management system and made available to NARS partners in east and Southern Africa to improve data capture and subsequent analysis (Vivek et al. 2007). The data management system should also be made available to the NARS partners in WCA.

Genetic studies conducted during the last decade have confirmed that the HSGCA, HGCMAT and molecular markers such as SNP and SSR are invaluable tools for assigning into heterotic groups the numerous tropical inbreds that have been developed in SSA (Badu-Apraku et al. 2015, 2016). The challenge now is to use these proven grouping methods to classify the numerous inbreds that are yet to be field tested to allow planned crosses to be made and to facilitate the development of superior hybrids. There is also a need to refine the heterotic groups that have been identified and to search for more efficient testers. Furthermore, research efforts need to be devoted to the development of proven testers for improving nutritional quality of maize in terms of provitamin A, quality protein content and micronutrients. Many technical, ecological, institutional and policy issues need to be addressed if maize is to contribute its expected share to the alleviation of poverty in SSA. There is an urgent need for increased enhancement of NARS capacity to develop and transfer technology through training, infrastructure development, use of biotechnology and geographic information systems. Other new tools and innovative approaches that should be employed include participatory plant breeding, use of the mother – baby approach on on-farm testing, and the development of micro-nutrient dense and quality protein maize varieties with multiple stress tolerance (to drought, low soil nitrogen, *Striga*, stem borers and nutrient depletion). These will help to improve the income generating capabilities and nutritional status of farmers, as well as increase the capacity of farmers and farmers' organizations to drive the research agenda and use the products of research to develop sustainable farming systems that increase the productivity of land and labour in the face of climate change and land degradation. There is also the need to improve the efficiency and effectiveness of the varietal release processes across countries and agroecological zones, developing effective partnerships between the public and private sectors in the development of effective seed production and distribution systems, ensuring that input and output markets and the rural infrastructure that support them are capable of providing farmers with access to efficient services, developing market information systems to guide and link farmers with traders, consumers, input dealers, and policymakers. In addition, other areas to explore are development new maize technologies for new products, specific markets, production areas, and processing methods to diversify the utilization of maize, stimulation of regional trade by promoting appropriate agro-enterprise and advocacy in the development of appropriate policy reforms for sustained maize production and productivity.

9 Where to look for further information

Additional information may be obtained may be obtained from the following publications:

Badu-Apraku, B., Jim, E.-J., Abebe, M., Samuel, A. and Fakorede, M.A.B. 2012. Building partnerships and encouraging innovation for sustainable maize production: the West and Central Africa Collaborative Maize Research Network, Achievements and Impact. IITA, Ibadan, Nigeria. 82 p.

Badu-Apraku, B., Fakorede, M. A. B., Gedil, M., Annor, B., Talabi, A. O., Akaogu, I. C., Oyekunle, M., Akinwale, R. O. and Fasanmade, T. Y. (2016). Heterotic patterns of IITA

and CIMMYT early-maturing yellow maize inbreds under contrasting environments. *Agronomy Journal* 108: 1321–36.
Badu-Apraku, B., Fakorede, M. A. B., Talabi, A. O., Oyekunle, M., Akaogu, I. C., Akinwale, R. O., Annor, B., Melaku, G., Fasanmade, Y. and Aderounmu, M. (2015).Gene Action and Heterotic Groups of Early White Quality Protein Maize Inbreds under Multiple Stress Environments. *Crop Science* 56: 183–99 (2016). doi:10.2135/cropsci2015.05.0276.
Badu--Apraku, B., Fakorede, M. A. B. Talabi, A. O. Akinwale, R. O. Oyekunle, M. Annor, B., Akaogu, I. C., Melaku, G. and Fasanmade, Y. 2015. Grouping of early quality protein maize inbreds based on combining ability and SNP markers under multiple stress environments. *Field Crops Research* 183: 169–83.

10 References

Agbaje, S. A., Badu-Apraku, B. and Fakorede, M. A. B. (2008). Heterotic patterns of early maturing maize inbred lines in *Striga*-free and *Striga* infested environments. *Maydica* 53: 87–96.
Ajayi, S. A. and Fakorede, M. A. B. (2001). Effect of storage environments and duration of equilibration on maize seed testing and seedling evaluation. *Maydica* 46: 267–75.
Ajmone-Marsan, P., Castiglioni, P., Fusari, F., Kuiper, M. and Motto, M. (1998). Genetic diversity and its relationship to hybrid performance in maize as revealed by RFLP and AFLP markers. *Theoretical and Applied Genetics* 96: 219–27.
Akinwale, R. O. (2012). Evaluation of the heterotic groups of *Striga* resistant early maize *(Zea mays* L.) inbred lines using diallel analysis and molecular markers. Ph.D. Thesis, Obafemi Awolowo University, Ile-Ife. Nigeria, 125 p.
Annor, B. and Badu-Apraku, B. (2016). Gene action controlling grain yield and other agronomic traits of extra-early quality protein maize under stress and non-stress conditions. *Euphytica* doi:10.1007/s10681–016-1757–4.
Babu, R., Rojas, N. P., Gao, S., Yan, J. and Pixley, K. (2012). Validation of the effects of molecular marker polymorphisms in LcyE and CrtRB1 on provitamin-A concentrations for 26 tropical maize populations. *Theoretical and Applied Genetics* (Published Online DOI 10.1007/s00122–012-1987–3).
Badu-Apraku, B., Fakorede, M. A. B. and Ajala, S. O. (2004). Enhancing human resources for maize research and development in West and Central Africa: A networking approach. *Jour. Natural Resources and Life Sciences Education* 33: 77–84.
Badu-Apraku, B., Menkir, A., Fakorede, M. A. B., Lum, A. F. and Obeng-Antwi, K. (2006). Multivariate analyses of the genetic diversity of forty-seven *Striga* resistant tropical early maturing maize inbred lines. *Maydica* 51: 551–9.
Badu-Apraku, B., Fakorede, M. A. B. and Lum, A. F. (2007) Evaluation of experimental varieties from recurrent selection for *Striga* resistance in two extra-early maize populations in the savannas of West and Central Africa. *Experimental Agriculture* 43: 183–200.
Badu-Apraku, B., Lum, A. F., Akinwale, R. O. and Oyekunle, M. (2011a). Biplot analysis of diallel crosses of early maturing tropical yellow maize inbreds in stress and nonstress environments. *Crop Science* 51: 173–88.
Badu-Apraku, B., Fakorede, M. A. B., Oyekunle, M. and Akinwale, R. O. (2011b). Selection of extra-early maize inbreds under low N and drought at flowering and grain- filling for hybrid production. *Maydica* 56: 1721–35.
Badu-Apraku, B., Fakorede, M. A. B., Menkir, A. and Sanogo, D. eds (2012). *Conduct and Management of Maize Field Trials*. IITA, Ibadan, Nigeria. 59p.
Badu-Apraku, B., Fakorede, M. A. B., Gedil, M., Talabi, A. O., Annor, B., Oyekunle, M., Akinwale, R. O., Fasanmade, T. Y., Akaogu, I. C. and Aderounmu, M. (2015). Heterotic responses among crosses

of IITA and CIMMYT early white maize inbred lines under multiple stress environments. *Euphytica* 206: 245–62.

Badu-Apraku, B., Fakorede, M. A. B., Gedil, M., Annor, B., Talabi, A. O., Akaogu, I. C., Oyekunle, M., Akinwale, R. O. and Fasanmade, T. Y. (2016). Heterotic patterns of IITA and CIMMYT early-maturing yellow maize inbreds under contrasting environments. *Agronomy Journal* 108: 1321–36.

Banziger, M. and de Meyer, J. (2000). Farmers' voices are heard here. CIMMYT Annual Report 99–2000.

Barata, C. and Carena, M. J. (2006). Classification of North Dakota maize inbred lines into heterotic groups based on molecular and testcross data. *Euphytica* 151: 339–49.

Betrán, F. J., Ribaut, J. M., Beck, D. and Gonzalez de Leon, D. (2003). Genetic diversity, specific combining ability, and heterosis in tropical maize under stress and nonstress environments. *Crop Science* 43: 797–806.

Bouchez, A. and Gallais, A. (2000). Efficiency of the use of doubled haploids in recurrent selection for combining ability. *Crop Science* 40: 23–9.

Bracco, M., Lia, V. V., Gottlieb, A. M., Cámara Hernández, J. and Poggio, L. (2009) Genetic diversity in maize landraces from indigenous settlements of Northeastern Argentina. *Genetica* 135(1): 39–49.

Cairns, J. E., Hellin, J., Sonder, K., Araus, J. L., MacRobert, J. F., Thierfelder, C. and Prasanna, B. M. (2013). Adapting maize production to climate change in sub-Saharan Africa. *Food Security* 5: 345–60.

Chaikam, V., Nair, S. K., Babu, R., Martinez, L., Tejomrtula, J. and Prasanna, B. M. (2015). Analysis of effectiveness of *R1-nj* anthocyanin marker for *in vivo* haploid identification in maize and molecular markers for predicting the inhibition of *R1-nj* expression. *Theoretical and Applied Genetics* 128: 159–71.

CIMMYT. (2014). Maize Doubled Haploid Facility for Africa. http://www.cimmyt.org/en/component/docman/doc_download/62-maize-doubled-haploid-facility-for-africa Accessed 14th April 2015.
Dhliwayo, T., Pixley, K., Menkir, A. and Marilyn, M. (2009). Combining ability, genetic distance and heterosis among elite CIMMYT and IITA tropical maize inbred lines. *Crop Science* 49: 1201–10

Edmeades, G. O., Banziger, M., Chapman, S. C., Ribaut, J. M. and Bolanos, J. (1995). Recent advances in breeding for drought tolerance in maize. In Badu-Apraku, B., Akoroda, M. O., Ouedraogo, M. and Quin, F. M. (eds), *Contributing to Food Self-Sufficiency: Maize Research and Development in West and Central Africa*. Proceedings of a Regional Maize Workshop, 29 May to 2 June 1995, IITA, Cotonou, Benin Republic. WECAMAN-IITA, pp. 22–41.

Fakorede, M. A. B. and Akinyemiju, O. A. (2003). Climatic change: effects on maize production in a tropical rain forest location. In Badu-Apraku, B., Fakorede, M. A. B., Ouedraogo, M., Carsky, R. J. and Menkir, A. (eds), *Maize revolution in West and Central Africa*. Proceedings of a Regional Maize Workshop, 14–18 May 2001, IITA, Cotonou, Benin Republic. WECAMAN-IITA, pp. 272–82.

Fakorede, M. A. B., Alofe, C. O., Adenola, O. A., Kim, S. K., Nweke, F. I., Iken, J. E. and Akinwumi, J. A. (1995). Research priorities for the improvement of maize production in Nigeria: farmers' views. In Badu-Apraku, B., Akoroda, M. O., Ouedraogo, M. and Quin, F. M. (eds), *Contributing to Food Self-Sufficiency: Maize Research and Development in West and Central Africa*. Proceedings of a Regional Maize Workshop, 29 May to 2 June 1995, IITA, Cotonou, Benin Republic. WECAMAN-IITA, pp. 382–7.

Fakorede, M. A. B., Oluwaranti, A. and Akinwale, R. O. (2014). Teaching report writing in Crop Science to undergraduate agriculture students in a Nigerian University. In The Proceedings of the Fourth RUFORUM Biennial Regional Conference, 21–5 July 2014, Maputo, Mozambique, pp. 71–6.

Fakorede, M. A. B., Saka, A. and Ajayi, S. A. (2004). Drought probability during the cropping season in a rainforest location: Implications for maize production. *Paper presented at the First Annual Conference of the Nigeria Society for Environmental Management*, 12–13 May 2004. Obafemi Awolowo University, Ile-Ife, Nigeria.

Fan, X. M., Chen, H. M., Tan, J., Xu, C. X., Zhang, Y. M., Huang, Y. X. and Kang, M. S. (2008). A new maize heterotic pattern between temperate and tropical germplasms. *Agronomy Journal* 100: 917–23.

Fan, X. M., Tan, J., Yang, J. Y., Liu, F., Huang, B. H. and Huang, Y. X. (2002). Study on combining ability for yield and genetic relationship between exotic tropical, subtropical maize inbreds and domestic temperate maize inbreds. *Science Agriculture Sin*. 35: 743–9.

FARA (Forum for Agricultural Research in Africa) (2006). Assessment of the requirements for efficient, effective and productive national agricultural research systems in Africa. Main report and strategic recommendations, 6 March 2006, p. 12.

Geiger, H. H. (2009). Doubled haploids. In Bennetzen, J. L. and Hake, S. (eds), *Maize Handbook – Volume II: Genetics and Genomics*, Springer Science and Business Media LLC, pp. 641–57.

George, M. L. C., Salazar, F., Warburton, M., Narro, L. and Vallejo, F. A. (2010). Genetic distance and hybrid value in tropical maize under P stress and non-stress conditions in acid soils. *Euphytica* doi:10.1007/s10681–010-0273–1.

Godshalk, E. B., Lee, M. and Lamkey, K. R. (1990). Relationship of restriction length polymorphisms to single cross hybrid performance of maize. *Theoretical and Applied Genetics* 80: 273–80.

Goodman, M. M. (1985). Exotic maize germplasm: status, prospects and remedies. *Iowa StateJournal of Research* 59: 497–527.

Gönül, C., Faheem, S. B., Benjamin, K., Ahmet, C. U. and Hakan, O. (2012). Diversity assessment of Turkish maize landraces based on fluorescent labelled SSR markers. *Plant Molecular Biology Reporter* 30: 261–74

Gordillo, G. A. and Geiger, H,H. (2008). Optimization of DH-line based recurrent selection procedures in maize under a restricted annual loss of genetic variance. *Euphytica* 161: 141–54.

Gowda, M., Das, B., Makumbi, D., Babu, R., Semagn, K., Mahuku, G., Olsen, M. S., Bright, J. M., Beyene, Y. and Prasanna, B. M. (2015). Genome-wide association and genomic prediction of resistance to maize lethal necrosis disease in tropical maize germplasm. *Theoretical and Applied Genetics* 128: 1957–68.

Harjes, C. E., Rocheford, T. R., Bai, L., Brutnell, T. P., Kandianis, C. B., Sowinski, S. G., Stapleton, A. E., Vallabhaneni, R., Williams, M., Wurtzel, E. T., Yan, J. and Buckler, E. S. (2008). Natural genetic variation in lycopene epsilon cyclase tapped for maize biofortification. *Science* 319: 330–3.

Hartings, H., Berardo, N., Mazzinelli, G. F., Valoti, P., Verderio, A. and Motto, M. (2008) Assessment of genetic diversity and relationships among maize (Zea mays L.) Italian landraces by morphological traits and AFLP profiling. *Theoretical and Applied Genetics* 117(6): 831–42.

IPGRI, (1993). Diversity and development. The strategy of International Plant Genetic Resources Institute, IPGRI, Rome, Italy.

Jagtap, S. S. (1995). Handbook for use of resource information system (RIS): Interactive and user-friendly geographic data analysis on personal computers. Agroecological Studies Unit, IITA, Ibadan, Nigeria.

Kamara, A., Defoer, T. and Coulibaly, N. (1997). Evaluation par la méthod participative de variétés de maïs en milieu paysan au Mali. In Badu-Apraku, B., Akoroda, M. O., Ouedraogo, M. and Quin, F. M. (eds), *Contributing to Food Self-Sufficiency: Maize Research and Development in West and Central Africa*. Proceedings of a Regional Maize Workshop, 29 May to 2 June 1995, IITA, Cotonou, Benin Republic. WECAMAN-IITA, pp. 388–98.

Kim, S. K. and Ajala, S. O. (1996). Combining ability of tropical maize germplasm in West Africa II: Tropical vs temperate × tropical origins. *Maydica* 41: 135–41.

Kim, S. K., Ajala, S. O. and Brewbaker, J. L. (1999). Combining ability of tropical maize germplasm in West Africa III. Tropical maize inbreds. *Maydica* 44: 285–91.

Lee, M. (1995). DNA markers and plant breeding programmes. *Advances in Agronomy* 55: 265–344.

Longin, C. F. H., Utz, H. F., Reif, J. C., Schipprack, W. and Melchinger, A. E. (2006). Hybrid maize breeding with doubled haploids: I. One-stage versus two-stage selection for testcross performance. *Theoretical and Applied Genetics* 112: 903–12.

Manyong, V. M., Kling, J. G., Makinde, K. O., Ajala, S. O. and Menkir, A. (2003) Impact of IITA Germplasm Improvement on Maize Production in West and Central Africa. In Evenson, R. E. and Gollin, D. (eds), *Crop Variety Improvement and its Effect on Productivity*, CABI, Wallingford, UK.

Masuka, B., Magorokosho, C., Olsen, M., Atlin, G. N., Bänziger, M., Pixley, K., Vivek, B., Labuschagne, M., Matemba-Mutasa, R., Burgenõ, J., Macrobert, J., Prasanna, B. M., Makumbi, D., Tarekegne, A., Crossa, J., Zaman-Allah, M. and Cairns, J. E. (2016). Gains in maize genetic improvement in Eastern and Southern Africa ii) CIMMYT open pollinated varieties (OPVs). *Crop Science submitted.*

Melani, M. D. and Carena, M. J. (2005) Alternative maize heterotic patterns for the northern corn belt. *Crop Science* 45: 2186–94.

Melchinger, A. E. (1993). Use of RFLP markers for analyses of genetic relationships among breeding materials and prediction of hybrid performance. In Buxton, D. R. et al. (ed.), *International Crop Science I.* CSSA, Madison, WI, pp. 621–8.

Melchinger, A. E. and Gumber, R. K. (1998). Overview of heterosis and heterotic groups in agronomic crops. In Lamkey, K. R. and Staub, J. E. (eds), *Concept and Breeding of Heterosis in Crop Plants.* CSSSA-SP, Publication no. 25, Madison, WI, pp. 29–44.

Menkir, A., Melake-Berhan, A., Thé, C., Ingelbrecht, I. and Adepoju, A. (2004). Grouping of tropical mid-altitude maize inbred lines on the basis of yield data and molecular markers. *Theoretical and Applied Genetics* 108: 1582–90.

Menkir, A., Kling, J. G., Badu-Apraku, B. and Ingelbrecht, I. (2005). Molecular marker-based genetic diversity assessment of *Striga*-resistant maize inbred lines. *Theoretical and Applied Genetics* 110(6): 1145–53.

Menkir, A., Badu-Apraku, B., Thé, C. and Adepoju, A. (2003). Evaluation of heterotic patterns of IITA's lowland white maize inbred lines. *Maydica* 48: 161–70.

Meseka, S., Menkir, A. and Obeng-Antwi, K. (2015). Exploitation of beneficial alleles from maize (*Zea mays* L.) landraces to enhance performance of an elite variety in water stress environments. *Euphytica* 201: 149–60.

Mohammadi, S. A., Prassana, B. M., Sudan, C. and Singh, N. N. (2008). SSR heterogenic patterns of maize parental lines and prediction of hybrid performance. *Biotechnology and Biotechnological Equipment* 22: 541–7.

Olosunde, A. A. (2015). Agronomic performance and seed potential storability of tropical maize landraces and improved varieties. Ph.D. Thesis. Department of Crop Production and Protection, Obafemi Awolowo University, Ile-Ife. Nigeria. 144p.

Paterniani, E. (1990). Maize breeding in the tropics. *CRC Critical Review in Plant Science* 9: 125–54.

Prasanna, B. M., Chaikam, V. and Mahuku, G. (eds), (2012). *Doubled Haploid Technology in Maize Breeding: Theory and Practice.* Mexico D. F.: CIMMYT, 50p.

Prasanna, B. M., Pixley, K. V., Warburton, M. and Xie, C. (2010). Molecular marker-assisted breeding for maize improvement in Asia. *Molecular Breeding* 26: 339–56.

Prigge, V., Sanchez, C., Dhillon, B. S., Schipprack, W., Araus, J. L., Bänziger, M. and Melchinger, A. E. (2011). Doubled haploids in tropical maize: 1. Effects of inducers and source germplasm on in vivo haploid induction rate. *Crop Science* 51: 1498–506.

Prigge, V., Xu, X., Li, L., Babu, R., Chen, S., Atlin, G. N. and Melchinger, A. E. (2012). New insights into the genetics of in vivo induction of maternal haploids, the backbone of doubled haploid technology in maize. *Genetics* 190: 781–93.

Rai, M. (2002) Genetic resources and intellectual property rights in agriculture perspective. *Indian Journal of Pulses Research* 15: 1–18.

Reif, J. C., Warburton, M. L., Xia, X. C., Hoisington, D. A., Crossa, J., Taba, S., Muminovic, J., Bohn, M., Frisch, M. and Melchinger, A. E. (2006). Grouping of accessions of Mexican races of maize revisited with SSR markers. *Theoretical Applied Genetics* 113: 177–85.

Ruiz de Galeretta, J. I. and Alvarez, A. (2001). Morphological classification of maize landraces from Northern Spain. *Genetic Resources and Crop Evolution* 48: 391–400.

Saville, D. J. and Rowarth, J. S. (2008). Statistical measures, hypotheses, and tests in applied research. *Natural Science Education* 37: 74–82.

Seitz, G. (2005). The use of doubled haploids in corn breeding. In Proc. of the 41st Annual Illinois Corn Breeders' School 2005. Urbana-Champaign, Illinois, USA, pp. 1–7.

Sieglinde, S. (2002). Quantifying farmer evaluation of technologies: The Mother and baby trial design. In Bellon, M. R. and Reeves, J. (eds), *Quantitative Analysis of Data from Participatory Methods in Plant Breeding*. Mexico, DF: CIMMYT.

Shaw, R. H. (1977).Climatic requirements. In Sprague, G. F. (ed.), *Corn and corn Improvement*, 2nd edition. American Society of Agronomy, Madison, WI, USA, pp. 591–623.

Tanksley, S. D. and Couch, S. R. (1997). Seed banks and molecular maps: Unlocking genetic potential from the wild. *Science* 277: 1063–6.

Van Deynze, A. and Bradford, K. J. (2006). Academic training of plant breeders no longer meets demand. *Seed World* January 2006: 20–1.

Vivek, B., Kasango, J., Chisoro, S. and Magorokosho, C. (2007). Fieldbook: Software for managing a maize breeding program: A cookbook for handling field experiments, data, stocks and pedigree information. Mexico, DF (Mexico): CIMMYT. 144 p.

Wu, J. C., Xu, C. X., Chen, H. M., Tan, J., Han, X. R., Huang, B. H. and Fan, X. M. (2007). Studies on combining ability and heterotic grouping of 27 quality protein maize inbreds and four temperate representative inbreds of Chinese major heterotic groups. *Science in Agriculture Sin* 40: 1288–96.

Xu, Y., Lu,Y., Gao, S., Wan, J. and Prasanna, B. M. (2012). Whole genome strategies for molecular marker-assisted plant breeding. *Molecular Breeding* 29: 833–54.

Yan, J., Kandianis, C., Harjes, C. and Bai, L. (2010). Rare genetic variation at *Zea mays* crtrb1 increases carotene in maize grain. *Nature Genetics* 42: 322–7.

Chapter 3

Developing maize-breeding methods and cultivars to meet the challenge of climate change

Marcelo J. Carena, North Dakota State University, USA

1 Introduction

The United Nations, through its Food and Agriculture Organization, has stated numerous times that people have the right to access adequate food. The agricultural sector has the potential to produce more food for a growing population, but the challenge is to produce it in a sustainable way. Many public and private institutions promote the agricultural sector, including breeders and farmers, to 'feed the world'. But there is still a long way to go, and developing the next generation of sustainable maize-breeding methods and cultivars is part of the solution. Breeders can develop improved cultivars for both extensive and intensive production systems in a sustainable manner that can reach all people, including farmers in the world's marginal environments growing maize to sustain their families.

http://dx.doi.org/10.19103/AS.2016.0001.05

Sustainable breeding is a viable scientific alternative to maintain enough food supply under the environmental challenges facing our planet. It includes responsible breeding with the most efficient use of resources, genetic diversity, maximizing conservation and minimizing waste. In order to achieve a more sustainable maize production, a commitment from independent scientists is needed for new ideas to solve breeding paradigms, improving the quality of life for all human beings, not just a few. The problems posed by climate change warrant innovative solutions to tackle the challenges that lie ahead.

This chapter addresses ideas to develop not only unique cultivars but also breeding methodologies to assist breeders selecting most desirable genes in genetically complex traits in an efficient and sustainable way. Climate change calls for increasing efforts in the integration of germplasm improvement with cultivar development, adapting germplasm carrying unique genetic properties and diversity not available in sequenced genomes or current cultivars available in expensive research projects (e.g. NAM, G2P, among others). It is also an opportunity to move breeding stations from traditional production centres to strategic marginal locations. Scientists are encouraged to be creative in their search for needed research to address current and future challenges. World maize breeders, with the support of their national leaderships and international centres, have the task of developing the next generation of sustainable cultivars and cost-effective breeding methodologies that will facilitate successful and sustainable results for the common good of all world citizens and their environments.

Sustainable maize production will be possible only with the development of the next generation of maize products carrying genetic diversity. Sustainable maize breeders face current and future climate challenges that can be addressed only by germplasm diversity. Only those maize-breeding programmes utilizing large samples of genetically broad-based germplasm have the potential to contribute useful and unique alleles to face climate challenges. These programmes need to integrate germplasm adaptation and improvement with cultivar development adapting and improving exotic and unique germplasm. Developing new products from these genetically broad-based germplasm sources should supplement the typical elite-by-elite line pedigree selection cultivar development process often used in breeding programmes focused on genetically narrow-based germplasm.

In order to obtain elite germplasm to develop new sustainable maize cultivars, breeders need to adapt and improve diverse genotypes to their target environments. However, the choice of germplasm determines the success of a breeding programme. No breeding methodology or technology can provide successful cultivars with a poor germplasm choice. However, once elite germplasm is chosen, breeders should be able to develop not only new cultivars but also new breeding methodologies for efficient selection of quantitative traits. There is a need for breeders to develop not only products but also methodologies that can save time and cost in measuring economically important traits that are genetically complex, difficult to screen and largely influenced by the environment. Breeders will need to develop not only new products but also new breeding methodologies to increase the high-throughput screening efficiency of quantitative traits of economic importance. Rate of dry down (Yang et al. 2010), cold tolerance (Carena 2013b) and drought tolerance (Sharma and Carena 2016) are examples of new and successful inventions on breeding methodologies that have facilitated the screening of thousands of genotypes for traits that are difficult to measure and significantly influenced by the environment. These are breeding techniques that avoid the destruction of plots and save millions of dollars in maize-breeding and production costs. Ideas to overcome the evaluation of genetically complex traits will continue to be a priority for developing the next generation of sustainable maize products. Molecular techniques were initially targeted at economically important traits, which could be the reason why this technology is now utilized in genetically simple traits. Technology, if used wisely, however, should target traits according to their genetic complexity. Flowering time is of the more simple traits for breeders to move maize to higher latitudes at a very low cost. On the other hand, drought tolerance and non-destructive root-screening systems (Sharma and Carena 2016) and the rate of dry down (Yang et al. 2010) are successful examples of genetically complex traits targeted with new breeding methodologies that can facilitate a high-throughput screening of large samples. Breeders should focus on new breeding methodologies for genetically complex traits, including the use of unmanned aircraft systems (UAS).

2 Early developments in breeding

Maize (*Zea mays* L.) is considered one of the important natural resources of the Western Hemisphere. It appeared as a wild species about 7000 to 10 000 years ago. N. I. Vavilov proposed that maize was derived from teosinte and that its centre of origin was in Central America (Fig. 1 and 2). After domestication, maize was widely grown by Native Americans in the United States before the arrival of Columbus in 1492. These first maize breeders are a successful example of breeder-directed evolution. As cultivation of maize became

Figure 1 'Nikolai Vavilov NYWTS' by World Telegram staff photographer – Library of Congress. New York World-Telegram & Sun Collection. http://hdl.loc.gov/loc.pnp/cph.3c18109. Licensed under Public Domain via Commons – https://commons.wikimedia.org/wiki/File:Nikolai_Vavilov_NYWTS.jpg#/media/File:Nikolai_Vavilov_NYWTS.jpg.

Figure 2 'Maize diversity in Vavilov's office (3421259242)' by Luigi Guarino from Suva, Fiji – Maize diversity in Vavilov's office. Licensed under CC BY 2.0 via Commons – https://commons.wikimedia.org/wiki/File:Maize_diversity_in_Vavilovs_office_(3421259242).jpg#/media/File:Maize_diversity_in_Vavilovs_office_(3421259242).jpg.

more extensive, selected open-pollinated strains were developed. By the end of the 1800s, U.S. farmers had 800–1000 unique landrace options to grow (Sturtevant 1899), some of them showing yields of over 6 t ha^{-1} when properly cultivated (Leaming 1883). During this time, seed cost was barely a concern for farmers who save their own seed for sowing the following year. Still this is a very low-cost and sustainable system for maize production worldwide. In short-season dryland North Dakota (ND) environments, unique varieties improved by recurrent selection methods have often provided grain yield values of 8–10 t ha^{-1} per se in experiment trials evaluated across years.

Improved populations are also used for the development of genetically broad-based inbred lines and hybrids with unique grain quality and stress-tolerant properties not available in commercial hybrids. They should be freely available for farmers all over the world to take advantage of this system if backed up by active public breeding programmes that improve varieties. It is desirable that improved populations be stored at the U.S. National Plant Genetic System (NPGS) in Fort Collins, CO for future generations as local university cold storage conditions are not often appropriate for not only the long term but also, in many cases, for the short- and mid-terms. In addition to the significantly low seed costs, the following year's crop of improved populations is expected to deliver, genetically, a similar crop. On the other hand, if farmers plant the seed harvested from a single-cross hybrid (e.g. where all plants are genetically identical) typically grown on extensive farms today, the following year's crop is expected to deliver, genetically, a poor crop. The system, therefore, makes farmers buy seed annually.

3 Hybrid breeding: heterosis

The use of landraces, and later on improved varieties, for production per se was discouraged with the discovery of heterosis. During the early 1900s, extension and research maize programmes of land-grant universities focused on demonstrating the yield advantages of double- and single-cross hybrids over unimproved landraces. There was a technical and economical transition from the production of maize landraces to single-cross maize hybrids. The practical value of hybrid vigor traced back to the controlled hybridization of different maize races by U.S. farmers in the 1800s (Hallauer and Carena 2009). Moreover, controlled crosses of open-pollinated cultivars showed potential for increasing maize yields well before the inbred–hybrid concept was developed (Beal 1878). At that time, genetic divergence among selected strains was sufficient to make variety hybrids superior in grain yield performance to selected varieties per se, but there was a limit of maize genetic divergence when utilizing very wide crosses (Moll et al. 1965). Heterosis, however, was mostly measured on crosses from unimproved varieties.

The population–hybrid concept (Carena 2005) is another low-cost system that can be extensively used worldwide with the advantage of exploiting heterosis (Dowswell et al. 1996). Although this system requires the use of isolations and detasseling for maize production, the seed production cost by the female population is cost-effective due to the lack of inbreeding depression (e.g. no inbred lines are involved). In addition, once the desired population–hybrid combination is found, there is no need for self-pollination, as it is needed in the inbred–hybrid system. These programmes can increase the efficiency of breeding and selection, saving time significantly.

The economic potential of population hybrids, however, is also dependent on genetic improvement programmes like intra- and inter-populations recurrent selection programmes. These alternative maize cultivars are possible only if maize-breeding programmes with continuous population improvement programmes are active (Hallauer et al. 2010). Unfortunately, the rate of development and improvement of new varieties and populations was not desirable for achieving maximum genetic gains. The emphasis laid on the inbred–hybrid system and the shift in research emphasis, both nationally and internationally (Sonino et al. 2006), reduced population genetic improvement efforts to just a few maize-breeding programmes integrating pre-breeding with cultivar development. Nevertheless, royalties were generated from this alternative maize production system and there are still effective maize-breeding programmes that can provide this type of cultivars. North Dakota State University (NDSU) and local cooperatives have produced and commercialized two of the top population hybrids developed at the NDSU maize-breeding program. This can serve as an example to any institution or cooperative system worldwide. It is never too late for this system, especially with the wise and extensive use of winter nurseries. Top population hybrids, including varieties improved by recurrent selection methods, have often provided yields of 10–12 t ha^{-1} in trials across years in short-season dryland ND environments, in addition to unique trait properties not found in commercial cultivars.

4 Inbred–hybrid breeding

In the early 1900s, by the time the first airplane was being tested, public scientists created the inbred–hybrid concept for maize (Shull 1908; East 1908), that is, inbreeding for several generations and restoring hybrid vigor through crossing. This research, however, was abandoned, due to the fact that seed production was challenged on inbred parents affected by severe inbreeding depression after self-pollinating within unimproved landraces. Nevertheless, the private sector exploited this research idea, even though it was considered impractical due to poor seed production on female hybrid parents. The heterosis expressed in farmer fields justified the increased breeding and production costs on hybrid providers. Most of this cost was transferred to farmers who bared a large part of the production risk and had to buy new seed every year.

Inbred lines or hybrid parents became the trade secret for hybrid providers who led a new competitive market worth billions of dollars. Land-grant universities provided an important contribution before intellectual property restrictions. B73, an inbred line developed at Iowa State University (ISU) through a long-term population improvement programme, is the most successful example. Progressive farmers wanting to utilize this system had to buy seed every year and depended on the price established in the market. As long as unit prices were stable and insurance could be obtained in eventual environmental emergencies, higher grain yields allowed farmers to annually pay for hybrid seed and be debt free. However, this was not true for many areas in the world and climate change increased this risk. Genetically diverse open-pollinated varieties utilized locally by farmers were replaced by genetically narrow-based hybrid maize cultivars, often, genetically identical hybrids with a combination of different Genetically modified organism (GMO) trait options. As a consequence, genetic vulnerability to abiotic and biotic stresses became a general concern.

The inbred–hybrid system is still the main method utilized for maize breeding and production today. However, low-cost alternative systems of maize production are still present worldwide (Dowswell et al. 1996; Carena 2005). Identifying the best hybrid combination of top-producing male and female lines continues to be the major challenge, though, and many of the seed business financial resources are geared towards predicting best cultivars. Genetically narrow-based maize hybrids have proved to be susceptible to climate change. Therefore, it would seem desirable to develop more sustainable breeding and production systems in maize.

5 Limitations in current breeding techniques

The inbred–hybrid system is over a hundred years old, and has proved economical in the developed world, but it had its consequences:

- Genetic diversity and sample sizes have declined (Brown 1953; Goodman 1985, 1999b). One type of cultivar, the single-cross hybrid, has been utilized in most farmer fields (Pollak 2003). This cultivar is made of heterozygous identical individuals extensively grown in homogeneous farmer fields. Farmer options have been reduced with retailers offering genetically identical hybrids with a combination of different transgenic events, thus increasing genetic vulnerability. Genetically uniform farmer fields have a high risk of disease epidemics and abiotic stresses.
- Maize cultivars have shown less stability and more susceptibility to climate change. National breeding programmes are still centred in ideal environmental conditions. New products need to be bred under marginal conditions for increasing abiotic and biotic stress tolerance and decreasing input needs associated with the excessive use of natural resources. The excessive focus on grain yield as the primary trait of maize hybrids has caused a significant reduction in grain quality and agronomic attributes, especially important for environments with climate changes.

- Public maize-breeding programmes integrating germplasm improvement with cultivar development have either been abandoned or replaced due to shift in research emphasis. Long-term maize improvement programmes on genetically broad-based germplasm utilizing recurrent selection programmes are barely conducted and these are sources of unique and more diverse products.
- The confidential nature of the maize business limits breeding rights to develop better industry products, which currently reduces the breeding efficiency to identify the most outstanding hybrids for farmers across regions. The ideal hybrid cultivar for farmers might not even be made due to intellectual property restrictions on improving identified weaknesses of industry products. Ex-Plant Variety Protection (PVP) lines are 20 years old, often narrow genetically, and are not expected to compete in the current market (Bari and Carena 2014, 2015). Seed prices and the trait integration business have been the priority over genetic diversity and breeding. As a consequence, genetic gains have slowed down.

World maize-breeding programmes have the opportunity to not only take advantage of the positive aspects of this system but also to address many of the problems raised above with new ideas for the development of the next generation of maize-breeding methods and cultivars tolerant to climate change. New ideas are needed to sustain maize production under climate change (Carena 2011).

Sustainable maize breeding requires active and continuous maize-breeding programmes integrating genetically broad-based germplasm improvement with the development of sustainable cultivars. These programmes focus not only on short-term decisions but also on long-term goals, and can provide different types of cultivars, each with a different need. They require a wise and cost-effective use of maize genetic resources and methods to meet the growing demands of food security, food quality and environmental conservation contributing to a minimum ecological footprint. Breeders of these programmes can enhance maize production and value through extensive germplasm adaptation, increased genetic diversity and maximization of genetic improvement.

6 Exploiting genetic diversity

Genetic diversity is one of the key factors in sustainable maize production. Ever since hybrid maize became a reality, Brown (1953) emphasized that approximately 98% of the world's maize germplasm was being ignored. Only recently a consortium of public and private organizations was formed to increase the genetic diversity of industry hybrids. This programme was expanded to the northern United States through the EarlyGEM program (Carena et al. 2009a,b; Carena 2012; Sharma and Carena 2012).

Results thus far have shown that germplasm carrying unique genes can break environmental margins for maize production (Carena 2011). Tropical germplasm can contribute useful and unique genes, especially after the expansion of new diseases (e.g. Goss's Wilt). Tropical maize possesses better grain quality and greater resistance to major pests (Holley and Goodman 1988; Carena et al. 2009a,b; Sharma and Carena 2012). However, incorporation of tropical germplasm has been slow. Industry needs to be ready to broaden its genetically narrow-based germplasm. Unfortunately, most farmer fields have increased the number of genetically identical hybrids with different single-gene events. This increases the probability of extensive epidemics and crop failures and, as a consequence, affects sustainability.

Long-term activities to adapt and maximize genetic improvement of elite diverse germplasm under climate change are required. Public breeding programmes have the luxury of applying long-term useful goals and ideas with continuous research, discriminating between innovative and 'cutting-edge' research. The defined cutting-edge research, often, is what most scientists do, which is not necessarily innovative. Therefore, sustainable breeding research needs to be innovative, unique and different from most work done. Maize-breeding programmes integrating genetically broad-based germplasm adaptation and improvement programmes with cultivar development have been scarce. They are the source of new knowledge, methods and sustainable maize cultivars. These unique programmes will provide national labs with genetic materials representing unique and diverse alleles increasing the sample sizes, currently limited by B73, NAM and G2F programmes. They will also help increase the sample sizes utilized in genetic mapping studies through classic mating designs (Carena and Wicks 2006; Bernardo 2008; Hallauer and Carena 2009) and validate models for genome-wide selection (Bernardo and Yu 2007) and elite cultivar prediction.

Sustainable maize breeding and production includes preserving maize genotypes carrying unique genetic diversity for future generations. This should be a wake-up call to all institutions, especially those housing maize-breeding programmes. Long-term cold storage and equipment infrastructure addressing long-term goals of adaptation and improvement are not very popular among administrative leaders. Greenhouses without adequate cold storage facilities for seed viability do not work even if they are more appealing investments. In a time of strong intellectual property policies, preserving germplasm is essential for future generations. In addition, exclusive agreements are more popular nowadays between university research foundations and private entities. As a consequence, many university official releases are not being registered nationally, that is, they are not preserved in national germplasm banks. This is a major concern to be addressed as these unique cultivars depend on local short- and long-term cold storage facilities if available or inadequately maintained. Exclusivity includes conditions. One of them is to provide private institutions years of advantage before a product could be registered publicly. Some of these products, probably the most successful ones for unique conditions, will never be registered. Therefore, public institutions have become genetic providers of unique maize inbred lines and germplasm to foundation seed companies who sell or license their products to retailer companies across regions. These retailer companies are the ones selling seed directly to farmers in exchange of licenses and royalties. Cooperation is desirable. Private and public institutions, however, currently suffer from their own intellectual property limitations and controlling systems. Royalties are believed to be the solution for basic and applied scientific products developed in the public sector in a confidential market. But, controlling breeding rights and breeding access limits not only the preservation but also the development of the ultimate outstanding product for farmers. Moreover, industry products (e.g. inbred lines) with known local weaknesses cannot be improved, often for 20 years. Once this time limit expires, ex-PVP products are not competitive (Bari and Carena 2015). Therefore, public–private breeding initiatives are encouraged to develop better industry versions for a more sustainable maize production.

There is a significant amount of genetic diversity in locally improved germplasm to be exploited within sustainable breeding programmes (Carena and Wicks III 2006), which may or may not represent the typical heterotic group classifications. These local germplasm sources often are not well preserved under local cold storage conditions either, which is also a call to preserve local germplasm carrying unique genetic diversity. Local sources of germplasm do not need additional time for adaptation and can directly be incorporated into genetic improvement programmes. The creation of synthetic varieties is also an option to be considered (Table 3).

7 Breeding for marginal environments

Climate change is a grim reality and national maize-breeding programmes need to factor in climate change by devising appropriate strategies. There is a need for applied breeding projects that will meet present and future demands of superior genotypes tolerant to climate change (Carena 2011). New ideas are needed to increase maize productivity and sustainability while taking advantage of breeding under increasing changes taking place in the climate. Breeding research programmes are essential for the future sustainability of maize production not only in current production areas but also in areas where maize production is headed (Carena 2013c). National governments have traditionally located breeding programmes in most favourable production areas (Fig. 3).

Millions of dollars are being wasted annually in growing hybrids that have been bred under ideal environmental conditions. In the northern United States, costs increase when hybrids with drought and cold susceptibility, later maturity, slow rate of dry down, low test weight and low quality for ethanol and feedstock utilization are grown.

Lowering farmer fertilizer and seed costs to farmers, for instance, has not been a very popular approach among scientists, but would be desirable for sustainable maize breeding. Breeding hybrids with low nitrogen (N) and plant density needs could be beneficial not only to farmers but also to breeders managing low seed amounts for testing across locations. Farmers, for instance, need to apply maximum N levels due to the fact that the market offers only hybrids bred under high N levels. Therefore, breeding research focused on developing maize cultivars under marginal conditions is essential to increase the efficiency of N uptake and use by maize plants. State and federal funds have been wasted with production research conducted on hybrids bred under high N levels. Many N levels are used in extension research trials to recommend hybrids tolerant to low N levels. But most of these trials include commercial hybrids bred only under high N levels. This does not help.

Another example includes the use of different plan populations in farmer fields. It has been extensively shown that grain yield of many single-cross maize hybrids has increased with the use of high plant densities (Troyer and Rosenbrook 1983). Indeed, most breeding programmes have increased selection under high plant densities over time to develop cultivars with better tolerance to those conditions. But selection under only high plant densities might not provide cultivars that are tolerant to low densities or even independent to plant densities (Carena et al. 1998; Carena and Cross 2003; Hyrkas and Carena 2005). One of the undesirable consequences of breeding under high densities, however, is that more seed is needed for breeding and production purposes. Seed is sold by the unit, which has increased overall seed costs. As with N research, scientists continue to spend state and federal funds to evaluate maize cultivars under different plant density levels (Edwards 2014) when knowingly these genotypes have all been bred under high plant densities. Carena and Cross (2003) and Hyrkas and Carena (2005) have confirmed the

Figure 3 Identifying desirable breeding locations for the development of the next generation of sustainable maize cultivars tolerant to climate changes.

availability of density independent maize populations that could be bred under low plant densities to reduce seed production needs in breeding programmes and annual seed costs to industry and farmers. Further research is needed to identify genotypes under different input needs. Like with N levels, researchers need to first breed under various plant density levels before evaluating genotypes at those different levels, especially if they are known to be bred solely under high plant densities.

Sustainable maize-breeding programmes need to be placed in challenging environments. Although non-stress environments are desirable to farmer fields and breeder seed production needs, most breeding work focuses on screening thousands of genetic materials and discarding weak genotypes. Thus, breeders need strategic and challenging environments to separate strong from weak genotypes fast under the pressure of climate changes. As such, breeders need marginal to extreme environments unlike the ideal conditions farmers need for production purposes. Marginal conditions can be used as an opportunity to direct evolution faster as breeding can be much more effective in marginal areas. These opportunities have often been neglected by simply trying to adapt hybrids developed under non-marginal conditions. As a consequence, adaptation of genotypes has been slower and stability of commercial hybrids weak. Therefore, public breeding programmes utilizing challenging environmental conditions for sustainable cultivar development could serve both marginal and non-marginal maize farmers independent from industry investment levels across years.

Even though maize yield productivity has increased, the stability of maize hybrids has decreased over the years (Carena 2013c). The lack of stable stress-tolerant maize hybrids reduces the options for farmers in marginal conditions and those that will be affected by

Table 1 Sample of temperate maize varieties developed and improved to short-season environments through reciprocal full-sib (FR), reciprocal half-sib with tester (R-T), modified ear-to-row (MER), inbred progeny (S) and intra-full sib (FS) recurrent selection programmes

Designation	Heterotic group	Genetic background	Tropical (%)
NDSS(FR)	SS (density and N)	North Dakota SS Synthetic	0
NDL(FR)	non-SS (density and N)	North Dakota Lancaster	0
NDSM(FS)	non-SS	North Dakota Synthetic M	0
NDSCD(FS)	non-SS	North Dakota Synthetic CD	0
NDBS21(R-T)	UR or non-SS	North Dakota Synthetic 21	0
NDBS22(R-T)	SS	North Dakota Synthetic 22	0
NDCG(FS) and (FR)	SS	North Dakota Synthetic CG	0
NDSAB(MER-FS)C15	non-SS	North Dakota Synthetic AB	0
NDLEAM(FR)	non-SS	Leaming(S-FS)C6	0

Source: Adapted from Hallauer and Carena (2013).

climate change in the future. Local maize adaptation to marginal environmental conditions is essential. Unfortunately, the lack of these programmes makes many commercial maize cultivars to suffer stability problems due to the fact of being often bred in non-marginal areas. The utilization of controlled winter nursery environments for abiotic stresses has been a solution for developing cold- and drought-tolerant maize products (Carena et al. 2009a).

Breeding under marginal or low input environments is needed for developing the next generation of sustainable maize cultivars. Maize research emphasizing the study of genotype × environment interactions has been necessary to find out if breeding under target conditions is required (Carena et al. 2009c). Finding genotype by N and genotype by density interactions would seem to be first necessary to find out if breeding under target conditions is needed. Selection under different N and plant density levels in North Dakota Stiff Stalk (NDSS), North Dakota Lancaster (NDL) and EarlyGEM populations has already shown that interactions are important. Other local or exotic temperate populations have either been incorporated or improved under short-season environments through intra- and inter-population recurrent selection programmes (Carena and Hallauer 2001a,b; Laude and Carena 2014) (Table 1).

8 High-throughput phenotyping

Breeders have the opportunity to develop not only products but also new breeding methodologies for non-destructive and better high-throughput phenotyping of genetically complex traits. Adaptation to climate change requires improvement of traits quantitatively inherited, often difficult to phenotype, and largely influenced by the environment. Molecular scientists initially developed labs and methods targeted at making breeding work easier or not needed at all, but unfortunately, the same scientists ended targeting genetically simple traits instead. Genetically simple traits are traits that are either easy to measure or not significantly biased by the environment or both. As a consequence, breeders can easily separate genetic from environmental effects for these traits by either simple visual observation or simple breeding techniques without being fooled by environmental bias. Molecular markers have had a limiting effect on quantitative traits (Sonnino et al. 2007) and shift in research emphasis has had a negative effect on sustainable breeding programmes worldwide (Guimaraes et al. 2006). The U.S. Science Foundation (NSF) and the U.S. Department of Agriculture (USDA) have spent millions of dollars targeting genetically simple traits such as flowering time in maize (Buckler et al. 2009).

Phenotyping is costly and wise decisions on how to target breeding technologies for trait screening are desirable. Therefore, developing methods and tools for easier phenotyping genetically complex traits are the ones needed. These traits are often controlled by hundreds of genes and each hybrid has its own specific gene combinations (Hallauer and Carena 2009). Active cooperation with industry should provide access to latest technologies to apply ideas for method and product development without having to invest in obsolete academic labs (Guimaraes et al. 2006; Sonnino et al. 2007). Large sample sizes carrying unique alleles are also essential when targeting cost and effective methodologies on traits that are genetically complex and difficult to measure. These considerations will fill the gap between geneticists, physiologists and breeders to develop the next generation of sustainable products with maximum efficiency. Interdisciplinary tools such as electronic sensors and UAS to increase the size of data sets can help, but scientists will need to make a wise selection of the most effective tools for their specific target environments.

There is a need for developing new maize-breeding methodologies that can facilitate breeders their work with quantitative traits in sustainable maize-breeding programmes. The development of breeding methodologies for quantitative traits is not common within breeding programmes, but it is needed for developing sustainable products. The NDSU maize-breeding program develops not only maize short-season products but also breeding methodologies for high-throughput screening of quantitative traits, especially those difficult to measure and largely influenced by the environment. Annually, the upper U.S. Midwest consumes more than $1.4 billion of fossil fuels to dry $19.7 billion of maize grain. Northern U.S. maize often needs to be harvested at moisture levels too high for safe storage and must be artificially dried for storage and transport.

Basic NDSU maize-breeding research on fast dry down (Yang et al. 2010) has generated a significant change in knowledge through the invention of a new breeding methodology (AUDDC-area under the dry down curve), which can save $ millions in artificial drying to farmers and screening costs to industry. The AUDDC was invented to better measure a genetically complex trait largely and historically challenged by the environment. The method has been validated to reduce production and industry costs related to artificial grain drying and hybrid development. Public and private institutions, nationally and internationally, have already taken advantage of this non-patented method developed at NDSU for breeding purposes to identify maize lines and hybrids with fast dry down. The NDSU maize-breeding program is the first programme to utilize a high-throughput methodology to develop short-season fast-drying maize lines and hybrids without the need to destroy plot samples for lab measurements.

As with the rate of dry down, drought tolerance and tolerance to acid soils (this topic is discussed in more detail in Chapters 21–23) are quantitatively inherited trait. These have challenged breeders for a long time due to their genetic complexity, screening difficulty and level of environmental influence. Growth of root systems and their lodging resistance capabilities have been important components of drought tolerance in maize genotypes. Phenotyping maize root systems, however, has been challenging because of its complex developmental patterns above and below the ground and the need for digging plots for screening purposes. The NDSU maize-breeding program proposed a new breeding methodology. Sharma and Carena (2016) developed BRACE, a non-destructive screening high-throughput breeding methodology for short-season drought tolerance involving visible brace root systems.

Several quantitative traits are still in need of breeding methodologies that could follow the examples described above. The maize scientific community will significantly advance existing knowledge if better targeting of traits with cost and efficient new phenotyping breeding ideas and technologies are followed.

9 Case study: use of exotic germplasm

Few maize-breeding programmes have had the long-term vision to incorporate exotic germplasm carrying unique alleles and the willingness to broaden the genetic basis of breeding programmes (Hallauer and Sears 1972; Goodman 1985; Pollak 2003; Carena 2007, 2013b; Carena et al. 2009a; Sharma and Carena 2012; Carena and Sharma 2016a; Hallauer and Carena 2016). The use of genetically diverse germplasm was encouraged after the Southern Corn Leaf Blight epidemic (*Bipolaris maydis*) in the 1970s. Tropical genetic materials have not only been continuously exposed to major pests for improvement (Holley and Goodman 1988) but also have shown useful genetic diversity for significant grain yield heterosis expression (Mungoma and Pollak 1988; Crossa et al. 1990; Echandi and Hallauer 1996). However, long-term programme efforts focused on the incorporation of tropical materials are still scarce.

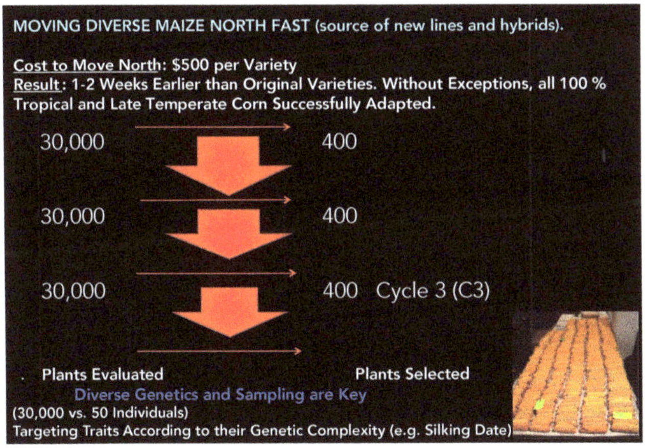

Figure 4 Use of stratified mass selection to adapt tropical and temperate maize varieties in the northern United States.

Source: Adapted from Carena et al. (2008); Hallauer and Carena (2014).

Mass selection has successfully been utilized to adapt exotic germplasm across latitudes (Hallauer 1999; Carena et al. 2008; Hallauer and Carena 2009, 2016; Hallauer et al. 2010). Even though mass selection is one of the oldest breeding methods, when properly conducted for genetically simple traits (e.g. flowering time), it has maximum efficiency and success across genotypes without exceptions. Stratified mass selection (Gardner 1961) for earliness, as primary selection target, has shown to be a cost-effective methodology to successfully adapt long-season maize to early maturing regions (Fig. 4). Unique tropical genetic materials have been effectively adapted to temperate environments (Hallauer and Carena 2014, 2016) and their agronomic performance has been comparable to adapted populations. This method requires good field isolations and, one of its major advantages, is that it allows largest sample sizes for an effective direct selection response per year. This method can generally be applied to any exotic population as it has shown to be effective across populations. Time of silk expression can reach an average of − 3 days per year with positive correlated responses on economically important and quantitative traits (Hallauer 1999; Hallauer and Carena 2009).

The adaptation of unique tropical and temperate genes to short-season environments has positive implications for farmers and industry in the long run. It provides a unique combination of agronomic and quality traits not available in industry hybrids and serves as the future sources of elite cultivars carrying diverse genetics.

The maize-breeding programme at the NDSU has worked towards increasing the genetic diversity from tropical and temperate regions while improving earliness, fast dry down, cold and drought tolerance, disease resistance and ethanol and feedstock quality. Summer maize-breeding nurseries include tropical varieties with sample sizes of up to 30000 individuals per pedigree. Several varieties are in the process of adaptation to ND, and the NDSU is the sole genetic provider (Table 2). They are the sources of new maize populations, lines and hybrids less vulnerable to pests, disease and climate change. One

Table 2 Sample of tropical and temperate maize varieties adapted to short-season environments through stratified mass selection, currently under genetic improvement through intra- and inter-population recurrent selection programmes

Designation	Heterotic group	Genetic background	Tropical (%)
NDSHLC	SS	ND Tropical Highland Synthetic	100
NDBSK(HI-M)C3	SS	ND Krug Yellow Dent	0
NDBS11(FR-M)C3	non-SS	ND Pioneer Two-Ear Synthetic	0
NDBS1011	UR	ND Pioneer/Iowa Synthetic	0
NDBSSS	SS	ND/IA Stiff Stalk Synthetic A	0
NDBS13	SS	ND/IA Stiff Stalk Synthetic B	0
NDBS39	non-SS	ND Tuson Composite(M)C11	100
NDBS16	non-SS	ND ETO Composite	100
NDBS28	SS	ND Tuxpeño Composite	100
NDBS29	non-SS	ND Suwan-1	100

SS = stiff stalk, non-SS = non-stiff stalk, UR = unrelated.

of the secrets to make these successful gains and desirable combination of traits is to choose the right maize germplasm and methodology in addition to utilizing large samples of genetically broad-based germplasm.

The genetic diversity from tropical maize contributes useful and unique genes that allow a combination of desirable trait properties. With enough genetic diversity, for instance, you can find the highest yielding hybrid with lowest grain moisture at harvest, a sustainable combination for northern U.S. environments that reduces risk. Adaptation and improvement of short-season tropical maize serve as new sources of unique public x public non-GMO and public x private GMO hybrid combinations that supplement the narrow genetic diversity present in most industry hybrids. This gene pool will offer unique early maturing competitive genes and products not available neither in the maize genome sequences available nor in the northern U.S. industry. These elite and uniquely diverse products will not only be tolerant to climate change with better abiotic and disease resistance with less input requirements but will also successfully impact the sustainability and stability of northern U.S. maize production and profitability. Similar procedures should be implemented to improve germplasm resources with greater tolerance to drought, acid soils and important pests that affect maize production worldwide.

In addition to the International Center for Maize and Wheat Breeding (CIMMYT), the Germplasm Enhancement Maize (GEM) project, ISU and NDSU EarlyGEM program efforts, North Carolina State University (NCSU) had one of the largest impacts for tropical maize in the United States (Goodman 1999a,b). It is, therefore, essential to continue supporting this type of programmes, which in this specific case, relies on decision making at NCSU. Decision making at NDSU and ISU, are also important as goals of the Iowa and North Dakota maize-breeding programmes include the typical elite-by-elite line pedigree selection cultivar development process. However, goals also include adaptation of exotic and unique germplasm, maximization of their genetic improvement, and development of unique products from genetically broad-based tropical and temperate germplasm for breeding and industry uses (Hallauer and Carena 2009, 2013; Hallauer et al. 2010; Carena and Sharma 2016a; Laude and Carena 2014, 2015). Therefore, national efforts to prevent local decisions against sustainable breeding programmes including germplasm preservation efforts would be encouraged (e.g. American Society of Agronomy (ASA) and American Seed Trade Association (ASTA) Germplasm Committee efforts could step in, among others).

The NDSU maize-breeding program leads the U.S. EarlyGEM project targeted at increasing the genetic diversity of short-season maize hybrids. With support of the ND and Minnesota (MN) Corn Grower Organizations, 22 maize unique EarlyGEM products have been adapted to ND and released to industry so far, which were extensively tested in industry and public network trials. These elite products originated from Argentina, Chile, Uruguay, Brazil, Mexico, Cuba, St. Croix and southern United States, and carry several percentages of tropical and temperate maize.

The plant materials introduced in ND were S_3 (or F_5) lines the GEM programme made available to network co-operators. The original linwwes trace back to the Latin American Maize Program (LAMP), in which 12 113 accessions were evaluated for multiple traits across 12 countries in Latin America (Sevilla and Salhuana 1997). From this extensive evaluation, 268 accessions were identified as elite germplasm sources (Pollak 2003). The GEM programme was created with U.S. Congress funds with the intention of broadening the genetic base of the hybrids grown by the U.S. producers. Different sets of accessions were identified and assigned to seed companies to make crosses with elite inbred lines. In some

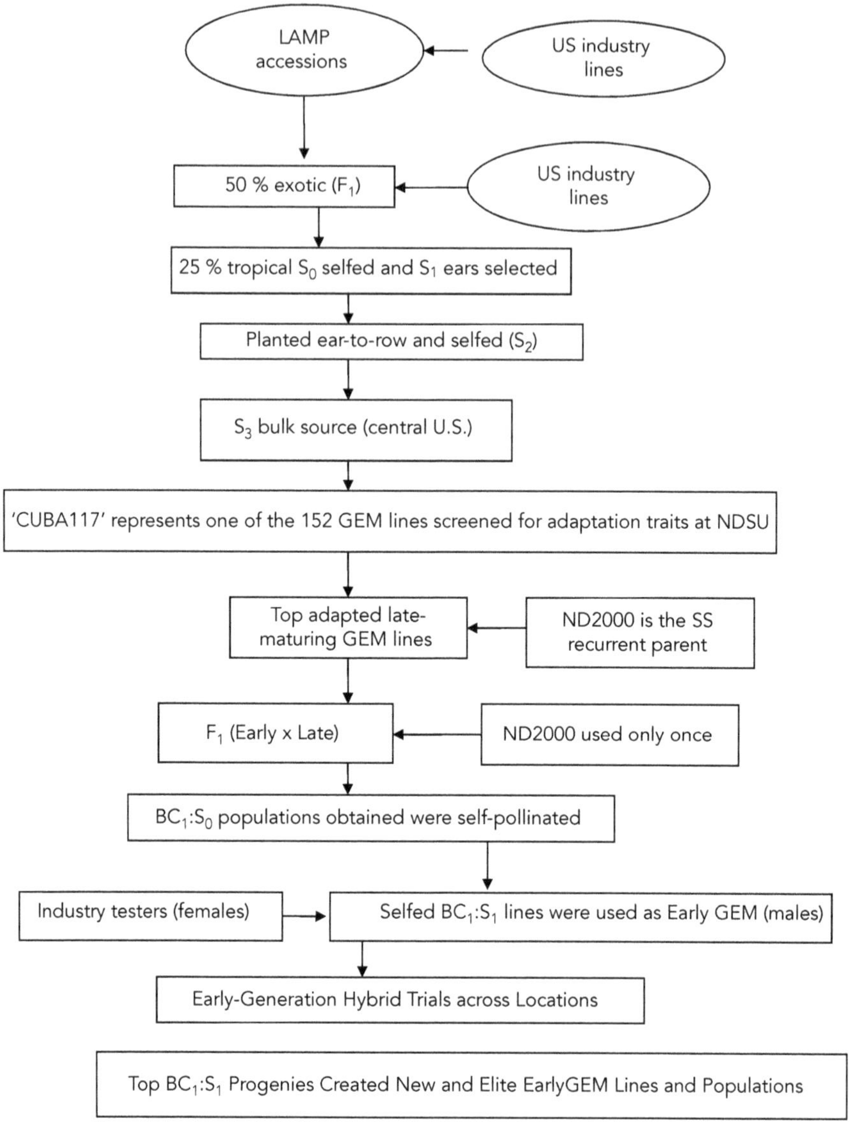

Figure 5 Long-term breeding process followed to develop NDSU EarlyGEM lines and populations. Source: Adapted from Hallauer et al. (2010).

cases, these breeding crosses were then assigned to seed companies to make additional crosses to adapted inbred lines (Salhuana 1997). Company lines were assumed to be the top ones industry had at that moment and their names were coded, but heterotic group information was preserved. The GEM programme later initiated inbred line development with early-generation testing to provide elite S_3 lines to network co-operators. These

Table 3 EarlyGEM maize short-season populations created after testing BC1:S1 early-generation lines in GMO and non-GMO hybrid combinations with industry and public maize inbred testers, respectively

Pedigree name	Heterotic group	Number of lines used for recombination	Accession origin
NDSU EarlyGEM22a	Non-SS	21	Chile
NDSU EarlyGEM22b	Non-SS	20	Chile
NDSU EarlyGEM4	Non-SS	22	St. Croix
NDSU EarlyGEM5	Non-SS	26	Brazil
NDSU EarlyGEM10	SS	20	Mexico
NDSU EarlyGEM3	SS	20	Cuba
NDSU EarlyGEM26	Non-SS	20	USA (Tropical)
NDSU EarlyGEM21a	SS	16	Argentina
NDSU EarlyGEM21b	SS	9	Argentina
NDSU EarlyGEM21c	SS	18	Argentina
NDSU EarlyGEM32	SS	45	Argentina
NDSU EarlyGEM33	Non-SS	22	St. Croix
NDSU EarlyGEM34	Non-SS	26	Brazil

SS = stiff stalk, non-SS = non-stiff stalk.

Figures 6 NDSU EarlyGEM maize before (left) and after (right) selection increased in the summer breeding nursery after a significant genetic progress. Selection has made maize one week earlier per year.

lines were still segregating to allow for further selection. The original efforts, however, concentrated on central and southern U.S. areas and the NDSU maize-breeding program became an active co-operator in 1999.

The EarlyGEM program (Carena 2007; Carena et al. 2009a,b; Sharma and Carena 2012) was initiated, through the NDSU maize-breeding program, to expand initial GEM efforts in northern latitudes. It started as a long-term project to increase the genetic diversity of industry hybrids in the northern United States and southern Canada with the rapid incorporation of elite exotic late-maturing germplasm in the year 2000. Therefore, the NDSU EarlyGEM program targeted to expand GEM to higher latitudes, north and west in the northern hemisphere. The goals were to move elite tropical and temperate maize germplasm northwards. This was achieved by adapting LAMP-GEM elite germplasm to short-season drought and cold-prone environments (Carena et al. 2009a,b) and developing new and unique early maturing lines for industry use (Fig. 5).

Rinke and Sentz (1962) inspired the backcross programme utilized for adapting GEM germplasm, with the difference that recurrent parents were short-season cultivars (Carena and Wanner 2003; Carena et al. 2003; Barata and Carena 2006) and that only one backcross was conducted to speed up the process. Early-generation hybrid multi-environment and multi-trait trials, including EarlyGEM lines in combination with industry and public testers, were extensively conducted, and S_1 seed of best progenies was used for the double purpose of developing new early maturing lines and creating new short-season exotic populations (Carena and Sharma 2016b). A range from 9 to 45 top progenies were inter-mated utilizing the bulk-entry methodology (Table 3), which allowed not only the development of the new EarlyGEM populations but also self-pollination of early-generation lines to continue inbred line development of top progenies. EarlyGEM unique products showed outstanding characteristics when compared to industry top checks and have been released to exclusive and non-exclusive partners. The NDSU maize-breeding program is the sole genetic provider of these elite and exotic genetic materials. New genetic materials have demonstrated to be, when properly tested, outstanding sources for grain yield, grain quality, cold and drought tolerance, and, especially, disease resistance. In years of extensive disease infection affecting summer breeding nurseries, only germplasm derived from the EarlyGEM program has shown high levels of disease resistance up to complete immunity. This programme has been strategically located to develop germplasm for environmental margins and climate changes faster (Carena 2011). Results have been successful (Fig. 6).

10 Case study: short-season quality maize hybrids

Farmers and ranchers would be able to increase their profitability if value-added maize products were available. Maize is mainly grown for manufactured food products, feed and ethanol. A primary nutritional limitation of the maize grain is the insufficient amounts of essential amino acids (i.e. lysine, methionine and cysteine, among others) to satisfy the nutritional requirements of mono-gastric animals, including humans. Also, there is potential to increase the quality of maize by-products after processing. The content and quality of protein in maize has been increased through genetic improvement. However, adaptation of Quality Protein Maize (QPM) genotypes to temperate environments has been quite challenging (Worral et al. 2015).

There is a need for marketing sustainable hybrids adapted to very short seasons with enhanced quality for ethanol production, human nutrition and livestock feed products. NDSU leads the first national programme to adapt tropical QPM genotypes to the northern United States. The maize-breeding programme created the NDSU EarlyQPM (Early QPM) and the NDSU EarlyQPMF (Early QPM for Feedstock) programs to increase the utilization of short-season products, adding value to the crop, and balancing the nutritional needs in the livestock market. It provides genetic diversity for added value to U.S. northern farmers and ranchers. The breeding programme is the first national programme to initiate efforts to adapt QPM genotypes to the northern United States and to extensively evaluate elite and adapted populations, lines and hybrids for amino acid composition. New high-quality populations, lines and hybrids are available for the first time in short-season environments. These particular programmes have the potential to increase the value for feedstock operations before and after ethanol utilization by producing not only high-yielding fast-drying maize hybrids but also those with maximum grain protein quality.

The NDSU maize-breeding quality programmes have already established a continuous grain quality lab for the full evaluation of short-season maize for high-quality protein through screening of grain quality components and essential amino acids across breeding plots. These efforts complement the programme's extensive partnership with the seed industry to screen the same genotypes for ethanol utilization. Each NDSU maize-breeding plot from all experiments is sampled for the NDSU EarlyQPM and QPMF programs, which have already shown the best performance on grain protein content, including the essential amino acids, lysine, methionine and cysteine. These genetic materials have been adapted for the first time to short seasons through a system similar to the one used for the NDSU EarlyGEM program (Dong and Carena 2016).

11 Case study: cold- and drought-resistant varieties

Genetically narrow-based maize hybrids have shown to be susceptible to climate change (Carena 2013c). Growing genetically diverse and shorter-season vigorous maize varieties solves most of the agronomic problems facing farmers in early maturing environments. Still, in northern U.S. environments, the market offers hybrids that are not stable to short-season environmental conditions. Influencing industry to move maize north (or south in the southern hemisphere) will be important to the future of short-season maize producers. Millions of dollars are wasted annually to growing and drying hybrids that have been bred under non-marginal conditions with hidden susceptibilities. Costs increase when hybrids with drought and cold susceptibility, later maturity, slow rate of dry down, low test weight and low quality for ethanol and feedstock utilization are grown.

There is the need for new product development in marginal areas. Breeding programmes have traditionally not been strategically located to work faster under the pressure of marginal environments. Industry is currently not offering stable stress-tolerant hybrids and northern U.S. farmers still do not have the same opportunities as their southern counterparts have to enhance their quality of life. A long-term solution for improving profitability in these vast regions is to develop the next generation of short-season stable and diverse products for a sustainable maize production. Breeding for adaptation is best done under target-challenging but uniform-controlled environmental conditions where strengths and weaknesses are quickly identified and stable genotypes succeed.

Figures 7 (a) Earliest (65RM) and most vigorous NDSU maize line (left) and most drought-tolerant line identified in southern New Zealand and southern Argentina winter nurseries with controlled conditions for short-season cold tolerance (right).

The ND maize-breeding program is currently known for being the most northern maize-breeding programme in North America, moving maize north to cooler seasons and west to dry areas fast. With the support of industry, growers and, especially, maize grower organizations, the programme annually screens approximately three million genotypes in summer and winter breeding nurseries and extensive population, inbred and hybrid multi-location trials serving as a unique genetic provider to foundation seed companies, retailer seed companies, processing industry and national and international breeders.

Since 1999, winter nurseries have been essential to the NDSU maize-breeding program not only to develop products fast but also to manage controlled environmental conditions (no rain and extreme cold when wanted and needed during ND winters). The purpose is to screen thousands of maize lines for drought and cold tolerance as well as for hybrid production of tolerant products identified during the process. One of the positive consequences for utilizing winter nurseries for screening purposes is that 94–100% of NDSU and industry experiments produced lines that in hybrid combinations have beaten top commercial industry checks in marginal environments, while only 54% of them were successful before controlled winter nurseries were initiated for abiotic stress screening purposes. Also, short-season drought-tolerant hybrids have increased their grain yield up to 194% relative to top industry checks in western ND irrigated and dryland trials while showing stable ranks across environments. Overall, advanced NDSU lines have shown higher yield and quality under drought stress and higher cold tolerance in hybrid combinations when compared to commercial checks. Three improved populations were directly utilized in high-starch hybrid cultivars for commercial production under drought conditions in central ND. Cold- and drought-tolerant lines are currently in pre-commercial industry trials as well. NDSU experimental hybrids showed larger yield and grain moisture differences as test locations were in higher latitudes. Therefore, a significant genetic improvement was achieved after utilizing our maize-breeding methodology for screening cold tolerance lines and hybrids in controlled winter nurseries (Fig. 7) complemented with progeny testing programmes at a rate of one year per selection cycle (Sezegen and Carena 2009).

12 Summary and future trends

Sustainable maize production will be possible only with the development of the next generation of maize products carrying genetic diversity. Sustainable maize breeders face current and future challenges (e.g. climate change) that can be addressed only by germplasm diversity (Carena 2011, 2013a) and the utilization of marginal environments for faster genetic improvement (Carena 2013b,c). Significant genetic gains can be obtained fast with the utilization of useful genetic diversity in breeding centres located in strategic marginal environments. Unlike the ideal conditions desired by farmers, challenging environmental conditions will be essential to sustainable breeders to facilitate the identification of strengths and weaknesses in genotypes for breeding purposes. Only those maize-breeding programmes utilizing large samples of genetically broad-based germplasm have the potential to contribute useful and unique alleles to face climate challenges (Fig. 8). These programmes need to integrate germplasm adaptation and improvement with cultivar development (Carena et al. 2010; Hallauer et al. 2010) adapting and improving exotic and unique germplasm. Developing new products from these genetically broad-based germplasm sources should supplement the typical elite-by-elite line pedigree selection cultivar development process often used in breeding programmes focused on genetically narrow-based germplasm.

The next generation of successful and sustainable maize cultivars is dependent upon screening diverse and large samples of elite germplasm under controlled environmental conditions (Carena 2013b). Adequate sampling includes all possible genotypes, not just the most elite ones (e.g. genome sequence of 'NAM', genotyping and phenotyping of 'B73 × Mo17', limited sample of the 'G2F' project, among others proposed). Maize germplasm ranging from tropical to very short-season genetic materials, including indigenous maize, ensures discovery, validation and utilization of all possible genes affecting economically important traits across environments, each with different needs.

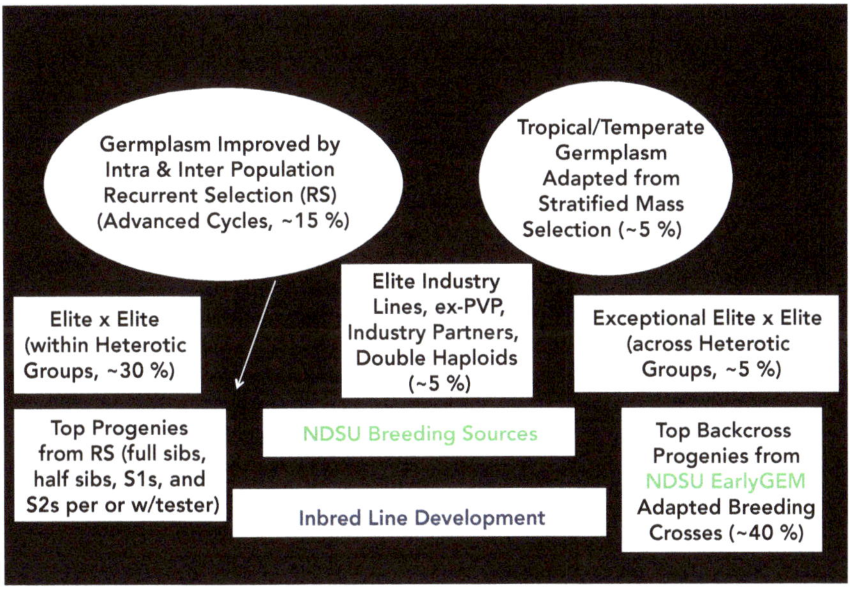

Figure 8 Current germplasm breeding sources for NDSU maize inbred line and hybrid development. Source: Adapted from Hallauer et al. (2010).

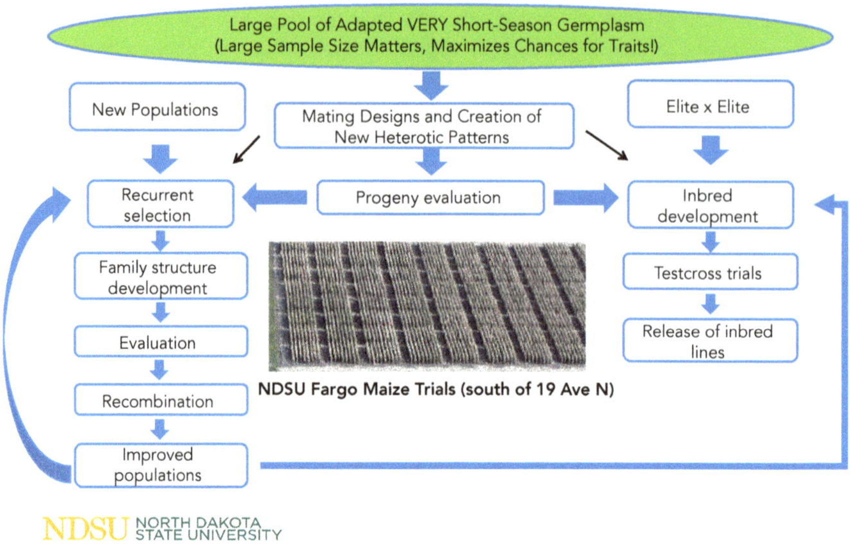

Figure 9 Integration of pre-breeding with cultivar development.
Source: Adapted from Hallauer et al. (2010).

Only this kind of sampling can find desirable combinations of traits for a sustainable maize production. The use of genetically narrow-based germplasm is one of the reasons why farmers and agronomists tend to believe that only later-maturing hybrids provide top yields. Desirable trait combinations can be found only in programmes utilizing large sampling in genetically broad-based germplasm (Fig. 9).

Farmer maize options have been significantly reduced with seed companies offering identical hybrids with only different single-gene events (e.g. RR, Bt, and combinations of events). The wide use of narrow genetics in developing industry hybrids is a real concern and prevention is needed to avoid another genetic vulnerability similar to the widespread disease epidemic that occurred in the United States in the 1970s, affecting over 15% of the national maize crop. Diverse elite gene pools are needed for the development of the next generation of maize cultivars with reduced risk and added value to farmers and ranchers. New genetically diverse maize populations and alternative heterotic patterns need to be created to provide the next generation of sustainable maize lines, populations and their hybrids. The NDSU maize-breeding program leads a genetic diversity programme carrying tropical germplasm at a range of 12.5–100%, and these are being successfully adapted to the northern United States. These have increased the maize diverse gene pool that could address climate change, especially for northern and southern short-season climates. The adaptation of unique tropical genes to short seasons has provided a combination of traits not available in industry hybrids (e.g. disease resistance, low grain moisture, high yield). Similar genes have been found for new uses and markets. These diverse products carry genes not identified in multi-million genome sequence projects (Fig. 10). Germplasm expressing these genes have been adapted at less than 1% of the total cost needed for finding minor genes on the same trait used for selection (Hallauer and Carena 2014). Results have shown that germplasm carrying unique genes can break environmental margins for

Figures 10 New short-season tropical maize lines adapted to the northern United States.

maize production. NDSU tropical-adapted products (e.g. including those adapted from South America) expressed unique genes for cold and drought tolerance already and we expect similar genes for above-average grain quality for new uses and markets. It is hoped that the NDSU will continue to advance existing knowledge by becoming a leader in developing the next generation of healthier short-season hybrids for achieving better livestock and human nutrition (Carena 2013b). The NDSU is the only genetic provider of these new products carrying unique genes and will be essential in the development of the next generation of sustainable cultivars tolerant to climate changes as long as the programme remains active with future breeders.

In order to obtain elite germplasm to develop new sustainable maize cultivars, breeders need to adapt and improve diverse genotypes to their target environments. However, choice of germplasm determines the success of a breeding programme. No breeding methodology or technology can provide successful cultivars with a poor germplasm choice. However, once elite germplasm is chosen, breeders should be able to develop not only new cultivars but also new breeding methodologies for efficient selection of quantitative traits. There is a need for breeders to develop not only products but also methodologies that can save time and cost in measuring economically important traits, which are genetically complex, difficult to screen, and largely influenced by the environment. Invention of new and practical breeding methodologies will increase high throughput screening efficiency of quantitative traits of economic importance. Rate of dry down (Yang et al. 2010), cold tolerance (Carena 2013b), and drought tolerance (Sharma and Carena 2016) are examples of new and successful inventions on breeding methodologies that have facilitated the screening of thousands of genotypes for traits that are difficult to measure and significantly influenced by the environment. These are breeding techniques that avoid the destruction of plots and save $ millions in maize breeding and production costs. The next generation of maize breeders will need to target technology according to the genetic complexity of the trait. Ideas to overcome the evaluation of genetically complex traits will continue to be a priority within our programme (Carena 2013c) to develop the next generation of sustainable maize products.

There are still many challenges ahead. Molecular techniques were initially targeted at economically important traits, which could be the reason why this technology is now utilized in genetically simple traits. Technology, if used wisely, however, should target traits according to their genetic complexity. For instance, flowering time is a genetically simple trait that breeders can utilize to move maize to higher latitudes at a very low cost. Drought tolerance through the use of non-destructive root-screening systems (Sharma and Carena 2016) and rate of dry down (Yang et al. 2010) are successful examples of genetically complex traits targeted with new breeding methodologies that can facilitate high-throughput screening of large samples. Breeders should focus on new breeding methodologies for genetically complex traits, including the use of UAS. Investment in breeders with cost-efficient practical ideas is desirable for innovation.

Public–private partnerships to develop the next generation of sustainable maize products are encouraged, especially to overcome industry product weaknesses under environmental stresses. The ideal hybrid for northern U.S. farms is still missing due to intellectual property limitations on breeding access. The development of better industry versions is needed. In addition, there is interest to utilized old versions of industry lines. However, ex-PVP lines have shown limited usefulness due to their established protection for 20 years. Despite the interest data have shown ex-PVP products to be non-competitive and genetically narrow old products (Bari and Carena 2014, 2015). Breeding efforts should focus on new and unique genetically diverse products.

National institutes and international centres should emphasize the potential of alternative sustainable systems of maize production (e.g. improved populations per and in hybrid combinations, lines × populations, among others). Also, scientists should focus on ideas to improve the current single-cross hybrid maize methodology which is old. The future relies on maize products carrying diverse genetics. This will prevent vulnerabilities to epidemics and climate change. Breeding goals of adaptation, genetic improvement and integration of pre-breeding with product and method development should help expand maize production in a sustainable fashion.

The NDSU maize-breeding program leads a large network of public and private co-operators assisting the programme in developing new products and methods without the need to spend federal and state funds on costly academic labs becoming obsolete, intellectual property, testers, checks, GMO transformation, grain quality equipment, plots and locations. The programme is a unique provider of products, methods and trained employees to seed companies and to national and international institutions. Adaptation and improvement of unique short-season and tropical germplasm has served as new sources of unique public × public non-GMO and public × private GMO hybrid combinations that supplement the narrow genetic diversity present in most industry hybrids. Maize research, development and production in marginal regions will continue to lead breeding research for climate change worldwide. These will be the source of unique sustainable cultivars that will meet the twenty-first-century demands of food, feed, fuel and fibre.

13 References

Barata, C. and Carena, M. J. (2006). Classification of North Dakota Maize Inbred Lines into Heterotic Groups Based on Molecular and Testcross Data. *Euphytica* 151: 349.

Bari, M. A. A. and Carena, M. J. (2014). Can Expired Proprietary Maize (*Zea mays* L.) Industry Lines be Useful for Short-Season Breeding Programs? I. Grain Quality and Nutritional Traits. *Euphytica* 202: 157–71.

Bari, M. A. A. and Carena, M. J. (2015). Can Expired Proprietary Maize (*Zea mays* L.) Industry Lines be Useful for Short-Season Breeding Programs? II. Agronomic Traits. *Euphytica* 207: 69–77.

Beal, W. J. (1878). *Report of Michigan State Board Agr.* 17: 445–7.

Bernardo, R. (2008). Molecular Markers and Selection for Complex Traits in Plants: Learning from the last 20 years. *Crop Sci.* 48: 1649–64.

Bernardo, B. and Yu, J. (2007). Prospects for Genome-Wide Selection for Quantitative Traits in Maize. *Crop Sci.* 47: 1082–90.

Buckler, E. S., Holland, J. B., Bradbury, P. J., Acharya, C. B., Brown, P. J., Browne, C., Ersoz, E., Flint-Garcia, S., Garcia, A., Glaubitz, J. C., Goodman, M. M., Harjes, C. G., Guill, K., Kroon, D. E., Larsson, S., Lepak, N. K., Li, H., Mitchell, S. E., Pressoir, G., Peiffer, J. A., Rosas, M., Rocherford, T. R., Romay, M., Romero, S., Salvo, S., Villeda, H. S., da Silva, H., Sun, Q., Tian, F., Upadyayula, N., Ware, D., Yates, H., Yu, J., Zhang, Z., Kresovich, S. and McMullen, M. D. (2009). The Genetic Architecture of Maize Flowering Time. *Science* 325: 714–18.

Brown, W. L. (1953). Maize of the West Indies. *Trop. Agric.* 30: 141–70.

Carena, M. J. (2005). Maize Commercial Hybrids Compared to Improved Population Hybrids for Grain Yield and Agronomic Performance. *Euphytica* 141: 201–8.

Carena, M. J. (2007). Moving Corn Belt GEM germplasm northward. In Agronomy Abstracts. ASA, Madison, WI.

Carena, M. J. (2011). Germplasm Enhancement for Adaptation to Climate Changes. *Crop Breed. Appl. Biotech.* 11: 56–65.

Carena, M. J. (2012). Increasing Genetic Diversity with Exotic Germplasm. *48th Illinois Corn Breeders School Proceedings.*

Carena, M. J. (2013a). Challenges and Opportunities for Developing Maize Cultivars in the Public Sector. *Euphytica* 191: 165–71.

Carena, M. J. (2013b). Developing Cold and Drought Tolerant Short-Season Maize Products for Fuel and Feed Utilization. *Crop Breed. Appl. Biotech.* 13: 1–8.

Carena, M. J. (2013c). Developing the Next Generation of Diverse and Healthier Maize Cultivars Tolerant to Climate Changes. *Euphytica* 190: 471–9.

Carena, M. J. and Cross, H. Z. (2003). Plant Density and Maize Germplasm Improvement in the Northern Corn Belt. *Maydica* 48: 105–11.

Carena, M. J. and Wicks III, Z. W. (2006). Maize Early Maturing Hybrids: An exploitation of U.S. Temperate Public Genetic Diversity in Reserve. *Maydica* 51: 201–8.

Carena, M. J. and Hallauer, A. R. (2001a). Expression of Heterosis in Leaming and Midland Yellow Dent Populations. *J. Iowa Acad. Sci.* 108: 73–8.

Carena, M. J. and Hallauer, A. R. (2001b). Response to Inbred Progeny Recurrent Selection in Leaming and Midland Yellow Dent Populations. *Maydica* 46: 1–10.

Carena, M. J. and Wanner, D. W. (2003). Registration of ND2000 Inbred Line of Maize. *Crop Sci.* 43: 1568–9.

Carena, M. J. and Sharma, S. (2016a). Genetic Variability of Grain Quality in Maize (*Zea mays* L.): Breeding Implications for Short-Season Drought Environments. *Euphytica* (under review).

Carena, M. J. and Sharma, S. (2016b). Registration of Five Short-Season Stiff Stalk (SS) EarlyGEM Maize Germplasms. *J. Plant Registrations* (accepted).

Carena, M. J., Wanner, D. W. and Cross, H. Z. (2003). Registration of ND291 Inbred Line of Maize. *Crop Sci.* 43: 1568.

Carena, M. J., Santiago, I. and Ordas, A. (1998). Direct and Correlated Response to Selection for Prolificacy in Maize at Two Planting Densities. *Maydica* 43: 95–102.

Carena, M. J., Eno, C. and Wanner, D. W. (2008). Registration of NDBS11(FR-M)C3, NDBS1011, and NDBSK(HI-M)C3 maize germplasms. *J. Plant Reg.* 2: 132–6.

Carena, M. J., Wanner, D. W. and Yang, J. (2010). Linking Prebreeding for Local Germplasm Improvement with Cultivar Development in Maize Breeding for Short-Season (85–95 RM) Hybrids. *J. Plant Reg.* 4: 86–92.

Carena, M. J., Pollak, L., Salhuana, W. and Denuc, M. (2009a). Development of Unique Lines for Early-Maturing Hybrids: Moving GEM Germplasm Northward and Westward. *Euphytica* 170: 87–97.

Carena, M. J., Bergman, G., Riveland, N., Eriksmoen, E. and Halvorson, M. (2009b). Breeding Maize for Higher Yield and Quality under Drought Stress. *Maydica* 54: 287–96.

Carena, M. J., Yang, J., Caffarel, J. C., Mergoum, M. and Hallauer, A. R. (2009c). Do Different Production Environments Justify Separate Maize Breeding Programs? *Euphytica* 169: 141–50.

Crossa, J., Taba, S. and Wellhausen, E. J. (1990). Heterotic Patterns Among Mexican Races of Maize. *Crop Sci.* 30: 1182–90.

Dong, N. and Carena, M. J. (2016). NDSU EarlyQPM and NDSU Early QPMF Programs: Developing the Next Generation of Corn Nutritional Products. *Euphytica* (under review).

Dowswell, C. R., Palival, R. L. and Cantrell, R. P. (1996). Maize in the Third World. West View Press, Boulder, CO, U.S.

East, E. M. (1908). Inbreeding in Corn. *Ann. Rept. Conn. Agr. Expt. Sta.* 1907–08: 419–28.

Echandi, C. R. and Hallauer, A. R. (1996). Evaluation of U.S. Corn Belt and Adapted Maize Cultivars and their Diallel Crosses. *Maydica* 41: 317–24.

Edwards, J. (2014). Genotype by Environment Interaction for Plant Density Response in Maize. *Crop Sci.* http://www.ars.usda.gov/research/publications/publications.htm?SEQ_NO_115=298838.

Gardner, C. O. (1961). An Evaluation of Effects of Mass Selection and Seed Irradiation with Thermal Neutrons on Yield of Corn. *Crop Sci.* 1: 241–5.

Goodman, M. M. (1985). Exotic Maize Germplasm: Status, Prospects, and Remedies. *Iowa State J. Res.* 59: 497–527.

Goodman, M. M. (1999a). Developing Temperate Inbreds from Tropical Germplasm: Rationale, Results and Conclusions. *Illinois Corn Breeders' School* 35: 1–19.

Goodman, M. M. (1999b). Broadening the Genetic Diversity in Maize Breeding by Use of Exotic Germplasm. In J. G. Coors and S. Pandey (eds), *Genetics and Exploitation of Heterosis in Crops*, pp. 139–48. ASA, CSSA, and SSSA, Madison, WI.

Guimaraes, E., Kueneman, E. and Carena, M. J. (2006). Assessment of National Plant Breeding and Biotechnology Capacity in Africa and Recommendations for Future Capacity Building. *Hort. Sci.* 41: 50–2.

Hallauer, A. R. (1999). Conversion of Tropical Germplasm for Temperate Area Use. *Illinois Corn Breeders' School* 35: 20–36.

Hallauer, A. R. and Carena, M. J. (2009). Maize Breeding. In: M. J. Carena (ed.), *Handbook of Plant Breeding: Cereals*. Springer, New York.

Hallauer, A. R. and Carena, M. J. (2013). Recurrent Selection in Maize. *Maydica* 57: 266–83.

Hallauer, A. R. and Carena, M. J. (2014). Adaptation of Tropical Maize Germplasm to Temperate Environments. *Euphytica* 196: 1–11.

Hallauer, A. R. and Carena, M. J. (2016). Registration of BS39 Maize Germplasm. *J. Plant Reg.* (in press).

Hallauer, A. R. and Sears, J. H. (1972). Integrating Exotic Germplasm into Corn Belt Breeding Programs. *Crop Sci.* 12: 203–6.

Hallauer, A. R., Carena, M. J. and Miranda Fo, J. B. (2010). Quantitative Genetics in Maize Breeding. 3rd Ed., Springer, New York.

Holley, R. N. and Goodman, M. M. (1988). Stalk Quality and Stalk Rot Resistance of Tropical Hybrid Maize Derivatives. *Plant Dis.* 72: 321–4.

Hyrkas, A. and Carena, M. J. (2005). Response to Long-Term Selection in Early Maturing Maize Synthetic Varieties. *Euphytica* 143: 43–9.

Laude, T. P. and Carena, M. J. (2014). Diallel Analysis Among 16 Maize Populations adapted to the U.S. Northern Corn Belt. Euphytica 200: 29–44.

Laude, T. P. and Carena, M. J. (2015). Genetic Diversity and Heterotic Grouping of Tropical and Temperate Maize Populations Adapted to the Northern U. S. Corn Belt. *Euphytica* 204: 661–77.

Leaming, J. S. (1883). Corn and Its Culture. *J. Steam Printing*. Wilmington, OH.

Moll, R. H., Lonnquist, J. H., Fortuno, J. V. and Johnson, E. J. (1965). The Relationship of Heterosis and Genetic Divergence in Maize. *Genetics* 42: 139–44.

Mungoma, C. and Pollak, L. M. (1988). Heterotic Patterns Among 10 Corn Belt and Exotic Maize Populations. *Crop Sci.* 28: 500–4.

Pollak, L. M. (2003). The History and Success of the Public-Private Project on Germplasm Enhancement of Maize (GEM). *Adv. Agron.* 78: 45–87.

Rinke, E. H. and Sentz, J. C. (1962). Moving Corn Belt Dent Germplasm Northward. *Minnesota Agr. Exp. Stn.* 1110: 53.

Salhuana, W. (1997). Proposal of International Maize Germplasm Enhancement Project (IMGEP). In W. Salhuana, R. Sevilla and S. A. Eberhart (eds), LAMP Final Report 1997, pp. 125–39.
Sevilla, R. and Salhuana, W. (1997). General Description of the Plan and Execution of Latin American Maize Project (LAMP), In W. Salhuana, R. Sevilla and S. A. Eberhart (eds), LAMP Final Report 1997, pp. 1–35.
Sezegen, B. and Carena, M. J. (2009). Divergent Recurrent Selection for Cold Tolerance in Two Improved Maize Populations. *Euphytica* 167: 237–44.
Sharma, S. and Carena, M. J. (2012). NDSU EarlyGEM: Increasing the Genetic Diversity of Northern U.S. Hybrids through the Development of Unique Exotic Elite Lines. *Maydica* 57: 34–42.
Sharma, S. and Carena, M. J. (2016). BRACE: A Method for High Throughput Maize Phenotyping of Root Traits for Short-Season Drought Tolerance. *Crop Sci.* (in press).
Shull, G. H. (1908). The Composition of a Field of Maize. *Am. Breeders' Assoc. Rep.* 4: 296.
Sonnino, A., Carena, M. J., Guimaraes, E. P., Baumung, R., Pilling, D. and Rischkowsky, B. (2007). An Assessment of the Use of Molecular Markers in Developing Countries. In E. P. Guimaraes, J. Ruane, B. D. Scherf, A. Sonino and J. D. Dargie (eds), *Marker-Assisted Selection: Current Status and Future Perspectives in Crops, Livestock, Forestry, and Fish*. Food and Agriculture Organization of the United Nations, Rome, IT, pp. 15–26.
Sturtevant, E. L. (1899). Varieties of Corn. *USDA Off. Exp. Stn. Bull.* 57: 1–108.
Troyer, A. F. and Rosenbrook, R. W. (1983). Utility of Higher Plant Densities for Corn Performance Testing. *Crop Sci.* 23: 863–7.
Worral, H. M, Scott, P. and Hallauer, A. R. (2015). Registration of Temperate Quality Protein Maize (QPM) Lines BQPM9, BQPM10, BQPM11, BQPM12, BQPM13, BQPM14, BQPM15, BQPM16, and BQPM17. *J. Plant Reg.* 9: 371–5.
Yang, J., Carena, M. J. and Uphaus, J. (2010). AUDDC: A New Method to Evaluate Rate of Dry Down in Maize. *Crop Sci.* 50: 1–8.

Chapter 4

Understanding and improving protein traits in maize

Yongrui Wu, National Key Laboratory of Plant Molecular Genetics, CAS Center for Excellence in Molecular Plant Sciences, Institute of Plant Physiology & Ecology, Shanghai Institutes for Biological Sciences, Chinese Academy of Sciences, China; and Joachim Messing, Waksman Institute of Microbiology, Rutgers University, USA

1 Introduction

Global maize (*Zea mays*, commonly known as corn) production yields approximately 1 billion tons of seeds per year. Given that seeds comprise about 10% protein, corn supplies 100 million tons of protein for feed and food annually. Here, we review several parameters that contribute to the supply and quality of corn protein. For instance, kernel hardness is critical for mechanic strength during harvest, transportation and storage. On the other hand, kernel hardness depends on the composition of storage proteins, with zeins as the major component. Zeins are extremely biased in amino acid composition. In contrast to α-zeins, β-zein, γ-zeins and δ-zeins are rich in sulphur-rich amino acids, but all of them are deficient in the essential amino acids lysine and tryptophan. Because of the preponderance of zeins, which account for 60% of total seed protein, maize is nutritionally inferior and needs costly dietary supplements. Three transcription factors (TFs), opaque2 (O2), prolamin-box binding factor (PBF) and opaque2 heterodimerizing proteins (OHPs), regulate 90% of zein transcription. Misfolded or low levels of proteins result in irregular protein body (PB) shapes and opaque, non-vitreous seed phenotypes. Reduction of α-zeins causes an increase in lysine levels because of the compensatory up-regulation of other proteins in the seed. Moreover, *o2* modifiers can overcome the unfavourable

http://dx.doi.org/10.19103/AS.2016.0001.04

concomitant defect in the loss of kernel hardness, which is known today as Quality Protein Maize (QPM). Still, QPM alone cannot alleviate the methionine deficiency in maize seed, which is mainly caused by limited availability of reduced sulphur. Therefore, increasing the capacity of sulphur reduction during photosynthesis should enhance the methionine level in addition to the overexpression of the methionine-rich 10-kD δ-zein gene.

2 Storage proteins in maize seed

The major parts of corn seed are embryo and endosperm, which constitute approximately 10% and 90% of dry weight, respectively. Both are storage organs, the former mainly accumulating oil and protein and the latter starch and protein. Based on the solubility in different solvents, the storage proteins are classified into (Osborne 1908; Shewry and Casey 1999):

- albumins (soluble in water),
- globulins (soluble in dilute saline),
- prolamins (soluble in 60–70% alcohol, also called zeins in maize) and
- glutelins (soluble in alkali)

In maize seed, they account for approximately 3%, 3%, 60% and 34% of total proteins, respectively. In the embryo, the primary storage proteins are globulins (GLBs) (Khavkin et al. 1978), containing two subgroups, GLB1 and GLB2, which are encoded by two single genes *globulin1* (*glb1*) and *globulin2* (*glb2*), respectively, and constitute 10% and 20% of total proteins in the mature maize embryo (Belanger and Kriz 1989; Kriz 1989; Wallace and Kriz 1991). In the endosperm, the major storage proteins are zeins, which are heterogeneous mixtures of alcohol-soluble proteins. Based on their structure and sequence similarity, zein proteins can be classified into four families, that is, α (19 and 22 kD), γ (50, 27 and 16 kD), β (15 kD) and δ (18 and 10 kD) (Esen 1987; Swarup et al. 1995; Coleman and Larkins 1999). Based on DNA sequence similarity, 19-kD α-zeins can be further classified into z1A, z1B and z1D subgroups and 22-kD α-zeins into z1C (Heidecker and Messing 1986). The α-zeins are the most predominant proteins in maize endosperm and account for 60–70% of the total zein fraction. These proteins are encoded by a large number of copies arranged in tandem duplications and distributed over several chromosomal loci (Song et al. 2001; Song and Messing 2002, 2003; Miclaus et al. 2011b), whereas other zein members are encoded by only one or two copies (Das et al. 1990, 1991a, 1991b; Swarup et al. 1995; Woo et al. 2001; Wu et al. 2012). The γ-zeins are the second predominant proteins, amounting to about 20–25% of total zeins (Thompson and Larkins 1994). The 27- and 16-kD γ-zeins originated from a common progenitor by allotetraploidization and share a high DNA sequence similarity (Xu and Messing 2008), whereas the 50-kD γ-zein, although very closely linked to the 27-kD γ-zein gene (Holding et al. 2008), is far more diverged and its expression level is lower than the other two γ-zeins (Yuan et al. 2014). The β- and δ-zeins represent just minor zein fractions (Thompson and Larkins 1994; Swarup et al. 1995).

All zein proteins are deficient in lysine and tryptophan and this is detrimental to the nutritional quality of corn seeds. This leads to protein deficiency when corn is used as the sole protein source (Osborne et al. 1914a, 1914b). The 27- and 16-kD γ-zeins are rich in cysteine and represent the main sink for cysteine, whereas the δ-zeins represent the

Table 1 Amino acid composition of zein proteins

Zeins	MW (kDa)	Accession no.	Mature peptide	Met%	Cys%	(Met + Cys)%	Lys%
	22 C (*fl2*)	L34340.1	241	0.00	0.41	0.41	0.00
α	19 A	BT034568.1	219	0.94	0.94	1.88	0.00
	19 B	DQ244961.1	220	0.46	0.91	1.37	0.00
	19 D	BT061340.1	219	0.46	0.46	0.92	0.00
	50	AF371263	278	1.08	5.40	6.48	2.52
γ	27	AF371261	204	0.49	7.35	7.84	0.00
	16	AF371262	163	1.84	7.36	9.20	0.00
β	15	AF371264	160	11.25	4.38	15.63	0.00
δ	18	AF371265	190	25.26	1.58	26.84	0.53
	10	AF371266	129	22.48	3.88	26.36	0.00
Total protein[a]	/	/	/	2.30	2.40	4.70	2.54

[a]Data were measured from B73 seed flour. It was calculated by the percentage of an amino acid in total seed protein.
Source: Adapted from Wu et al. (2012).

reservoir for methionine (Kirihara et al. 1988; Swarup et al. 1995; Wu et al. 2009) and the β-zein is abundant in both cysteine and methionine (Table 1) (Wu et al. 2012).

3 Regulation of storage protein genes

The expression of storage protein genes is described by three characteristics: levels of expression, tissue specificity and strict temporal regulation (Wu and Messing 2012a). In angiosperm plants, a common motif in the promoter region of seed storage protein genes seems to be a 'G' box, which contains the ACGT core element, which is recognized by bZIP transcription factors (Ueda et al. 1992; Alonso et al. 2009; Zhang et al. 2015). For instance, in *Arabidopsis thaliana*, the model plant for dicots, the sole storage organ is the embryo because the endosperm is degraded and absorbed by the embryo during the early seed development. Its main storage proteins are 2S albumins and 12S cruciferins (Bäumlein et al. 1994). They are transcriptionally regulated by bZIPs (AtbZIP10 and AtbZIP25) (Lara et al. 2003) and B3s (ABSCISIC ACID INSENSITIVE3 (AtABI3), FUSCA3 (AtFUS3) and LEAFY COTYLEDON2 (AtLEC2)) (Giraudat et al. 1992; Bäumlein et al. 1994; Parcy et al. 1994; Stone et al. 2001). In the maize embryo, the mutation in *Viviparous1* (*Vp1*), the *AtABI3* homolog in maize, causes a viviparous phenotype and a dramatic reduction in the synthesis of GLBs (Kriz et al. 1990), indicating that the embryo storage protein genes in dicot and monocot are regulated by similar trans factors. However, the bZIP TF that interacts with Vp1 to regulate *Glbs* has not yet been identified (Fig. 1a) (Zhang et al. 2015).

In maize endosperm, zein genes are expressed starting about ten days after pollination (Woo et al. 2001). Consistent with their highly abundant protein accumulation, zein RNA

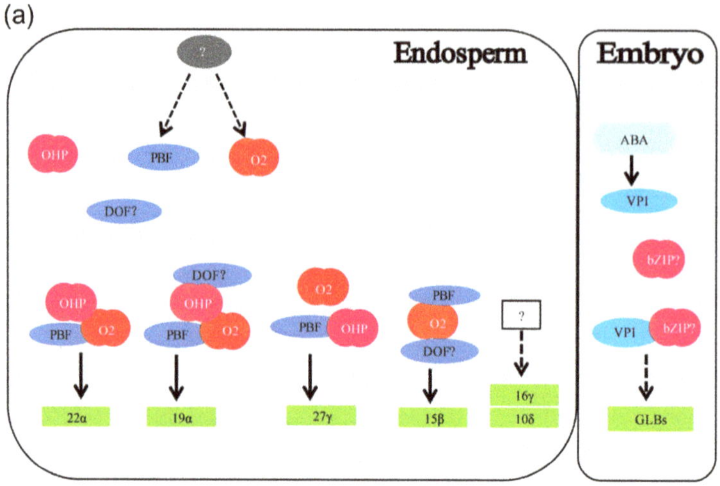

(a) Transcriptional regulation of zein genes. The interaction of bZIP and DOF transcription factors with their target genes in the endosperm and embryo is shown. In the embryo, it has been suggested that *Glb1* and *Glb2* are regulated by VP1 and an unidentified bZIP TF. In the endosperm, zein genes are regulated by the interaction of DOF and bZIP TFs. O2 is the main bZIP TF regulating α- and β-zein genes, whereas OHPs are the main regulators for the 27-kD γ-zein gene. PBF regulates 22-kD α- and 27-kD γ-zein genes through interactions with O2 and OPHs, respectively. In the presence of *PbfRNAi*, their expression is decreased, whereas the 15-kD β- and the 19-kD α-zein genes are not significantly affected, suggesting redundancy of another unidentified DOF TF with PBF. Transcriptional regulation of 16-kD γ- and δ-zein genes is unknown. Also, the upstream factors that activated *O2* and *Pbf* at 10 DAP, which in turn activate zein gene expression, have not been identified.

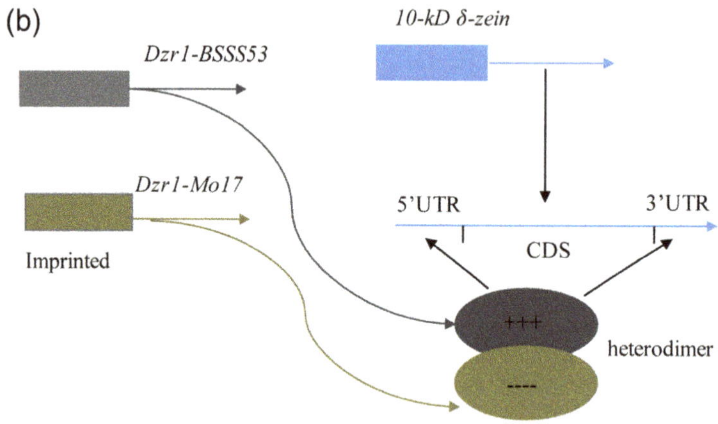

(b) Post-transcriptional regulation of the 10-kD δ-zein gene. The regulatory factor Dzr1 controls the accumulation of the 10-kD δ-zein mRNA at the post-transcriptional level, hypothetically as a dimer to explain the dominant negative function of the *dzr1-Mo17* allele over the *dzr1-BSSS53* allele. An interesting aspect of this regulation is that the *dzr1-Mo17* allele is imprinted. During female gametogenesis this imprint is removed after the egg nucleus has been formed so that the active Dzr1-Mo17 is not inherited and acts on the accumulation of the 10-kD δ-zein mRNA in a dominant negative fashion independent of grandparentage in violation of Mendelian rule.

Figure 1 Summary of known regulatory mechanisms of seed storage protein genes in maize endosperm and embryo.

sequences account for nearly 50% of total endosperm transcripts in a non-normalized cDNA library, of which α- and γ-zeins comprise around 30% and 15%, respectively (Woo et al. 2001). RNA-seq analysis of gene expression in the middle phase of endosperm development revealed that three-quarters of zein genes are ranked among the 100 most highly expressed genes in the endosperm, with α- and γ-zein subfamilies contributing the most abundant transcripts (Chen et al. 2014).

Most of the zein genes are regulated by the additive or synergistic action of bZIP and DNA-binding one zinc finger (DOF) transcription factors (Wang and Messing 1998; Zhang et al. 2015). Opaque2 (O2) is an endosperm-specific bZIP transcription factor, recognizing the O2 box (CCACGT), whose mutation causes a dramatic reduction in 22-kD α- and 15-kD β-zein accumulation (Schmidt et al. 1992; Ueda et al. 1992; Neto et al. 1995). However, RNA-seq analysis of transcriptome alterations in the *o2* mutant reveals that the transcript levels of all zein genes, except the 16-kD γ-zein, are reduced (Li et al. 2015). However, at the protein level, there is no significant reduction of 27-kD γ-zein, indicating that O2 might be a minor regulator for 27-kD γ-zein gene, and this amount of decrease of its transcripts is insufficient to lead to a translational reduction. In addition, the transcript and protein levels of 19-kD α-zeins are not reduced as much as those of 22-kD α-zeins in *o2*, indicating that another transcription factor(s) may function along with O2 in regulating zein genes (Zhang et al. 2015). For instance, O2 heterodimerizing proteins (OHP1 and OHP2) are bZIP TFs that are constitutively expressed and could form heterodimers with O2 (Pysh et al. 1993; Zhang et al. 2015). OHPs can recognize both the CCACGT and TTACGT motif in the 22-kD α- and 27-kD γ-zein promoters, respectively. RNA interference (RNAi) silencing of OHPs dramatically decreases the expression of 27-kD γ-zein, indicating that OHPs have been recruited as the main bZIP TFs for regulating the 27-kD γ-zein gene (Fig. 1a) (Zhang et al. 2015).

Prolamin-box binding factor (PBF) is an endosperm-specific DOF TF, which recognizes the P box (TGTAAAG) that is conserved in most of the endosperm storage protein gene promoters in cereals (Vicente-Carbajosa et al. 1997; Wu and Messing 2012a). Silencing of *Pbf* causes a dramatic reduction in the transcript levels of 27-kD γ- and 22-kD α-zein genes. In the double mutants of *OhpRNAi;o2* and *PbfRNAi;o2*, the mRNA and protein levels of 22- and 19-kDa α-zeins are further reduced compared to those of the single mutants, indicating that *PbfRNAi* and *OhpRNAi* have additive or synergistic effects on α-zein gene expression in the absence of O2. The zein content of double mutants, *OhpRNAi;o2* and *PbfRNAi;o2*, and the triple mutant, *PbfRNAi;OhpRNAi;o2*, is reduced by 83%, 89% and 90%, respectively, indicating that OHPs, O2 and PBF are the main TFs of zein genes (Fig. 1a) (Zhang et al. 2015). Still, the expression of OHPs, O2 and PBF is only a prerequisite but not sufficient for tissue-specific expression of zein genes. The ectopic expression of O2 and PBF in leaf tissue, where OHPs are constitutively expressed, fails to trans-activate endogenous zein gene transcription (Wu and Messing 2012a). On the other hand, the 16-kD γ-zein gene has a deletion in its promoter that eliminates the P- and O2-like boxes. Interestingly, it is still specifically expressed in endosperm, albeit at a lower level than 27-kD γ-zein gene, suggesting that endosperm-specificity of zein gene expression is directed by other factors possibly including epigenetic factors that are still unknown (Wu and Messing 2012a; Zhang et al. 2015). In this respect, an interesting case is the copy number variation of the 22-kD α-zein gene cluster on chromosome 4S between B73 and BSSS53, where a new copy at the 3′ end of the cluster in BSSS53 no longer requires O2 (Song et al. 2001). Moreover, subjecting maize endosperm to tissue culture treatment can change α-zein genes from a silenced state to an active state (Miclaus et al. 2011b; Xu et al. 2016).

The delta zein regulator (*dzr1*) controls the expression of the 10-kD δ-zein gene post-transcriptionally (Cruz-Alvarez et al. 1991; Schickler et al. 1993), involving the untranslated terminal region (UTRs) of the mRNA (Lai and Messing 2002). Expression of the *dzr1* allele of Mo17 is imprinted, but other genotypes exhibit a typical dosage effect in the triploid endosperm. For instance, in reciprocal crosses of BSSS53 and W64A, the amount of 10-kD δ-zein is determined by the dosage of the *Dzr1-BSSS53* allele. However, in hybrids of BSSS53 and Mo17, the levels of 10-kD δ-zein depend on the parental origin of the allele (Fig. 1b). When the *Dzr1-Mo17* allele is maternally transmitted, the level of the 10-kD δ-zein is as low as that in Mo17. In contrast, when it is paternally inherited, the level of the 10-kD δ-zein is as high as in BSSS53 (Chaudhuri and Messing 1994; Wu et al. 2009). However, the lack of an apparent visible phenotype in seeds expressing low levels of 10-kD δ-zein has greatly impeded the cloning of the *dzr1* gene and the breeding of high methionine, in contrast to the QPM corn.

4 Synthesis and deposition of zein proteins

In maize endosperm, zein proteins are synthesized on the polyribosomes of the rough endoplasmic reticulum (RER) and then co-translocated into the lumen of the RER, where they are deposited into PBs (Larkins and Dalby 1975; Burr and Burr 1976; Larkins and Hurkman 1978). The γ- and β-zeins are expressed before the α-zeins and δ-zeins and located in the peripheral region of PBs, whereas α- and δ-zeins are subsequently positioned in the central area to fill PBs to their full size (Lending and Larkins 1989). Their earlier expression and their cysteine-rich nature are consistent with the role of γ- and β-zeins in cross-linking via disulphide bonds and in the initiation and stabilization of PBs (Coleman et al. 1996, 2004; Bagga et al. 1997; Hinchliffe and Kemp 2002). Whereas silencing of γ- and β-zein expression results in irregular shapes of PBs and an opaque phenotype, the combination of *γRNAi*, which silences the expression of 27-kD and 16-kD γ-zeins, and a null mutant of 10-kD and 18-kD δ-zeins causes only a slight alteration in the morphology of PBs, indicating that β-zeins can compensate in part for the lack of γ-zeins, thereby maintaining the normal shape of PBs during their development (Wu and Messing 2010b). Recent analysis also showed that suppression of different zein subclasses by RNAi could be used to further define the non-redundant roles of zeins in PB initiation and expansion (Guo et al. 2013). These results are consistently shown, with 27-kD γ-zein playing a critical role in PB initiation, whereas α-zeins are mainly responsible for PB expansion (Messing and Holding 2013).

Mutations interfering with the processing of the signal peptide or the folding of zein proteins result in lack of protein deposition and amorphous PBs. The *floury2* (*fl2*), *De-B30** and *floury4* (*fl4*) alleles bear a point mutation in the signal peptide of a 22-kD and 19-kD α-zein copy, which prevents the correct processing of zein proteins at maturity (Coleman et al. 1997; Gillikin et al. 1997; Kim et al. 2004; Wang et al. 2014a). *Mucronate1* (*Mc1*) contains a 38-bp deletion in the coding sequence of 16-kD γ-zein gene, which leads to a frame shift and produces an abnormal 16-kD γ-zein. As a consequence, PBs are irregularly shaped in *Mc1* mutant background (Kim et al. 2006). Using a unique genetic approach, one can demonstrate that a *γ-RNAi* transgene is dominant over this dominant negative mutation (Wu and Messing 2010a). Interestingly, the sorghum highly digestible high-lysine mutant *hdhl*, which develops invaginated PBs and soft endosperm, is caused by a mutation in a 22-kD α-kafirin copy, the homolog of 22-kD α-zein in sorghum (Wu et al. 2013). This mutation coincidently changed the same amino acid residue as in the maize *fl2* allele, and the nucleotide substitution is identical to the one in *fl4* (Wu et al. 2013; Wang et al. 2014a). Floury1 (Fl1) is located in the PB membrane and is required for the proper deposition of 22-kD α-zeins in PBs (Holding et al. 2007). All zein proteins are abundant in the amino acid proline. When the biogenesis of proline is blocked in *proline responding1* (*pro1*), the synthesis of zein proteins is dramatically inhibited (Wang et al. 2014b). Opaque7 (O7) encodes an acyl-CoA synthetase-like protein. Lack of this enzyme causes a preferential reduction of 19-kD α-zeins compared to other zeins by a yet unknown mechanism (Miclaus et al. 2011a; Wang et al. 2011). Yet another mutation *opaque1* (*o1*) lacks a myosin XI motor protein, which is associated with ER and PBs. O1 was shown to play an important role in ER streaming and PB biogenesis. Although zein synthesis is not reduced, the ER structures are affected, and the number of smaller and misshapen PBs is increased in *o1* compared to the normal endosperm (Wang et al. 2012).

5 Improving protein quality in maize seed

A well-known long-term selection for protein traits at the University of Illinois yielded an Illinois High Protein (IHP) line, in which the protein level has been raised to 30% of the seed dry weight (Moose et al. 2004; Dudley 2007). However, the increase is mainly due to the enhanced expression of α-zeins, which further lowers the lysine level of the total protein. Therefore, IHP contributed little value to feeding monogastric animals (Wu and Messing 2012b). The same long-term selection was also applied to lower the protein content, resulting in the Illinois Low Protein (ILP) line, which has an opaque kernel phenotype. In ILP, the level of zeins is more decreased than that of non-zeins, and therefore, the absolute lysine content is lower than that in normal corn, but the relative lysine content is significantly higher. However, ILP is not a suitable candidate for breeding of a high-lysine corn, because the total protein concentration is too low (Wu and Messing 2012b). The *o2* mutation was found to create a doubling of lysine level (Mertz et al. 1964; Nelson et al. 1965; Nelson 2001; Crow and Kermicle 2002). The elevated lysine content in *o2* comes from the compensatory synthesis of non-zein proteins and the total protein level is maintained at a relatively constant level (Holding and Larkins 2009; Wu and Messing 2012b; Zhang et al. 2015). However, as described above, O2 is a transcriptional factor for the expression of zein genes, where a single copy is sufficient, which makes the *o2* mutation recessive. A recessive mutation complicates hybrid seed production, but this can be overcome with the expression of an α-zein RNAi, which is dominant (Segal et al. 2003; Huang et al. 2006; Wu and Messing 2012b).

A significant problem arises with high-lysine corn because of its soft kernel texture that causes susceptibility to disease, leading to a drop in yield, and subsequently damage during storage in elevators and transportation in large vessels (Vasal et al. 1980). These unfavourable facets could be overcome by selecting for *o2* modifiers, which are quantitative trait loci (QTLs) that convert soft kernels into hard ones without losing the high-lysine trait, yielding the modified *o2* mutant called QPM (Paez et al. 1969; Vasal et al. 1980). Moreover, the *o2* modifiers are also dominant over *α-zeinRNAi* mutants. Therefore, unlike the recessive *o2* mutation, the dominant *α-zeinRNAi* transgene can accelerate breeding of QPM and can be more reliable (Wu and Messing 2011).

Breeding of QPM has greatly alleviated malnutrition in developing countries, where corn is often consumed as the main protein resource. However, neither the mechanism nor the genetic components controlling endosperm modification were well understood. In total, seven *o2 modifiers* have been mapped to six chromosomes. Among them the one close to the 27-kDa γ-zein locus provides the major contribution, whereas the other six only have smaller effects (Holding et al. 2008). The linkage of one of the QTLs to the 27-kD γ-zein locus also correlates with elevated 27-kD γ-zein transcript and protein levels in QPM that are two- to three-fold higher than those in the isogenic normal and *o2* mutant lines (Geetha et al. 1991; Lopes and Larkins 1991, 1995; Holding et al. 2008). Direct evidence of the role of the 27-kDa γ-zein gene in kernel hardness comes from its suppression with RNA interference or a deletion of the 27-kDa γ-zein, caused by γ-irradiation, which completely destroys the hard endosperm texture and restores the *o2* phenotype (Wu et al. 2010; Guo et al. 2013; Yuan et al. 2014).

Despite these contradictory properties of high lysine and soft endosperm texture, a solution could be achieved independent of the *o2* mutation and modifiers. For instance, as described above, the IHP line contains a protein level more than twice that in normal corn. When the most prominent fraction of α-zeins in IHP is suppressed with RNA interference, the high protein content remains the same because other proteins with higher lysine content are elevated to make up the difference. However, RNA interference is only a reduction of gene expression levels and leaves enough residual zein levels that are sufficient to allow for a vitreous endosperm; the resulting seed is high-lysine, high-protein and hard-kernel phenotype (Wu and Messing 2012b).

6 Maintaining sulphur storage in maize seed

The strategy for lysine enhancement had been built on rebalancing the proteome because about 57% of the predicted maize proteins have a lysine content of more than 4% (Wu et al. 2012). One can hypothesize that during photosynthesis nitrogen is reduced and subsequently stored in the seed in the form of proteins, driving an up-regulation of protein synthesis in the absence of zeins. However, only 8% of the predicted maize proteome contains methionine above 4%, requiring the expression of specialized proteins to store reduced sulphur in the seed. In fact, there are maize storage proteins like the 15-kD β-zein, 10-kD and 18-kD δ-zeins that contain 11.25%, 22.48% and 25.26% methionine residues, respectively (Table 1). In some genetic backgrounds, expression of the δ-zeins can be elevated to a level that can overcome methionine deficiencies in animal feed without supplementation (Messing and Fisher 1991), which constitutes today $3 billion in annual sales of synthetic methionine. Genetic variation of δ-zein levels in different maize inbred lines can be overcome with the overexpression of chimeric genes (Lai and Messing 2002). However, it was found that overexpression of the 10-kD δ-zein in maize seed caused a compensatory decrease of β-zein, 27-kDa and 16-kD γ-zeins accumulation. Consistent with this observation, selecting high-methionine variants from a maize population leads to low accumulation of 27-kDa γ-zein (Newell et al. 2014), whereas reduction of β-zein and γ-zeins with RNA interference results in increased levels of 10-kD δ-zein (Wu et al. 2012). It appears that the flow of sulphur can alternate between different gene products. Given that sulphur is one of the essential elements for plant growth because it is involved in many fundamental cellular processes including protein synthesis, any limitation is expected to severely impact normal development. However, cysteine and methionine are the only two amino acids among the 20 L-amino acids that contain sulphur. Sulphate (SO_4^{2-}) absorbed by roots has to be reduced from an oxidation state +6 to −2 before utilization during plant growth, which is catalyzed by three enzymes, ATP sulphurylase (APS), APS reductase (APR) and sulphite reductase, in this pathway, yielding sulphide. Another critical enzyme is serine acetyl-transferase (SAT) that catalyzes serine and acetyl-CoA to form *O*-acetylserine (OAS), which then reacts with sulphide to produce the end assimilation product of cysteine catalyzed by OAS thiol-lyase (Leustek et al. 2000). Then, the enzymes cystathionine γ-synthase and cystathionine β-synthase transfer the sulphur moiety from cysteine to methionine.

The energy required for the reduction of sulphur is produced in leaf cells during photosynthesis. Sulphur amino acids are then transported from leaves to seeds. During maize seed development, the γ- and δ-zeins are the main reservoir for cysteine and methionine in the endosperm because of the preponderance of these two amino acids. In addition, β-zein is rich in cysteine and methionine and acts as a sink for both. Because the capacity of sulphur absorption and reduction is limited, the increased expression of 10-kD δ-zein is at least in part at the expense of β-zein and γ-zeins (Wu et al. 2012; Wu and Messing 2014). If this is true, then the methionine level in maize seeds should be increased by the overexpression of the committing enzymes APR and SAT during photosynthesis.

7 Future trends

Improvement of nutritional quality represents a significant strategy to address the world's limited agricultural resources of arable land and irrigation because the same amount of crops would nourish more people. For instance, proper nourishment is currently driven by the world protein market, which depends mainly on soybean yields and the chemical production of methionine. Human and animal diet requires carbohydrates and oil as major calories balanced with essential amino acids derived from protein. This formulation applies both to vegetarians and to the production of animal protein. Given that corn has evolved as the major calorie source, the question has always been whether breeders could also generate varieties with the same yield and sufficient quality protein so that the same size of arable land and amount of irrigation could deliver both the calories and the protein simultaneously. This question has prompted a long-term investigation of maize seed proteins. Interestingly, research on storage proteins spans nearly 200 years (Gorham 1821). Although a lot of progress have been made since the first maize storage protein sequence became known (Geraghty et al. 1981), there are still many details unattended, and there is no large commercial realization of a well-rounded corn variety. With the focus on nutritional quality of corn, we propose that three major questions should preferentially be addressed in the coming years. First, what are the regulatory networks of storage protein genes (*zeins* and *glbs*)? This includes identification of new transcription factors and their binding sites of downstream-regulated genes and the epigenetic factors that control spatial- and temporal-specific expression of these genes. Second, what are the components and mechanisms of endosperm modification in QPM? QPM has already benefitted countless undernourished people in the developing countries. Cloning the underlying QTL will enable breeding programmes to integrate these into elite lines by seed companies. Third, teosinte and IHP accumulate a similar high level of proteins and most of the increased portion results from the expression of α-zeins, but, when suppressed, protein levels are sustained by the expression of non-zein proteins. What are the QTLs and mechanisms that control protein levels in endosperm and what artificial selection takes place during maize domestication?

Advances in our understanding of these questions could also come from comparative genomics. Today, the sequencing of the genomes of many maize inbred lines will permit us to identify the gene products to give rise to QTLs. More than one thousand inbred lines from the US and China have already been genotyped by next-generation sequencing, providing the maize research community with extensive information on the genetic variation from natural populations. This depth of comparative analysis has explained the molecular signatures of quantitative traits underlying the nutritional quality of maize. Moreover, if this knowledge would require us to edit genomic information by design, we could take advantage of the recently developed CRSPR/cas9 system (Jinek et al. 2012) for maize nutritional improvement.

8 Acknowledgements

We thank the anonymous reviewer for critical reading and valuable editing of our manuscript. We also like to acknowledge the support of the Selman A. Waksman, Chair in Molecular Genetics to JM, of the National Natural Science Foundation of China to YW (31371630, 91335109 and 31422040), Chinese Academy of Sciences to YW (XDA08020107) and of Chinese Thousand Talents Program Grant to YW.

9 References

Alonso, R., L. Onate-Sanchez, F. Weltmeier, A. Ehlert, I. Diaz et al. 2009. A pivotal role of the basic leucine zipper transcription factor bZIP53 in the regulation of Arabidopsis seed maturation gene expression based on heterodimerization and protein complex formation. *Plant Cell* 21: 1747–61.

Bagga, S., H. P. Adams, F. D. Rodriguez, J. D. Kemp and C. Sengupta-Gopalan. 1997. Coexpression of the maize delta-zein and beta-zein genes results in stable accumulation of delta-zein in endoplasmic reticulum-derived protein bodies formed by beta-zein. *Plant Cell* 9: 1683–96.

Bäumlein, H., S. Miséra, H. Luerßen, K. Kölle, C. Horstmann et al. 1994. The FUS3 gene of Arabidopsis thaliana is a regulator of gene expression during late embryogenesis. *Plant J.* 6: 379–87.

Belanger, F. C. and A. L. Kriz. 1989. Molecular characterization of the major maize embryo globulin encoded by the glb1 gene. *Plant Physiol.* 91: 636–43.

Burr, B. and F. A. Burr. 1976. Zein synthesis in maize endosperm by polyribosomes attached to protein bodies. *Proc. Natl. Acad. Sci. USA* 73: 515–19.

Chaudhuri, S. and J. Messing. 1994. Allele-specific parental imprinting of *dzr1*, a posttranscriptional regulator of zein accumulation. *Proc. Natl. Acad. Sci. USA* 91: 4867–71.

Chen, J., B. Zeng, M. Zhang, S. Xie, G. Wang et al. 2014. Dynamic transcriptome landscape of maize embryo and endosperm development. *Plant Physiol.* 166: 252–64.

Coleman, C. E. and B. Larkins. 1999. The Prolamins of Maize. In *Seed Proteins*, edited by P. Shewry and R. Casey, 109–39. Springer, Netherlands.

Coleman, C. E., A. M. Clore, J. P. Ranch, R. Higgins, M. A. Lopes et al. 1997. Expression of a mutant alpha-zein creates the floury2 phenotype in transgenic maize. *Proc. Natl. Acad. Sci. USA* 94: 7094–7.

Coleman, C. E., E. M. Herman, K. Takasaki and B. A. Larkins. 1996. The maize gamma-zein sequesters alpha-zein and stabilizes its accumulation in protein bodies of transgenic tobacco endosperm. *Plant Cell* 8: 2335–45.

Coleman, C. E., P. R. Yoho, S. Escobar and M. Ogawa. 2004. The accumulation of alpha-zein in transgenic tobacco endosperm is stabilized by co-expression of beta-zein. *Plant Cell Physiol.* 45: 864–71.

Crow, J. F. and J. Kermicle. 2002. Oliver Nelson and quality protein maize. *Genetics* 160: 819–21.

Cruz-Alvarez, M., J. A. Kirihara and J. Messing. 1991. Post-transcriptional regulation of methionine content in maize kernels. *Mol. Gen. Genet.* 225: 331–9.

Das, O. P., S. Levi-Minzi, M. Koury, M. Benner and J. Messing. 1990. A somatic gene rearrangement contributing to genetic diversity in maize. *Proc. Natl. Acad. Sci. USA* 87: 7809–13.

Das, O. P., E. Poliak, K. Ward and J. Messing. 1991a. A new allele of the duplicated 27kD zein locus of maize generated by homologous recombination. *Nucleic Acids Res.* 19: 3325–30.

Das, O. P., K. Ward, S. Ray and J. Messing. 1991b. Sequence variation between alleles reveals two types of copy correction at the 27-kDa zein locus of maize. *Genomics* 11: 849–56.

Dudley, J. W. 2007. From means to QTL: the Illinois long-term selection experiment as a case study in quantitative genetics. *Crop Sci.* 47.

Esen, A. 1987. A proposed nomenclature for the alcohol-soluble proteins (zeins) of maize (*Zea mays* L.). *J. Cereal Sci.* 5: 117–28.

Geetha, K. B., C. R. Lending, M. A. Lopes, J. C. Wallace and B. A. Larkins. 1991. Opaque-2 modifiers increase gamma-zein synthesis and alter its spatial distribution in maize endosperm. *Plant Cell* 3: 1207–19.

Geraghty, D., M. A. Peifer, I. Rubenstein and J. Messing. 1981. The primary structure of a plant storage protein: zein. *Nuc. Acids Res.* 9: 5163–74.

Gillikin, J. W., F. Zhang, C. E. Coleman, H. W. Bass, B. A. Larkins et al. 1997. A defective signal peptide tethers the floury-2 zein to the endoplasmic reticulum membrane. *Plant Physiol.* 114: 345–52.

Giraudat, J., B. M. Hauge, C. Valon, J. Smalle, F. Parcy et al. 1992. Isolation of the Arabidopsis ABI3 gene by positional cloning. *Plant Cell* 4: 1251–61.

Gorham, J. 1821. Analysis of Indian Corn. *Quart. J. Sci. Lit. Art* 11: 206–8.

Guo, X., L. Yuan, H. Chen, S. J. Sato, T. E. Clemente et al. 2013. Nonredundant function of zeins and their correct stoichiometric ratio drive protein body formation in maize endosperm. *Plant Physiol.* 162: 1359–69.

Heidecker, G. and J. Messing. 1986. Structural analysis of plant genes. *Annu. Rev. Plant Physiol.* 37: 439–66.

Hinchliffe, D. J. and J. D. Kemp. 2002. β-zein protein bodies sequester and protect the 18-kDa δ-zein protein from degradation. *Plant Sci.* 163: 741–52.

Holding, D. R. and B. G. Larkins. 2009. Zein storage proteins. In *Molecular Genetic Approaches to Maize Improvement*, edited by A. Kriz and B. Larkins, 269–86. Springer, Berlin Heidelberg.

Holding, D. R., B. G. Hunter, T. Chung, B. C. Gibbon, C. F. Ford et al. 2008. Genetic analysis of Opaque2 modifier loci in quality protein maize. *Theor. Appl. Genet.* 117: 157–70.

Holding, D. R., M. S. Otegui, B. Li, R. B. Meeley, T. Dam et al. 2007. The maize floury1 gene encodes a novel endoplasmic reticulum protein involved in zein protein body formation. *Plant Cell* 19: 2569–82.

Huang, S., A. Frizzi, C. A. Florida, D. E. Kruger and M. H. Luethy. 2006. High lysine and high tryptophan transgenic maize resulting from the reduction of both 19- and 22-kD alpha-zeins. *Plant Mol. Biol.* 61: 525–35.

Jinek, M., K. Chylinski, I. Fonfara, M. Hauer, J. A. Doudna et al. 2012. A programmable dual-RNA-guided DNA endonuclease in adaptive bacterial immunity. *Science* 337: 816–21.

Khavkin, E. E., S. I. Misharin, Y. Y. Markov and A. A. Peshkova. 1978. Identification of embryonal antigens of maize: globulins as primary reserve proteins of the embryo. *Planta* 143: 11–20.

Kim, C. S., B. C. Gibbon, J. W. Gillikin, B. A. Larkins, R. S. Boston et al. 2006. The maize Mucronate mutation is a deletion in the 16-kDa gamma-zein gene that induces the unfolded protein response. *Plant J.* 48: 440–51.

Kim, C. S., B. G. Hunter, J. Kraft, R. S. Boston, S. Yans et al. 2004. A defective signal peptide in a 19-kD alpha-zein protein causes the unfolded protein response and an opaque endosperm phenotype in the maize De*-B30 mutant. *Plant Physiol.* 134: 380–7.

Kirihara, J. A., J. B. Petri and J. Messing. 1988. Isolation and sequence of a gene encoding a methionine-rich 10-kDa zein protein from maize. *Gene* 71: 359–70.

Kriz, A. L. 1989. Characterization of embryo globulins encoded by the maize Glb genes. *Biochem. Genet.* 27: 239–51.

Kriz, A. R., M. S. Wallace and R. Paiva. 1990. Globulin gene expression in embryos of maize viviparous mutants : evidence for regulation of the glb1 gene by abscissic acid. *Plant Physiol.* 92: 538–42.

Lai, J. and J. Messing. 2002. Increasing maize seed methionine by mRNA stability. *Plant J* 30: 395–402.

Lara, P., L. Onate-Sanchez, Z. Abraham, C. Ferrandiz, I. Diaz et al. 2003. Synergistic activation of seed storage protein gene expression in Arabidopsis by ABI3 and two bZIPs related to Opaque2. *J. Biol. Chem.* 278: 21003–11.

Larkins, B. A. and A. Dalby. 1975. In vitro synthesis of zein-like protein by maize polyribosomes. *Biochem. Biophys. Res. Commun.* 66: 1048–54.

Larkins, B. A. and W. J. Hurkman. 1978. Synthesis and deposition of zein in protein bodies of maize endosperm. *Plant Physiol.* 62: 256–63.

Lending, C. R. and B. A. Larkins. 1989. Changes in the zein composition of protein bodies during maize endosperm development. *Plant Cell* 1: 1011–23.

Leustek, T., M. N. Martin, J. A. Bick and J. P. Davies. 2000. Pathways and regulation of Sulfur metabolism revealed through molecular and genetic studies. *Annu. Rev. Plant Physiol. Plant Mol. Biol.* 51: 141–65.

Li, C., Z. Qiao, W. Qi, Q. Wang, Y. Yuan et al. 2015. Genome-wide characterization of cis-acting DNA targets reveals the transcriptional regulatory framework of Opaque2 in maize. *Plant Cell* 27: 532–45.

Lopes, M. A. and B. A. Larkins. 1991. Gamma-zein content is related to endosperm modification in quality protein maize. *Crop Sci.* 31: 1655–62.

Lopes, M. A. and B. A. Larkins. 1995. Genetic analysis of Opaque2 modifier gene activity in maize endosperm. *Theor. Appl. Genet.* 91: 274–81.

Mertz, E. T., L. S. Bates and O. E. Nelson. 1964. Mutant gene that changes protein composition and increases lysine content of maize endosperm. *Science* 145: 279–80.

Messing, J. and H. Fisher. 1991. Maternal effect on high methionine levels in hybrid corn. *J. Biotechnol.* 21: 229–37.

Messing, J. and D. Holding. 2013. Evolution, structure, and function of prolamin storage proteins. In *Seed Genomics*, edited by I. S. U. Phil Becraft and I. O. Ames, 139–58. Wiley-Blackwell Publishing, Hoboken, NJ USA.

Miclaus, M., Y. Wu, J. H. Xu, H. K. Dooner and J. Messing. 2011a. The maize high-lysine mutant opaque7 is defective in an acyl-CoA synthetase-like protein. *Genetics* 189: 1271–80.

Miclaus, M., J. H. Xu and J. Messing. 2011b. Differential gene expression and epiregulation of alpha zein gene copies in maize haplotypes. *PLoS Genet* 7: e1002131.

Moose, S. P., J. W. Dudley and T. R. Rocheford. 2004. Maize selection passes the century mark: a unique resource for 21st century genomics. *Trends Plant Sci.* 9: 358–64.

Nelson, O. E. 2001. Maize: the long trail to QPM. In *Encyclopedia of Genetics*, edited by E. C. R. Reeve, 657–60. Fitzroy Dearborn, London/Chicago.

Nelson, O. E., E. T. Mertz and L. S. Bates. 1965. Second mutant gene affecting the amino acid pattern of maize endosperm proteins. *Science* 150: 1469–70.

Neto, G., J. Yunes, M. da Silva, A. Vettore, P. Arruda et al. 1995. The involvement of Opaque 2 on β-prolamin gene regulation in maize and Coix suggests a more general role for this transcriptional activator. *Plant Mol. Biol.* 27: 1015–29.

Newell, M. A., K. E. Vogel, M. Adams, N. Aydin, A. L. Bodnar et al. 2014. Genetic and biochemical differences in populations bred for extremes in maize grain methionine concentration. *BMC Plant Biol.* 14: 49.

Osborne, T. B. 1908. Our present knowledge of plant proteins. *Science* 28: 417–27.

Osborne, T. B., L. B. Mendel, With the coöperation of L. F. Edna and J. W. Alfred. 1914a. Amino-acids in nutrition and growth. *J. Biol. Chem.* 17: 325–49.

Osborne, T. B., L. B. Mendel, With the coöperation of L. F. Edna and J. W. Alfred. 1914b. Nutritive properties of proteins of the maize kernel. *J. Biol. Chem.* 18: 1–16.

Paez, A. V., J. L. Helm and M. S. Zuber. 1969. Lysine content of Opaque-2 maize kernels having different phenotypes. *Crop Sci.* 9: 251–2.

Parcy, F., C. Valon, M. Raynal, P. Gaubier-Comella, M. Delseny et al. 1994. Regulation of gene expression programs during Arabidopsis seed development: roles of the ABI3 locus and of endogenous abscisic acid. *Plant Cell* 6: 1567–82.

Pysh, L. D., M. J. Aukerman and R. J. Schmidt. 1993. OHP1: a maize basic domain/leucine zipper protein that interacts with Opaque2. *Plant Cell* 5: 227–36.

Schickler, H., M. S. Benner and J. Messing. 1993. Repression of the high-methionine zein gene in the maize inbred line Mo17. *Plant J.* 3: 221–9.

Schmidt, R. J., M. Ketudat, M. J. Aukerman and G. Hoschek. 1992. Opaque-2 is a transcriptional activator that recognizes a specific target site in 22-kD zein genes. *Plant Cell* 4: 689–700.

Segal, G., R. Song and J. Messing. 2003. A new opaque variant of maize by a single dominant RNA-interference-inducing transgene. *Genetics* 165: 387–97.

Shewry, P. and R. Casey. 1999. Seed proteins. In *Seed Proteins*, edited by P. Shewry and R. Casey, 1–10. Springer, Netherlands.

Song, R., V. Llaca, E. Linton and J. Messing. 2001. Sequence, regulation, and evolution of the maize 22-kD alpha zein gene family. *Genome Res.* 11: 1817–25.

Song, R. and J. Messing. 2002. Contiguous genomic DNA sequence comprising the 19-kD zein gene family from maize. *Plant Physiol.* 130: 1626–35.

Song, R. and J. Messing. 2003. Gene expression of a gene family in maize based on noncollinear haplotypes. *Proc. Natl. Acad. Sci. USA* 100: 9055–60.

Stone, S. L., L. W. Kwong, K. M. Yee, J. Pelletier, L. Lepiniec et al. 2001. LEAFY COTYLEDON2 encodes a B3 domain transcription factor that induces embryo development. *Proc. Natl. Acad. Sci. USA* 98: 11806–11.

Swarup, S., M. C. Timmermans, S. Chaudhuri and J. Messing. 1995. Determinants of the high-methionine trait in wild and exotic germplasm may have escaped selection during early cultivation of maize. *Plant J.: For Cell and Mol. Biol.* 8: 359–68.

Thompson, G. and B. Larkins. 1994. Characterization of zein genes and their regulation in maize endosperm. In *The Maize Handbook*, edited by M. Freeling and V. Walbot, 639–47. Springer, New York.

Ueda, T., W. Waverczak, K. Ward, N. Sher, M. Ketudat et al. 1992. Mutations of the 22- and 27-kD zein promoters affect transactivation by the Opaque-2 protein. *The Plant Cell* 4: 701–9.

Vasal, S. K., E. Villegas, M. Bjarnason, B. Gelaw and P. Goertz. 1980. Genetic modifiers and breeding strategies in developing hard endosperm *Opaque-2* materials. In *Improvement of Quality Traits of Maize for Grain and Silage Use*, edited by W. G. Pollmer and R. H. Phillips, 37–73. Martinus Nijhoff, London.

Vicente-Carbajosa, J., S. P. Moose, R. L. Parsons and R. J. Schmidt. 1997. A maize zinc-finger protein binds the prolamin box in zein gene promoters and interacts with the basic leucine zipper transcriptional activator Opaque2. *Proc. Natl. Acad. Sci. USA* 94: 7685–90.

Wallace, N. H. and A. L. Kriz. 1991. Nucleotide sequence of a cDNA clone corresponding to the maize globulin-2 gene. *Plant Physiol.* 95: 973–5.

Wang, G., W. Qi, Q. Wu, D. Yao, J. Zhang et al., 2014a. Identification and characterization of maize floury4 as a novel semidominant opaque mutant that disrupts protein body assembly. *Plant Physiol.* 165: 582–94.

Wang, G., X. Sun, G. Wang, F. Wang, Q. Gao et al. 2011. Opaque7 encodes an acyl-activating enzyme-like protein that affects storage protein synthesis in maize endosperm. *Genetics* 189: 1281–95.

Wang, G., F. Wang, G. Wang, F. Wang, X. Zhang et al., 2012 Opaque1 encodes a myosin XI motor protein that is required for endoplasmic reticulum motility and protein body formation in maize endosperm. *Plant Cell* 24: 3447–62.

Wang, G., J. Zhang, G. Wang, X. Fan, X. Sun et al. 2014b. Proline responding1 plays a critical role in regulating general protein synthesis and the cell cycle in maize. *Plant Cell* 26: 2582–600.
Wang, Z. and J. Messing. 1998. Modulation of gene expression by DNA-protein and protein-protein interactions in the promoter region of the zein multigene family. *Gene* 223: 333–45.
Woo, Y. M., D. W. Hu, B. A. Larkins and R. Jung. 2001. Genomics analysis of genes expressed in maize endosperm identifies novel seed proteins and clarifies patterns of zein gene expression. *Plant Cell* 13: 2297–317.
Wu, Y., W. Goettel and J. Messing. 2009. Non-Mendelian regulation and allelic variation of methionine-rich delta-zein genes in maize. *Theor. Appl. Genet.* 119: 721–31.
Wu, Y., D. R. Holding and J. Messing. 2010. Gamma-zeins are essential for endosperm modification in quality protein maize. *Proc. Natl. Acad. Sci. USA* 107: 12810–15.
Wu, Y. and J. Messing. 2010a. Rescue of a dominant mutant with RNA interference. *Genetics* 186: 1493–6.
Wu, Y. and J. Messing. 2010b. RNA interference-mediated change in protein body morphology and seed opacity through loss of different zein proteins. *Plant Physiol.* 153: 337–47.
Wu, Y. and J. Messing. 2011. Novel genetic selection system for quantitative trait loci of quality protein maize. *Genetics* 188: 1019–22.
Wu, Y. and J. Messing. 2012a. Rapid divergence of prolamin gene promoters of maize after gene amplification and dispersal. *Genetics* 192: 507–19.
Wu, Y. and J. Messing. 2012b. RNA interference can rebalance the nitrogen sink of maize seeds without losing hard endosperm. *PLoS One* 7: e32850.
Wu, Y. and J. Messing. 2014. Proteome balancing of the maize seed for higher nutritional value. *Front Plant Sci.* 5: 240.
Wu, Y., W. Wang and J. Messing. 2012. Balancing of sulfur storage in maize seed. *BMC Plant Biol.* 12: 77.
Wu, Y., L. Yuan, X. Guo, D. R. Holding and J. Messing. 2013. Mutation in the seed storage protein kafirin creates a high-value food trait in sorghum. *Nat. Commun.* 4: 2217.
Xu, J. H. and J. Messing. 2008. Organization of the prolamin gene family provides insight into the evolution of the maize genome and gene duplications in grass species. *Proc. Natl. Acad. Sci. USA* 105: 14330–5.
Xu, J. H., R. Wang, X. Li, M. Miclaus and J. Messing. 2016. Locus- and site-specific DNA methylation of 19 kDa zein genes in maize. *PLoS One* 11: e0146416.
Yuan, L., Y. Dou, S. F. Kianian, C. Zhang and D. R. Holding. 2014. Deletion mutagenesis identifies a haploinsufficient role for gamma-zein in Opaque2 endosperm modification. *Plant Physiol.* 164: 119–30.
Zhang, Z., J. Yang and Y. Wu. 2015. Transcriptional regulation of zein gene expression in maize through the additive and synergistic action of Opaque2, prolamin-box binding factor, and O2 heterodimerizing proteins. *Plant Cell* 27: 1162–72.

Chapter 5

Advances in mycotoxin-resistant maize varieties

Marilyn L. Warburton and W. Paul Williams, USDA-ARS Corn Host Plant Research Resistance Unit, USA

1 Introduction

Maize (*Zea mays* L) is one of the most important food crops in the world, but the productivity and safety of grain produced by this plant is seriously decreased by a multitude of fungal pathogens causing ear rots. Depending on the growing environment, maize grains can be infected by one or more ear rot fungi, which may decrease grain yield and quality. Many of them also produce secondary metabolites, known as mycotoxins, which can have serious detrimental effects on humans and animals that consume the infected grain. The most commonly occurring ear rot fungi include Aspergillus, Fusarium and Gibberella species, all of which produce one or more mycotoxins (Table 1) and occur over very large geographic regions. Many other fungal species that are not as widespread also infect maize and may produce mycotoxins in the ears (Table 1). Mycotoxin content in maize is highly regulated in many countries of the world, leading to more of an economic than a health problem, as infected grain is destroyed before it can enter the food stream. In some countries, however, regulatory infrastructure is unequal to the task of inspection and enforcement; therefore, much of the infected grain is consumed, often within the household of the farmer who produced it. Immediate symptoms of large doses of mycotoxins may include abdominal pain, vomiting and diarrhoea, and may even lead to death, while long-term exposure to sub-lethal doses of mycotoxins has been linked to liver and oesophageal cancers, neural tube defects in newborns, stunted growth during childhood and depressed or severely

http://dx.doi.org/10.19103/AS.2016.0001.09

Table 1 Causal organism and associated mycotoxins of the fungal rots known to affect maize ears

Ear rot	Causal organism	Associated mycotoxin	Geographic spread
Commonly occurring ear rots:			
Aspergillus ear rot	*Aspergillus flavus* and *A. parasiticus*	aflatoxin	Worldwide but more common in tropical areas
Gibberella ear rot	*Gibberella zeae* (same organism that causes Fusarium head blight in wheat *Fusarium graminearum*), *G. fujikuroi* and *G. moniliformis*	deoxynivalenol (also known as DON or vomitoxin), zearalenone	Worldwide
Fusarium ear rot	*Fusarium verticillioides*, *F. moniliformis*, *F. proliferatum*, *F. sporotrichioides*, *F. subglutinans*, *F. avenaceum*, *F. cerealis*, and *F. poae*	Fumonisin, deoxynivalenol, trichothecene, zearalenones; minor toxins include beauvericin, fusaproliferin, nivalenol, fusarenone-X, moniliformin, T-2 toxin, and diacetoxyscirpenol	Worldwide
Less commonly occurring ear rots:			
Penicillium ear rot	*Penicillium oxalicum* or *P. roqueforti*	PR toxin, roquefortin C, patulin, mycophenolic acid	Worldwide
Diplodia or Stenocarpella ear rot	*Stenocarpella maydis*	no mycotoxins	Tropical maize-growing countries and states
Cladosporium ear rot	*Cladosporium herbarum* and *C. cladosporoides*	no mycotoxins	Widespread but rare, usually occurring under cool and humid conditions
Nigrospora ear rot	*Nigrospora oryzae*	no mycotoxins	Widespread but rare, usually occurring when plants are damaged
Trichoderma ear rot	*Trichoderma viride*	no mycotoxins	Widespread but rare, usually occurring when plants are damaged

compromised immune systems, among others (Eaton and Groopman, 1993; Bennett and Klich, 2003; Williams et al., 2004).

Many ear rot fungi grow more vigorously and produce higher levels of mycotoxins, under certain environmental conditions, which often include high temperatures and drought stress. It is thus suggested that increased temperatures and erratic rainfall due to global climate change may exacerbate the problem (Wu et al., 2011). Ear rot fungal spores may travel down the silk channel into the kernel or be introduced into the kernels by insects feeding on the ears. The spores may accumulate in the debris of previous cropping cycles. Thus, there are several potential remedial actions that can be taken to prevent the accumulation of mycotoxins in maize grain. These include crop rotation, tillage of crop residues, modification of planting date, use of biocontrol strains of fungi which do not produce toxins, optimal irrigation and fertilization, control of insects which spread the fungus, application of fungicides, adequate drying following harvest, good post-harvest storage conditions and treatment of infected grains with binders to render mycotoxins harmless. However, the most economically feasible solution for most farmers, and the simplest to implement once the resource is created, is the use of genetically resistant maize varieties. Efforts to reduce mycotoxin accumulation in maize have focused on identifying and selecting natural host resistance factors (Holley et al., 1989; King and Scott 1981; Gendloff et al., 1986). Nonetheless, to date, most commercial hybrid maize varieties still lack sufficient resistance to mycotoxin accumulation (Munkvold, 2003; Abbas et al., 2002), especially in years when environmental conditions favour mycotoxin production (Payne, 1992; Fountain et al., 2015).

2 Key challenges in developing new varieties

The development of maize breeding lines with resistance against toxigenic fungi or accumulation of the toxins they create has been slow to date. Breeding is hampered by low heritability, caused by difficulties in precise spore inoculation and measurement of toxin, high genotype by environment interaction (indicating the confounding effect of different environments on the expression of different genes) and the highly quantitative nature of inheritance, generally involving many genes of small phenotypic effects. In addition, the exact mechanisms of resistance are not known and may vary depending on the fungi involved, environmental conditions in which the plant is grown and even, perhaps, host genetic background. Despite the difficulties, stably resistant maize breeding lines, which are used to produce new cultivated maize hybrids, have been developed for Gibberella (Cullen et al., 1983; Gendloff et al., 1986), Aspergillus (McMillian et al., 1993; Williams and Windham 2001, 2006; Betran et al., 2002) (Fig. 1) and Fusarium ear rots (Eller et al., 2010; Mesterházy et al., 2012). These lines have demonstrated resistance to fungal infection or growth, decreased biomass and/or decreased toxin accumulation. However, there are other lines that have demonstrated high values for traits indirectly associated with lower toxin levels, including resistance to corn earworm, a tighter husk and upright ears, and harder grains (as reviewed in Munkvold 2003; Warburton and Williams, 2014; Williams et al., 2015). These indirect traits that correlate with lower mycotoxin levels may allow breeding to occur without directly selecting for lower fungal biomass or mycotoxin levels, both of which are expensive and time-consuming to measure.

Work has been done to identify genetic variation and, in some cases, genomic regions associated with resistance to the more minor ear rots (Cantone et al., 1983; Dorrance et al., 1998). In some cases, resistance to one fungal species will boost resistance to a related fungus. Thus, the possibility for genetic improvement via selection (with or without molecular markers) exists in maize. On the other hand, investment into breeding for resistance to these minor ear rots is not generally very high, as they are geographically less

Figure 1 *Aspergillus flavus* susceptible (*left*) and resistant (*right*) corn genotypes. Both genotypes were inoculated by injecting *A. flavus* spores under the husk into an immature ear. Clear spread from the single inoculation site by the fungus can be seen in the photo on the left; no growth is seen in the ear on the right.

widespread and of lower economic importance than the three major ear rots listed above. Thus, the creation of new resistant breeding lines and hybrids may not occur as rapidly as biologically possible, and work remains to be done.

3 Techniques for developing new varieties

During the late 1970s and 1980s, there was considerable interest in identifying maize germplasm with resistance to mycotoxin accumulation. Initial reports of germplasm with some resistance to the major ear rots were promising (Clements et al., 2004; Scott and Zummo, 1988; Reid et al., 1992, 1993), but resistance was generally not found in sufficiently high enough levels in released varieties. Thus, development of new resistant breeding lines began. Researchers quickly discovered that artificial inoculation of developing ears with target fungal spores would be necessary to successfully quantify resistance. A variety of screening techniques were developed and used at various locations (Scott and Zummo, 1987). Among the early techniques used were injection of a spore suspension into the silk channel or developing ear, dipping a knife tip into a spore suspension and inserting the tip into the side of the ear, and dipping a double row of pins imbedded in plastic into a spore suspension and inoculating a single row of kernels – the last of which is known as the pinbar technique. Its use led to the development of the first germplasm lines, Mp313E (Scott and Zummo, 1990) and Mp420 (Scott and Zummo, 1992), to be used as sources of resistance to *Aspergillus flavus* kernel infection and/or accumulation of aflatoxin. From the points of inoculation, infection spreads to neighbouring kernels and can cover significant portions of the ear by harvest (Fig. 1). Thus, selection in Mp313E and Mp420 was for reduced levels of kernel infection based on the percentage of infection of non-wounded kernels following inoculation of developing ears with the pinbar. Resistance to fungi in maize depends on resistance at the initial point of introduction and reducing the spread of the fungus in the developing ear, and both may help to alleviate the mycotoxin problem by decreasing fungal load in resistant ears.

Once breeding lines with stable resistance to toxigenic fungi are developed, the resistance must generally be transferred into more elite backgrounds – the parents of the commercial hybrids currently used. The most resistant germplasm developed to date tends to have poor agronomic performance in much of the temperate maize-growing world, because most of the natural resistance has been identified in tropical maize backgrounds (Holland et al., 1996; Warburton et al., 2013; Chiuraise et al., 2015). Transfer of resistance to new genetic backgrounds has been hampered by the highly quantitative nature of the trait, which occurs when many genes work together to create a phenotype (in this case, fungal resistance) and when the environment influences the final expression of the phenotype. Thus, marker-assisted selection or backcrossing, has been similarly slowed by a lack of single genes with large effects, validated linked markers and/or the need to move only the genes conferring resistance loci from unadapted genetic backgrounds into breeding pools with as little linkage drag as possible. On the other hand, for breeding programmes in the tropical world, currently available resistant breeding lines may perform very well, thus needing fewer rounds of backcrossing or improvement (Chiuraise et al., 2015).

Mapping studies are beginning to identify markers linked to Quantitative Trait Loci (QTL – the genes or genetic location of genes causing the expression of a quantitative trait) and loci linked to or associated with ear rot resistance in general (Xiang et al., 2012) and for specific fungi including Gibberella (Martin et al., 2012; Zhang et al., 2012), Aspergillus (Warburton et al., 2015 and reviewed in Mideros et al., 2014) and Fusarium (Zila et al., 2014; Lanubile et al., 2014; Chen et al., 2012; Robertson-Hoyt et al., 2006). However, identification of resistance factors is only the first step. Validation of the phenotypic variation accounted for by each marker must be done in multiple genetic backgrounds and environments, to ensure stable expression of these resistance factors in new resistant germplasm. Not all loci conferring resistance are going to be stable across backgrounds or environments or cause large enough phenotypic effect to justify the expense of marker development and use. Loci associated with dominant or epistatic gene effect may not be as easy to use in all crosses as loci associated with an additive gene effect. However, once markers with adequate stability, magnitude and type of effect have been validated, they can be adopted by breeding programmes to more quickly select for resistant maize among offspring of resistant x susceptible crosses, and more quickly return to the genetic background of a recurrent parent during conversion of susceptible lines to resistant ones via backcrossing. The lack of closely linked, validated markers for ear rot resistance is currently slowing resistance breeding.

Marker-assisted backcrossing to move one or more resistant QTL or loci into a cultivar with good agronomic performance will be precise, but time-consuming. Another method to increase mycotoxin resistance in maize involves population-level improvement. One would start with the creation of a population with multiple sources of resistance as parental material that had been intermated for several generations. Parents with excellent agronomic performance could also be included. Selection via phenotype, markers or both, for ear rot, fungal biomass or toxin accumulation resistance should increase the frequency of the positive alleles at target loci with each generation of selection. Selection can also be done for good agronomic performance and resistance to any other target disease or insect pests, although selection for multiple traits simultaneously will slow progress for any given trait. After several generations of selection and intermating, fixed lines could be selected from the population, which should possess high levels of resistance and good agronomic performance. Lines could be fixed very quickly using doubled haploid production, where a heterozygous plant is made homozygous in one generation by growing a haploid plant from a gamete, and then doubling the chromosome number (i.e. with colchicine). Further rounds of selection would improve the population further, and new and improved lines could be released every few years. The molecular markers needed for marker-assisted recurrent selection (MARS) in maize populations for fungal resistance are being developed and may soon be ready for such an endeavour.

A related technique for population-level improvement is the use of genomic selection to increase the quantitative traits, which should work well in breeding programmes that have existing populations with some level of mycotoxin or fungal resistance in them. Genomic selection uses a large number of markers to improve a trait, but these markers do not have to be linked to the trait of interest, thus allowing selection to occur without costly and time-consuming mapping studies. If a breeding programme has populations with no fungal resistance and sources of resistance have only been identified in germplasm very unrelated to the programme's populations, introgression of the resistance into the population via genomic selection may not be as efficient. It would take many rounds of intermating to remove the effect of population substructure caused by the mixing of the resistant germplasm into existing and unrelated susceptible populations; under these conditions, genomic selection is inefficient (Windhausen et al., 2012).

4 Case study: creating *Aspergillus flavus* resistant maize breeding lines

The successful creation of maize germplasm with resistance to *A. flavus* infection and aflatoxin accumulation requires the identification of genetic variation for resistance in maize germplasm, techniques to identify and quantify resistance, and breeding methods or procedures to develop resistant lines. An early priority of the USDA ARS Corn Host Plant Resistance Research Unit (CHPRRU) breeding programme at Mississippi State was to identify inbred lines and single-cross hybrids that could be used as resistant and susceptible checks. As the breeding programme evolved, these checks have served as benchmarks for resistance and susceptibility. Techniques for inoculating ears in the field and techniques for assessing resistance to kernel infection and aflatoxin accumulation are available (Zummo and Scott, 1989; Scott and Zummo, 1994; Windham et al., 2003). Field inoculation and evaluation is slow and costly, but heritability and repeatability can be improved by screening at least three replications of at least ten individual plants each. In addition, natural infection of developing ears by *A. flavus* is sporadic from year to year and can yield misleading data in aflatoxin resistance evaluations (Windham and Williams, 2002). Due to high genotype by environment interactions for resistance, multiple years and/or environments are strongly suggested. Measurement of fungal biomass via quantitative PCR (qPCR; Mideros et al., 2009), or the measurement of a primary metabolite, such as ergosterol (Bakan et al., 2002), created by growing fungus can be used to determine the amount of fungus present in each sample. The correlation between fungal biomass and amount of toxin is not strong enough to be predictive, so if the final trait to be measured is toxin level, it must be done independently. The use of affinity columns or spectrophotometry ensures an accurate quantification of toxin; faster or cheaper options must be tested for high correlation to these 'gold standard' measurements before being used routinely in a breeding or gene identification programme, but enzyme-linked immunosorbent assay (ELISA) correlate well with spectrophotometric results and are cheaper and faster to run. The use of UV light boxes to measure aflatoxin levels in corn grain is very simple and economical, but can greatly underestimate the amount of toxin present (Hara et al., 1974).

Because there is little resistance to *A. flavus* infection and aflatoxin accumulation in most publicly released varieties or in most temperate maize-breeding pools, germplasm from many sources have been screened to identify useful breeding material (Williams, 2006). Breeding crosses and lines must be evaluated repeatedly over many years and locations to ensure the resistance is stable. Resistance found in populations should be moved into a fixed (inbred) background to allow the study of heritability and gene action, and for ease of crossing. Thus, progress towards the creation of resistant breeding lines has been slow. However, fixed breeding lines showing stable resistance have been identified over the years from breeding programmes in Mississippi, Georgia, North Carolina and Louisiana (reviewed in Williams et al., 2015); Texas (Mayfield et al., 2012; Xu et al., 2004) and the International Institute for Tropical Agriculture (IITA) and the International Maize and Wheat Improvement Centre (known by its acronym in Spanish, CIMMYT), primarily working in Africa (Chiuraise et al., 2015; Menkir et al., 2006). Although correlation between disease score rated visually and total accumulation of aflatoxin is not phenotypically high enough to be predictive, they are genetically correlated (Lanubile et al., 2014); lines created via selection for low levels of either fungus or toxin have been shown to have reduced levels of the other trait as well. The most resistant germplasm has been identified within tropical maize (Warburton et al., 2013), and older breeding lines released for temperate growing areas have typically been very tall, late and low yielding, due to adaptation problems. However, the newest maize breeding lines released for temperate growing areas are significantly shorter and mature earlier (Mayfield et al., 2012; Williams and Windham, 2012), and many breeding lines with a higher percentage of tropical pedigree should do very well in many growing environments of Africa and Latin America.

Once resistant breeding lines are identified, they may be used as parents of new resistant hybrids, but because maize breeding programmes are usually dependent on finely tuned heterotic groups and with closed pedigrees, adding unadapted and unrelated breeding lines is not always successful. Therefore, a backcross breeding method to transfer resistance into an existing breeding pool, while transferring as little of the genetic background from the resistant donor, may be more productive. Although new aflatoxin-resistant breeding lines have been developed in this way, transfer of quantitative traits using phenotypic selection is often unsuccessful, as transfer is often incomplete and the resulting backcross lines are rarely as resistant as their resistant donor parents. Transfer of resistance via marker-assisted backcrossing would be much easier, if markers linked to genomic regions expressing resistance could be found. Multiple QTL mapping studies have found such markers for aflatoxin accumulation resistance from multiple donors (reviewed in Mideros et al., 2014), and newer Genome Wide Association Studies (GWAS)

(Warburton et al., 2015) and genomics studies including gene expression and proteomics studies (Huang et al., 1997; Chen et al., 2002; Pechanova et al., 2011; Kelley et al., 2012; Lanubile et al., 2015) are identifying more regions. Finally, biochemical and physiological studies can point to compounds that promote fungal resistance, and the genes encoding these proteins, hormones, enzymes and regulators will provide yet more markers for marker-assisted selection (Christensen and Kolomiets, 2011; Lanubile et al., 2015; Fountain et al., 2015).

Once these markers have been validated in multiple environments and genetic backgrounds, markers linked to the regions with the highest effect on resistance can be used to transfer resistance to new cultivars. Unfortunately, validation is also costly and time-consuming. In some cases, markers must be converted into a more user-friendly form, such as converting Single Nucleotide Polymorphisms (SNPs) identified via sequencing to stand-alone assays; or converting expensive chip or hybridization-based markers into simple PCR-based markers, such as sequence characterized amplified region (SCAR) markers. Multiple, closely linked markers must also be found and converted to user-friendly assays, because in any given cross between two unrelated maize inbred lines, more than half will not be polymorphic. Finally, the chromosomal regions being tagged by these markers must be tested in different genetic crosses and environmental conditions. The beneficial allele from each donor inbred line in which it was identified should be crossed with several different susceptible lines, and testcrossed or backcrossed progeny should be tested for increased resistance in replicated field trials in multiple fields and years. It is not surprising that many QTL and gene identification experiments have been published, but fewer publications detail the validation of these regions in independent genetic backgrounds. Once validated, however, these markers can speed selection for fungal resistance via marker-assisted selection or backcrossing, or population-level improvement.

5 Summary

Of all mycotoxin prevention or mitigation strategies, the use of host plant resistance and the release of stably resistant varieties is the easiest to transfer to the farmer's fields and is the most economical control measure for all markets. Implementation is as simple as the purchase of a bag of improved seed by the farmer, and no new equipment or know-how is necessary. Host plant resistance works well with other management strategies, including biocontrol, control of insect vectors, optimal agronomic practices, proper post-harvest storage conditions and post-harvest grain treatment such as nixtamalization, but it also decreases mycotoxin levels in the absence of any other strategy. Host plant resistance can be a non-transgenic method and thus not controversial in areas of the world which do not accept transgenic technology; conversely, it can be combined with transgenics to increase resistance even further.

With the use of resistant maize varieties, pre- and post-harvest grain yield losses will be reduced, leading to less waste and more food security. The health of primary consumers will benefit from resistant maize varieties enormously, particularly in countries where local infrastructure is unable to ensure proper testing of maize before consumption. The health and productivity of animals that are fed maize grain or distillers dried grains (DDG) will not be adversely affected either. Not only will the health of the farmer's families increase with the increased availability of the protein and nutrients provided by meat, egg and dairy animals, the economic opportunities available to farmers and their families will increase as well, as the wealth of the family is often tied up in livestock. In industrialized countries, the ability to sell DDG after the grain has been used for another purpose (e.g. ethanol) is necessary to make the whole industrial production system economically feasible. Because toxins concentrate in the DDG after ethanol is extracted, the use of the by-product would be precluded in mycotoxin-susceptible varieties, and farmers would be unable to sell their grain for this purpose, and possibly for any other purpose as well. Thus, with resistant varieties, industrialized country farmers' incomes will increase as well, as they face less rejection at the silo and when marketing their grain over state or country boundaries.

The availability of resistant maize breeding lines is no longer the limiting factor in the quest for mycotoxin-resistant maize. Resistant lines exist and have been released for use against all major toxin-producing fungal pathogens, and for fungi of lesser economic importance or geographic distribution as well. Although these lines are not currently well adapted to all maize-growing areas, the use of these lines as parents in new breeding populations and crosses will soon be expedited as molecular markers linked to genes increasing resistance are identified, validated and distributed. The identification of more genes and genomic regions contributing to resistance, better methods for selecting resistant germplasm and new methods for increasing resistance (including possible biotech-driven methods) will speed up the adoption of mycotoxin resistance in major maize breeding programmes.

6 Future trends

Current research is exploring many ways to tackle the problem of mycotoxin contamination of maize. Future efforts will need to tie together some of these and capitalize on synergies that different studies can bring. Gene identification studies will highlight proteins for future protein characterization studies, and proteomic studies will in turn give information to guide new candidate gene mapping analyses; both genomics and proteomics will provide input to metabolomic and pathway analyses and guide large-scale marker-assisted and genomic selection efforts. Identification of new classes of defence compounds, such as phytoalexins, phytoanticipins and phenylpropanoid derivatives will result in more targeted breeding efforts. Understanding the specific mechanisms and pathways involved in resistance will enable manipulation of the key genes in a pathway to increase the expression of the trait. This has been done with considerable success in maize with carotenoid content (Yan et al., 2010) and pro-Vitamin A level (Harjes et al., 2008) using natural variants present in the maize primary gene pool. With more complicated and quantitative mycotoxin resistance traits, such straightforward selection may not be possible, but key points in pathways will certainly become targets of selection or transgenic engineering. Once validation of some of the many markers identified to be linked to resistance to date has been completed, selection can move to a marker-assisted backcrossing and eventually to population-level improvement. Synergies between laboratories studying different fungal pathogens will build as well. While studies have shown that selection for resistance for one ear rot does not always give resistance to other ear rots (Henry et al., 2009), germplasm and some QTL enhancing resistance to multiple fungi have been found (Robertson-Hoyt et al., 2006; Wisser et al., 2006) and thus the creation of a maize breeding line with multiple fungal resistance is quite possible.

To increase heritability and repeatability of fungal resistance, which has traditionally been quite low, germplasm should be screened with the best phenotyping techniques available. Precision phenotyping and the control of variation introduced by non-genetic causes (i.e. the environment, measurement error, etc) will increase heritability by increasing the variation in the numerator of the heritability equation and reducing the variation in the denominator. In other words, it will be easier to assign the changes in phenotype to genes, and select for the effect of those genes without wasting effort in selecting lines that look resistant but are actually not, due to measurement errors or environmental conditions that did not allow for the full expression of the trait.

More work is needed to understand the reasons and conditions under which mycotoxin is produced. Developing maize that can block the environmental cues promoting toxin synthesis by the fungus may be one way to reduce toxin levels in the plant. More work is also needed to understand the multiple pathways by which resistance can be achieved in different maize germplasm, as there will be more than one mechanism even against the same fungus. Ensuring that multiple resistance mechanisms work simultaneously, rather than interfere with each other, will be necessary. Newest genetic techniques for increasing resistance may include the use of transgenics and RNAi or synthetic peptides that interfere with fungal colonization or spread, or transgenics that detoxify the toxin in the infected grain. The use of transgenics may spur interest in working on the trait in the private sector, which has historically found the possible return on investment of this recalcitrant and highly quantitative trait to be very low. When candidate genes and pathways have been identified, the use of gene editing, including the clustered regularly interspaced short palindromic repeats (CRISPR)/Cas-targeted gene editing system, may allow the creation of resistant lines in any genetic background, including the best elite breeding lines and commercial hybrids. Very soon, however, with education for producers and consumers and the deployment of resistant germplasm created with currently available inbred parents, the problem of mycotoxin in maize could be greatly reduced and over the long term, with newer resistant lines and technologies, it will be resolved.

7 Where to look for further information

Although plant breeding has not yet completely resolved the problem of mycotoxin contamination in maize, a significant body of work on the topic has been published over the years. An extensive reference section can be found at the end of this review. Information on ongoing aflatoxin research in maize can be found in the International Food Policy Research Institute (IFPRI) report entitled *Aflatoxins – Finding Solutions for Improved Food Safety* edited by Laurian Unnevehr and Delia Grace, published in 2013. Review articles and books that introduce the topic of host plant resistance breeding for mycotoxin reduction include Warburton and Williams (2014), Lanubile et al. (2014), Mesterházy et al. (2012), Menkir et al. (2006), Munkvold (2003) and Payne (1992). Articles covering fungal/plant interactions include Christensen and Kolomiets (2011), Dixon (2001), Balint-Kurti and Holland (2015) and Lo Presti et al. (2015).

Journals that cover the question of mycotoxin reduction via host plant resistance include Crop Science, Phytopathology, Plant Disease, Toxins, and the World Mycotoxin Journal. Finally, IFPRI, CIMMYT, IITA, the European MycoRED project, the USAID (particularly through its Feed the Future initiative) and the USDA post useful information on mycotoxins in maize in their websites, and then search for the keyword 'aflatoxin' in their blogs:

- www.cimmyt.org
- www.ifpri.org
- www.iita.org
- http://www.mycored.eu/d/1/The_Project/
- www.usda.gov
- www.usaid.gov
- http://www.feedthefuture.gov/
- http://www.ars.usda.gov/Research/docs.htm?docid=23012
- http://blog.cimmyt.org/tag/aflatoxin/
- http://www.cgiar.org/consortium-news/battling-the-deadly-aflatoxin-iita-scientist-wins-top-award/
- http://www.aflasafe.com/

8 References

Abbas, H. K., Williams, W. P., Windham, G. L., Pringle, H. C., Xie, W. and Shier, W. T. (2002), 'Aflatoxin and fumonisin contamination of commercial corn (*Zea mays*) hybrids in Mississippi'. *J. Agric. Food Chem., 50*, 5246–54.

Balint-Kurti, P. J. and Holland, J. B. (2015), 'New insight into a complex plant-fungal pathogen interaction'. *Nat. Genet., 47(2)*, 101–3.

Bakan, B., Melcion, D., Richard-Molard, D. and Cahagnier, B. (2002), 'Fungal growth and Fusarium mycotoxin content in isogenic traditional maize and genetically modified maize growing in France and Spain'. *J. Agric. Food Chem., 50*, 728–31.

Bennett, J. W. and Klich, M. (2003), 'Mycotoxins'. *Clin. Microbiol. Rev., 16(3)*, 497–516.

Betran, F. J., Isakeit, T. and Odvody, G. (2002), 'Aflatoxin accumulation of white and yellow maize inbreds in diallel crosses'. *Crop Sci., 42*, 1894–1901.

Cantone, F. A., Tuite, J., Bauman, L. F. and Stroshine, R. (1983), 'Genotypic differences in reaction of stored corn kernels to attack by selected *Aspergillus* and *Penicillium* spp.'. *Phytopathology, 73(9)*, 1250–5.

Chen, J., Ding, J., Li, H., Li, Z., Sun, X., Li, J., Wang, R., Dai, X., Dong, H., Song, W., Chen, W., Xia, Z., Wu, J. (2012), 'Detection and verification of quantitative trait loci for resistance to Fusarium ear rot in maize'. *Mol. Breed., 30(4)*, 1649–56.

Chen, Z. Y., Brown, R. L., Damann, K. E. and Cleveland, T. E. (2002), 'Identification of unique or elevated levels of kernel proteins in aflatoxin-resistant maize genotypes through proteome analysis'. *Phytopathology, 92(10)*, 1084–94.

Chiuraise, N., Derera, J., Yobo, K. S., Magorokosho, C., Nunkumar, A. and Qwabe, N. F. P. (2015), 'Progress in stacking aflatoxin and fumonisin contamination resistance genes in maize hybrids'. *Euphytica*, doi:10.1007/s10681-015-1529-6.

Christensen, S. A. and Kolomiets, M. V. (2011), 'The lipid language of plant-fungal interactions'. *Fungal Genet. Biol., 48*, 4–14.

Clements, M. J., Maragos, C. M., Pataky, J. K. and White, D. G. (2004), 'Sources of resistance to fumonisin accumulation in grain and Fusarium ear and kernel rot of corn'. *Phytopathology, 94(3)*, 251–60.

Cullen, D., Caldwell, R. W. and Smalley, E. B. (1983), 'Susceptibility of maize to Gibberella ear rot: relationship of host genotype, pathogen virulence, zearalenone contamination'. *Plant Dis., 67*, 89–91.

Dixon, R. A. (2001), 'Natural products and plant disease resistance'. *Nature, 411(6839)*, 843–7.

Dorrance, A. E., Hinkelmann, K. H. and Warren, H. L. (1998), 'Diallel analysis of diplodia ear rot resistance in maize'. *Plant Dis., 82(6)*, 699–703.

Eaton, D. L. and Groopman, J. D. (1993), *The Toxicology of Aflatoxins: Human Health, Veterinary and Agricultural Significance*. London: Academic Press.

Eller, M., Payne, G. A. and Holland, J. B. (2010), 'Selection for reduced *Fusarium* ear rot and fumonisin content in advanced backcross maize lines and their topcross hybrids'. *Crop Sci., 50*, 2249–60.

Fountain, J. C., Khera, P., Yang, L., Nayak, S. N., Scully, B. T., Lee, R. D., Chen, Z. Y., Kemerait, R. C., Varshney, R. K. and Guo, B. (2015), 'Resistance to *Aspergillus flavus* in maize and peanut: molecular biology, breeding, environmental stress, and future perspectives'. *Crop J.*, doi:10.1016/j.cj.2015.02.003.

Fountain, J. C., Scully, B. T., Chen, Z. Y., Gold, S. E., Glenn, A. E., Abbas, H. K., Lee, R. D., Kemerait, R. C. and Guo, B. (2015), 'Effects of hydrogen peroxide on different toxigenic and atoxigenic isolates of *Aspergillus flavus*'. *Toxins, 7*, 2985–99.

Gendloff, E. H., Rossman, E. C., Casale, W. L., Isleib, T. G. and Hart, L. P. (1986), 'Components of resistance to Fusarium ear rot in field corn'. *Phytopathology, 76*, 684–8.

Hara, S., Fennell, D. I. and Hesseltine, C. W. (1974), 'Aflatoxin-producing strains of *Aspergillus flavus* detected by fluorescence of agar medium under ultraviolet light'. *Appl. Microbiol., 27(6)*, 1118–23.

Harjes, Carlos E., Rocheford, T. R., Bai, L., Brutnell, T. P., Kandianis, C. B., Sowinski, S. G., Stapleton, A. E., Vallabhaneni, R., Williams, M., Wurtzel, E. T., Yan, J. and Buckler, E. S. (2008), 'Natural genetic variation in lycopene epsilon cyclase tapped for maize biofortification'. *Science, 319(5861)*, 330.

Henry, W. B., Williams, W. P., Windham, G. L. and Hawkins, L. K. (2009), 'Evaluation of maize inbred lines for resistance to Aspergillus and Fusarium ear rot and mycotoxin accumulation'. *Agron. J., 101(5)*, 1219–21.

Holland, J., Goodman, M. and Castillo-Gonzalez, F. (1996), 'Identification of agronomically superior Latin American maize accessions via multi-stage evaluations'. *Crop Sci., 36*, 778–84.

Holley, R. N., Hamilton, P. B. and Goodman, M. M. (1989), 'Evaluation of tropical maize germplasm for resistance to kernel colonization by *Fusarium moniliforme*'. *Plant Dis., 73*, 578–80.

Huang, Z., White, D. G. and Payne, G. A. (1997), 'Corn seed proteins inhibitory to *Aspergillus flavus* and aflatoxin biosynthesis'. *Phytopathology, 87(6)*, 622–7.

Kelley, R. Y., Williams, W. P., Mylroie, J. E., Boykin, D. L., Harper, J. W., Windham, G. L., Ankala, A. and Shan, X. (2012), 'Identification of maize genes associated with host plant resistance or susceptibility to *aspergillus flavus* infection and aflatoxin accumulation'. *PLoS ONE, 7(5)*, Article ID e36892, 2012.

King, S. B. and Scott, G. E. (1981), 'Genotypic differences in maize to kernel infection by *Fusarium moniliforme*'. *Phytopathology, 71*, 1245–7.

Lanubile, A., Machietto, V. and Marocco, A. (2014), 'Breeding maize for resistance to mycotoxins'. In J. F. Leslie and A. Logrieco (eds), *Mycotoxin Reduction in Grain Chains*, pp. 37–58. John Wiley and Sons, Iowa, USA.

Lanubile, A., Ferrarini, A., Maschietto, V., Delledonne, M., Marocco, A. and Bellin, D. (2015), 'Functional genomic analysis of constitutive and inducible defense responses to Fusarium verticillioides infection in maize genotypes with contrasting ear rot resistance'. *BMC genomics, 15(1)*, 710.

Lo Presti, L., Lanver, D., Schweizer, G., Tanaka, S., Liang, L., Tollot, M., Zuccaro, A., Reissmann, S. and Kahmann, R. (2015), 'Fungal effectors and plant susceptibility'. *Ann. Rev. Plant Biol., 66*, 513–45.

Martin, M., Miedaner, T., Schwegler, D. D., Kessel, B., Ouzunova, M., Dhillon, B. S., Schipprack, W., Utz, H. F. and Melchinger, A. E. (2012), 'Comparative quantitative trait loci mapping for gibberella ear rot resistance and reduced deoxynivalenol contamination across connected maize populations'. *Crop Sci., 52(1)*, 32–43.

Mayfield, K., Betran, F. J., Isakeit, T., Odvody, G., Murray, S. C. and Landivar, J. C. (2012), 'Registration of maize germplasm lines Tx736, Tx739, and Tx740 for reducing preharvest aflatoxin accumulation'. *Crop Sci., 6*, 88–94. doi:10.3198/jpr2010.12.0675crg.

McMillian, W. W., Widstrom, N. W. and Wilson, D. M. (1993), 'Registration of GT-MAS:gk maize germplasm'. *Crop Sci.*, *33*, 882.

Menkir, A., Brown, R. L., Bandyopadhyay, R., Chen, Z. Y. and Cleveland, T. E. (2006), 'Breeding maize for resistance to mycotoxins at IITA'. In J. F. Leslie, R. Bandyopadhyay and A. Visconti (eds), *Mycotoxins: Detection Methods, Management, Public Health, and Agricultural Trade*, pp. 277–86. CAB International, Oxfordshire, UK.

Mesterházy, A., Lemmens, M. and Reid, L. M. (2012), 'Breeding for resistance to ear rots caused by Fusarium spp. in maize – a review'. *J. Plant Breed.*, *131*, doi:10.1111/j.1439-0523.2011.01936.x.

Mideros, S. X., Warburton, M. L., Jamann, T. M., Windham, G. L., Williams, W. P. and Nelson, R. J. (2014), 'Quantitative trait loci for resistance to Aspergillus ear rot: analysis by linkage mapping, characterization of near-isogenic lines and meta-analysis'. *Crop Sci.*, *54*, 127–42.

Mideros, S. X., Windham, G. L., Williams, W. P. and Nelson, R. J. (2009), '*Aspergillus flavus* biomass in maize estimated by quantitative real-time polymerase chain reaction is strongly correlated with aflatoxin concentration'. *Plant Dis.*, *93(11)*, 1163–70.

Munkvold, G. P. (2003), 'Cultural and genetic approaches to managing mycotoxins in maize'. *Ann. Rev. Phytopathol.*, *41*, 99–116.

Payne, G. A. (1992), 'Aflatoxins in maize'. *Crit. Rev. Plant Sci.*, *10*, 423–40.

Pechanova, O., Pechan, T., Williams, W. P. and Luthe, D. S. (2011), 'Proteomic analysis of the maize rachis: Potential roles of constitutive and induced proteins in resistance to *Aspergillus flavus* infection and aflatoxin accumulation'. *Proteomics*, *11(1)*, 114–27.

Reid, L. M., Mather, D. E., Hamilton, R. I. and Bolton, A. T. (1992), 'Diallel analysis of resistance in maize to Fusarium graminaerum infection via the silk'. *Can. J. Plant Sci.*, *72*, 915–23.

Reid, L. M., Spaner, D., Mather, D. E., Bolton, A. T. and Hamilton, R. I. (1993), 'Resistance of maize hybrids and inbreds following silk inoculation with three isolates of *Fusarium graminearum*'. *Plant Dis.*, *77*, 1248–51.

Robertson-Hoyt, L. A., Jines, M. P., Balint-Kurti, P. J., Kleinschmidt, C. E., White, D. G., Payne, G. A., Maragos, C. M., Molnár, T. L. and Holland, J. B. (2006), 'QTL mapping for Fusarium ear rot and fumonisin contamination resistance in two maize populations'. *Crop Sci.*, *46*, 1734–43.

Scott, G. E. and Zummo, N. (1987), 'Host-plant resistance screening techniques'. In M. S. Zuber, E. B. Lillehoj and B. L. Renfro (eds), *Aflatoxin in Maize: A Proceedings of the Workshop*, pp. 221–33. CIMMYT, Mexico, DF.

Scott, G. E. and Zummo, N. (1988), 'Sources of resistance to maize kernel infection by *Aspergillus flavus* in the field'. *Crop Sci.*, *28*, 504–7.

Scott, G. E. and Zummo, N. (1990), 'Registration of Mp313E parental line of maize'. *Crop Sci.*, *30*, 1378.

Scott, G. E. and Zummo, N. (1992), 'Registration of Mp420 germplasm line of maize'. *Crop Sci.*, *32*, 1296.

Scott, G. E. and Zummo, N. (1994), 'Kernel infection and aflatoxin production in maize by Aspergillus flavus relative to inoculation and harvest dates'. *Plant Dis.*, *78*, 123–5.

Warburton, M. L. and Williams, W. P. (2014), 'Aflatoxin Resistance Genomics Review Paper Aflatoxin resistance in maize: What have we learned lately?'. *Adv. Botany*, Article ID 352831, http://dx.doi.org/10.1155/2014/352831.

Warburton, M. L., Williams, W. P., Windham, G., Murray, S., Xu, W., Hawkins, L. and Franco, J. (2013), 'Phenotypic and genetic characterization of a maize association mapping panel developed for the identification of new sources of resistance to *Aspergillus flavus* and aflatoxin accumulation'. *Crop Sci.*, *53*, 2374–83.

Warburton, M. L., Tang, J., Windham, G. L., Hawkins, L. K., Murray, S. C., Xu, W., Boykin, D., Perkins, A. and Williams, W. P. (2015), 'Genome wide association mapping of *Aspergillus flavus* and aflatoxin accumulation resistance in maize'. *Crop Sci.*, First Look. doi:10.2135/cropsci2014.06.0424.

Williams, J. H., Phillips, T. D., Jolly, P. E., Stiles, J. K., Jolly, C. M. and Aggarwal, D. (2004), 'Human aflatoxicosis in developing countries: a review of toxicology, exposure, potential health consequences, and interventions'. *Am. J. Clin. Nutr.*, *80*, 1106–22.

Williams, W. P. (2006), 'Breeding for resistance to aflatoxin accumulation in maize'. *Mycotoxin Res., 22(1)*, 27–32.

Williams, W. P. and Windham, G. L. (2001), 'Registration of maize germplasm line Mp715'. *Crop Sci., 41*, 1373–5.

Williams, W. P. and Windham, G. L. (2006), 'Registration of maize germplasm line Mp717'. *Crop Sci., 46*, 1407–8.

Williams, W. P. and Windham, G. L. (2012), 'Registration of Mp718 and Mp719 germplasm lines of maize'. *J. Plant Registrations, 6(2)*, 200–2.

Williams, W. P., Krakowsky, M. D., Scully, B. T., Brown, R. L., Menkir, A., Warburton, M. L. and Windham, G. L. (2015), 'Identifying and developing maize germplasm with resistance to accumulation of aflatoxins'. *World Mycotoxin J., 8(2)*, 193–209.

Windham, G. L. and Williams, W. P. (2002), 'Evaluation of corn inbreds and advanced breeding lines for resistance to aflatoxin contamination in the field'. *Plant Dis., 86(3)*, 232–4.

Windham, G. L., Williams, W. P., Buckley, P. and Abbas, H. (2003), 'Inoculation techniques used to quantify aflatoxin resistance in corn'. *J. Toxicol-Toxin Rev., 22*, 313–25.

Windhausen, V. S., Atlin, G. N., Hickey, J. M., Crossa, J., Jannink, J. L., Sorrells, M. E., Raman, B., Cairns, J. E., Tarekegne, A., Semagn, K., Beyene, Y., Grudloyma, P., Technow, F., Riedelsheimer, C. and Melchinger, A. E. (2012), 'Effectiveness of genomic prediction of maize hybrid performance in different breeding populations and environments'. *G3, 2(11)*, 1427–36.

Wisser, R. J., Balint-Kurti, P. J. and Nelson, R. J. (2006), 'The genetic architecture of disease resistance in maize: a synthesis of published studies'. *Phytopathology, 96(2)*, 120–9.

Wu, F., Bhatnagar, D., Bui-Klimke, T., Carbone, I., Hellmich, R., Munkvold, G., Paul, P., Payne, G. and Takle, E. (2011), 'Climate change impacts on mycotoxin risks in US maize'. *World Mycotoxin J., 4*, 79–93. doi:http://dx.doi.org/10.3920/WMJ2010.1246.

Xiang, K., Reid, L. M., Zhang, Z. M., Zhu, X. Y. and Pan, G. T. (2012), 'Characterization of correlation between grain moisture and ear rot resistance in maize by QTL meta-analysis'. *Euphytica, 183(2)*, 185–95.

Xu, W., Archer, T. L., Bynum, E. D. and Odvody, G. (2004), 'Registration of maize germplasm line Tx202'. *Crop Sci., 44*, 1883–4.

Yan, J., Kandianis, C. B., Harjes, C. E., Bai, L., Kim, E.-H., Yang, X., Skinner, D. J., Fu, Z., Mitchell, S., Li, Q., Salas Fernandez, M. G., Zaharieva, M., Babu, R., Fu, Y., Palacios, N., Li, J., DellaPenna, D., Brutnell, T., Buckler, E. S., Warburton, M. L. and Rocheford, T. (2010), 'Rare genetic variation at Zea mays crtRB1 increases [beta]-carotene in maize grain'. *Nat. Genet., 42(4)*, 322–7.

Zhang, D., Liu, Y., Guo, Y., Yang, Q., Ye, J., Chen, S., and Xu, M. (2012), 'Fine-mapping of qRfg2, a QTL for resistance to Gibberella stalk rot in maize'. *Theor. App. Genet., 124(3)*, 585–96.

Zila, C. T., Ogut, F., Romay, M. C., Gardner, C. A., Buckler, E. S. and Holland, J. B. (2014), 'Genome-wide association study of Fusarium ear rot disease in the USA maize inbred line collection'. *BMC Plant Biol., 14(1)*, 372.

Zummo, N. and Scott, G. (1989), 'Evaluation of field inoculation techniques for screening maize genotypes against kernel infection by *Aspergillus flavus* in Mississippi'. *Plant Dis., 73*, 313–16.

Chapter 6

Advances in cold-tolerant maize varieties

Csaba L. Marton and Zsuzsanna Tóthné Zsubori,
MTA Centre for Agricultural Research, Hungary

1 Introduction

Maize originates from subtropical regions, probably from the highlands of Mexico (Mangelsdorf 1986), and the newest molecular assays have confirmed this hypothesis (Matsuoka et al. 2002). On the basis of fossil studies, archaeologists also proved that teosinte, the ancestor of maize, was first domesticated in Oaxaca, Mexico (Benz 2005).

In pre-Columbian times, the cultivation of maize spread to Chile in the south and to southern Canada in the north, and to altitudes ranging from sea level to 3000 m in the Andes.

From 1493 onwards there was a further expansion of the maize cultivation area: (1) Columbus took samples of maize back to Europe on his first voyage to America, and these were successfully grown on the Iberian Peninsula. (2) Traders and sailors from Asia Minor swiftly introduced maize both to the southern countries of Europe and to the Middle East. (3) Maize first reached the Carpathian Basin via Italy, and it was only some 70 years later that it was again brought to the region by the Turks via the Balkans. The name 'Turkish wheat', however, is commonly used in the Carpathian Basin, which definitely points to the role of the Turks in its introduction.

http://dx.doi.org/10.19103/AS.2016.0001.11

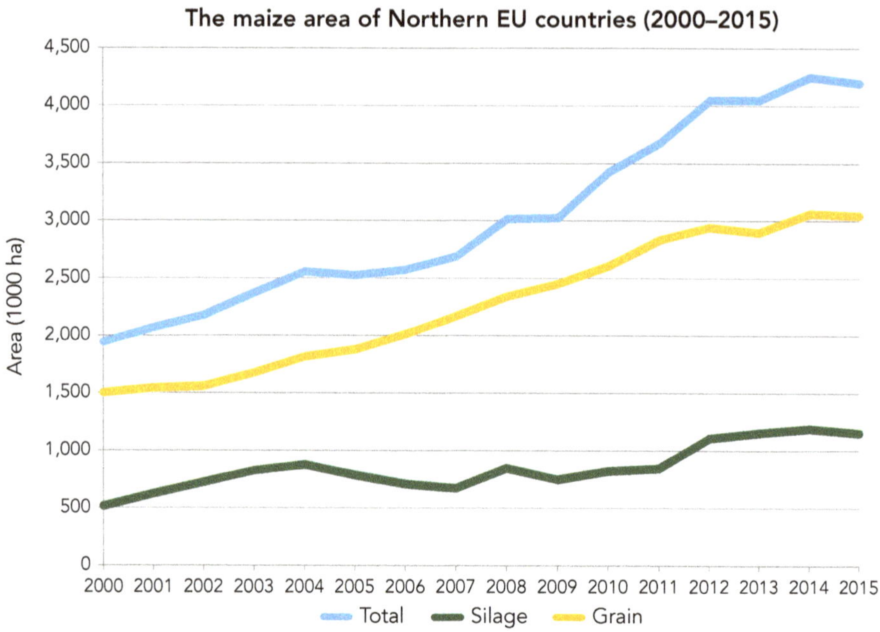

Figure 1 The maize-growing area of northern EU countries (2000–15).

Over the last 50 years the boundary of maize production in Europe has moved further northward than in almost 500 years previously, from a latitude of 50°N (Paris – Prague – Kiev) in 1960 to 55°N in 1990 and to an estimated 60°N at present (Fig. 1).

This intensive advance took place parallel to the spread of heterosis breeding, which is now a dominant factor in maize breeding. Heterosis for earliness has undoubtedly promoted the development of hybrids with shorter growing periods. The sowing area of maize has doubled in northern European countries over the last 15 years, increasing from 1.5 million to 3 million hectares for silage maize and from 500 000 ha to 1.2 million ha for grain maize. This northerly expansion of maize cultivation in Europe has probably been assisted by the global warming that has characterised recent decades. Today, maize is a leading crop in many temperate regions.

As maize is a thermophilic plant species, it characteristically needs a high optimum temperature for germination and growth. However, it can germinate at lower temperatures as well, but in this case the seedling vigour is weaker and the emergence period is longer.

Maize genotypes that are cultivated in cooler climates or at higher latitudes need to have a short vegetation period and the ability to adapt at low temperatures. This has been made possible by its high level of heterozygosity, which has facilitated natural and human selection for these traits (Lee and Estes 1982). Real adaptation to low temperature seems to be rather complex, requiring a number of features such as resistance to chilling, resistance to soil fungi during germination and the ability to grow at low temperatures.

2 The effect of temperature on maize growth

2.1 Suboptimal temperature ranges

Cold tolerance is defined as capability of surviving, growing and developing at low temperatures.

As the optimum temperature for maize is around 30°C, the term 'low temperature' covers a broad range of temperatures when interpreting chilling tolerance. Three ranges are generally distinguished in the literature:

Freezing:	below 0°C
Low, non-freezing:	0–6°C
Suboptimal:	6–20°C

Frost damage to maize kernels depends on the moisture content (Hall 2009). Kernels stored with a moisture content of below 15% can withstand even −20°C for more than six months; however, kernels with a moisture content greater than 25% lose their germination ability. Dry seeds in wet soil can withstand −5 to −10°C for 2 days, but they are killed if the freezing treatment is preceded by exposure to 20°C for 1–2 days. After emergence the aerial parts are easily killed by frost. Young seedlings, however, may recover if the shoot apical meristem, which is below soil level, is not killed. The freezing injury usually depends on the duration of temperature and time of exposure (Black et al. 2006). Buican (1969) showed that 50% of the plants were killed by −6°C for 2 h, −5°C for 3 h, −4°C for 3–6 h, −3°C for 24–36 h or −2°C for 48 h. Frost is also detrimental to plant development at the later stages. All maize genotypes are damaged if the temperature drops below −2°C. It can be concluded that maize is sensitive to frost in all phases of its growth cycle except as dry seed. Genetic variation seems to be very small, so tolerance to freezing temperature cannot be successfully improved by breeding. Freezing injury limits the expansion of the maize-growing area.

Low non-freezing temperatures (0–6°C) cause physiological damage to the seedlings. Wilting and discolouration of the leaves are symptoms of chilling injury; in the case of severe chilling, whole plants or plant parts are killed. The dysfunction of membranes at low temperatures could be the primary cause of chilling injury, and the severity of the damage depends on the temperature, the duration of the cold treatment, the stage of development, and to a certain extent, the genotype (Devries 2006, Burris and Knittle 1985). In farm practice, however, maize is sown in spring, when soil temperatures are seldom below 6°C for long periods of time, so it seems unlikely that survival and emergence are affected by chilling.

The temperature during emergence and the seedling stage frequently falls in the *suboptimal range* (6–20°C). These temperatures are not generally detrimental to seed or seedlings, but significantly retard growth and development. According to Wallace and Bressmann (1949), in the American Corn Belt temperatures of 13°C or lower in May not only delay emergence but also reduce the maize yield by around 15%. Therefore, breeders have long attempted to improve the chilling tolerance of maize.

By growing cold-tolerant hybrids it is possible to achieve an improvement in stand uniformity, the rate of plant development and the yield. In addition, cold-tolerant hybrids can be sown earlier, so that hybrids with longer vegetation periods and greater yield

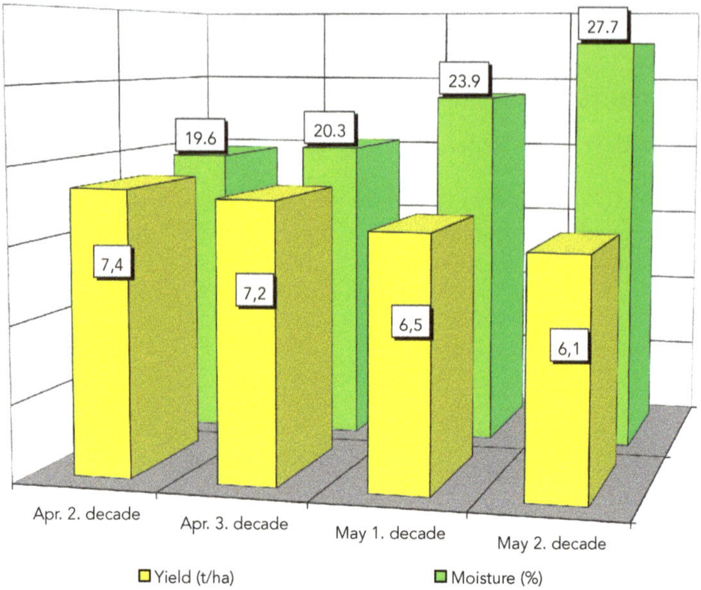

Figure 2 Effect of planting date on yield and moisture content of maize (mean of 4 years' field experiments, Martonvásár).

potential can be grown. Continental climate provides further advantage to the cold-tolerant hybrids during the flowering stage, wherein the most sensitive phenophase of maize from the point of view of water supplies occurs earlier when the crop is sown early, avoiding the summer heat and drought. Due to this escape mechanism, yields are higher and the grain moisture at harvest is lower, as proved by the results of sowing time experiments in Martonvásár (Hungary, Europe) (Fig. 2).

2.2 Cold tolerance and 'heat unit' methods

Réaumur (1735) was the first to point out that various plant development phases were reached each year at the same heat sum. Although the temperature alone does not determine the date of ripening or of individual phenophases, its use proved to be more accurate than the methods previously employed. Extensive experiments were initiated to examine the relationship between temperature and the various phenophases.

The daily mean temperatures accumulated above a certain threshold temperature proved to be useful for the prediction of growth and development, first in peas (Boswell 1929) and sweet corn (Lana and Haber 1952) and later in other crops, including fodder maize (Gilmore and Rogers 1958, Cross and Zuber 1972). An evaluation covering not only the flowering time but also the whole of the growing period was performed by Mederski et al. (1973).

The determination of heat thresholds is also the basis and problematic point of the 'heat unit' methods that serve to define the vegetation period. Growing degree days is the most widely used method to determine heat thresholds, which employs a temperature threshold of 10°C, but many authors (Arnold 1975, Bloc and Gouet 1976, Bunting 1976, Gilmore and Rogers 1958) suggested different temperatures. In Europe, threshold temperatures lower than 10°C usually give better estimations of the vegetation period (Derieux and Bonhomme 1982).

Herczegh and Marton (1986) investigated the minimum heat required for germination by hybrids and inbred lines using a gradient chamber with a temperature gradient of 4–16°C, rising in steps of 1.3°C. Germination began in two hybrids even at 4°C. Different genotypes have widely different temperature requirements, indicating that each genotype should be evaluated separately, rather than attempting to determine a single threshold temperature that is valid for maize in general. Stamp (1984) also reported differences in the temperature requirements of individual maize genotypes.

The period from sowing to emergence is very variable between genotypes and years, and is significantly modified not only by temperature but also by other factors, including the soil structure, the sowing depth and the soil moisture content. In addition to the temperature, the emergence and development of maize also depends on many other factors, such as genotype, soil, pathogen, seed quality and so on. It is the task of the breeder to develop chilling-tolerant genotypes, while the grower is responsible for ensuring satisfactory soil status and the seed producer for producing and processing physiologically mature, injury-free seed given satisfactory dressing.

Knowledge of the maize genotypes' chilling tolerance level is of practical importance not only in commercial maize production, but also to ensure the synchronised flowering of the parental lines in seed production. One aspect of the problem is the diverse heat threshold required by the genotypes for germination. If the temperature in spring is cooler than average, the differences in the emergence date may be so great that synchronised flowering is endangered. Marton (1990) studied the seedling growth of inbred lines over a temperature range of 9–18°C and found a significant genotype × temperature interaction, suggesting that the genotypes responded differently to various temperatures. This means that seed production technologies have varying success in ensuring synchronised flowering in different years. Improvements in the chilling tolerance of inbred lines, or at least better knowledge of their chilling tolerance, are also extremely important from this point of view.

3 Factors affecting seed quality and germination at low temperatures

Requirements from maize grown in cooler regions are the cold-tolerant genotype, good seed quality and resistance to fungal diseases by seed coating. Cold tolerance of the seeds depends on the seed quality and the conditions for germination.

3.1 Seed quality

Maturity

Despite its deficiencies, the use of the cold test has led to the accumulation of a substantial body of knowledge on maize chilling tolerance, as regards both seed quality and breeding/genetic aspects. Brown (1948), Livingston (1951) and Rush and Neal (1951) investigated the relationship between chilling tolerance and the biological maturity of the seed. The chilling tolerance of immature seed was found to be much poorer than that of mature seed. Kisselbach (1939) recommended that seed harvested with high moisture content should be dried in air with a temperature no higher than 40°C in order to preserve the biological value of the seed. According to Burris (1989), the chilling tolerance of maize kernels harvested at a moisture content of 30% or above did not deteriorate significantly after drying at 45°C, provided they were preliminarily dried at 35°C. In addition to the biological maturity of the seed, the grain moisture content at harvest, the drying temperature and the chilling tolerance also depends to a considerable extent on the external environmental conditions to which the mother plants were exposed (Isely 1957).

Injuries

Chilling tolerance is reduced by injuries suffered during seed processing and storage. There may be a substantial deterioration in chilling tolerance if the technology used in seed processing is too harsh (Wortman and Rinke 1951). Kaerwer (1953) reported a 47% rate of kernels with damaged pericarps after mechanical shelling and only 13% in the case of manual shelling. The ratio of pericarp injury was found to be closely correlated with the results of the cold test (Tatum 1942, Kovács 1961). Pericarp damage close to the germ made the seedlings especially prone to infection. Even in the absence of pathogens, pericarp injury may reduce the germination ability (Schulz and Bateman 1969).

Moisture content and uptake

Martinson (1971) confirmed that soaking maize seeds at low temperature could lead to damage and drew attention to the role of soil moisture in cold tests. Soaking is basically a physical process, as the water uptake of dead seeds was almost the same as that of live ones. In the course of vigour tests, Szirtes et al. (1982) found that water facilitated the diffusion of mineral salts, amino acids and sugars through cracks in the cell membrane arising due to chillness, thus resulting in greater damage than that occurring in consequence of soil chilling. The damage was manifested as a reduction in both the seedling length and the germination percentage. More permeable seeds are more readily penetrated by soil microorganisms and might be subject to greater leaching, which could serve as a substrate for soil pathogens (Tatum 1954).

At low temperature the seeds germinate at higher moisture content than at optimum temperature. At 12°C the embryo in particular imbibes a great deal of water (Blacklow 1972a,b). In maize, as in other chilling-sensitive plants, seed inhibition at low temperature resulted in exudation from the kernels (Miedema 1982), which could be expected as a consequence of membrane dysfunction.

Storage

The cold tolerance of seeds is greatly influenced by the method and duration of storage. Hoppe (1956) found that cold test value declined by 5% after one year of storage and by 20% after two years. Neal and Davis (1956) observed considerable differences between the cold test values of seeds stored at room temperature or at 8.3°C: higher storing temperature decreased the viability of seeds.

3.2 Germination

Soil moisture content

In cold test methods there is little difference in the temperature applied during the incubation period, but much greater deviations can be observed in the soil moisture content. Nagy et al. (1984) recommended using soil saturated to 50–60% of field capacity, while Svien (1951) advised 60%, Herczegh (1978) 65–70% and Szundy (1981) 70%. In cold soil, the soil moisture content determines the injuries suffered during seed soaking (Martinson 1971). As the duration of cold incubation increased, there was a steady deterioration in the germination percentage (Neptune 1953, Kovács 1961, Nagy et al. 1984). Despite the deficiencies of the cold test, a positive correlation was found by many authors (Tatum 1942, Pinell 1949, Hoppe 1958, Kovács 1961, Herczegh 1978) between the results of chilling tolerance experiments in the field (early sowing, staggered sowing) and laboratory field tests.

Pathogens

The pathogens present in the soil have a decisive effect on the results of the cold test. American authors generally consider *Pythium* species to be the most dangerous for germinating seeds, but the damage caused by *Fusarium* species is also frequently mentioned in the literature.

Dickson (1923), while researching on *Gibberella saubinettii* (Mont) Sacc., drew the following conclusions about the role of pathogens when maize germinates in the cold: (1) infection takes place in the early stages of germination; (2) the soil temperature is the most important factor determining the extent of germ death and (3) germ death occurs at temperatures unfavourable for maize growth, even if this is far from the optimum temperature for the vegetative growth of the pathogen (28°C for the pathogen examined). The results obtained by Dickson in sowing date experiments under field conditions agreed with those of laboratory experiments and showed that seedling blight was reduced in warm soils.

In severely infected soil the cold test percentage values are determined by the resistance level of the genotypes, while in mildly infected or pathogen-free soil the genetic chilling tolerance is decisive (Tatum and Zuber 1943, Maryam and Jones 1983a, Christeller 1984). Marton (1990), while evaluating the emergence of inbred lines inoculated with *Fusarium* species, found that the emergence percentage of inoculated treatments improved gradually as the temperature rose. The shoot dry mass, however, only reached 60% of the control value even at the highest temperature tested. The level of resistance to *Fusarium* species has a substantial modifying effect on the practical chilling tolerance of the genotypes. In order to avoid the problems caused by pathogens, many authors performed chilling tolerance tests in sterilised soil or in artificial media.

The pericarp plays an important role in defence against pathogens. If it suffers damage it is unable to prevent pathogen penetration, facilitating infection in the early phases of germ development. According to Hooker and Dickson (1952), the advantage of resistant lines was based on the more rapid development of resistance in the course of germination. The role of the maternal effect was demonstrated by Lunsford et al. (1975) in the inheritance of resistance to *Fusarium moniliforme* in the early stages of seedling development.

4 Cold test methods

Cold tests are an important part of breeding efforts aimed at improving chilling tolerance, although they are used more widely in seed certification as a way of testing vigour.

Various field and laboratory methods have been elaborated for the testing of cold tolerance. In most cases early sowing is applied in field experiments (Andrew 1954, Kovács 1958, Miedema 1979, Ivakhnenko 1979, Stamp 1984). Dressed or undressed seeds are sown at intervals and the results achieved for early and optimal sowing dates are compared.

Field selection is an effective way of improving chilling tolerance, but it depends to a great extent on the climatic factors at the given growing site in the given year, so it is very uncertain. To avoid this problem, breeders have elaborated laboratory testing methods, the greatest advantage of which is their speed and the fact that they can be performed regardless of the season. Tatum (1942) was the first to elaborate and use a laboratory method for the selection of chilling-tolerant plants. Seeds of the genotypes to be tested were incubated at low temperature and germinated at optimum temperature. This also became the basis for the cold tests developed later (Bochicchio 1985).

Cold tests are generally based on climatic programmes with a combination of two temperatures: (1) cold incubation for 5–10 days at 8–10°C and (2) germination at optimum temperature. Although programmes of this type can be used to demonstrate genetically determined chilling tolerance, they are really more suitable for estimating the biological value of the seed. It is not by chance that in many countries the cold test is applied as a seed testing method in the certification of seed.

Isely (1950) and Clark (1954) proposed that various cold test methods should be combined in order to make the results more easily comparable. Hoppe (1951) made a substantial contribution to the perfection of the method. The generally applied 'rolled towel method' is also linked with his name (Hoppe 1955). Kovács (1961) used the method recommended by Clark (1954) and developed it further by increasing the length of the cold treatment.

In accordance with the technical standard of the time, the experiments were carried out in an ice pit. The development of thermostats for cooling or warming was a great step forward, followed later by the use of phytotron chambers. In order to achieve a more precise demonstration of genotypic differences, Herczegh (1978) elaborated a new method for type G climate chambers. After 10 days of cold incubation at 8°C, germination was carried out not at optimum temperature, but at 13.5°C, which allowed information to be obtained not only on the emergence percentage, but also on the rate of emergence. This method was adapted by Szundy (1981) for type E 15 climate chambers. Herczegh and Marton (1986) used a temperature gradient chamber to investigate the emergence of maize (Fig. 3).

The susceptibility of maize to low temperatures during germination, emergence and early growth can be attributed to two main causes: (1) chilling injuries to the seed, embryo or seedling and (2) infection with microorganisms. The results of cold tests thus depend to a great extent on the quality of seed dressing (Grabe 1976). In breeding, where the experiments generally involve the use of undressed seeds, the cold test can be interpreted as a pathogenic test.

Figure 3 Study of cold tolerance in a temperature gradient chamber.

As the cold test does not provide sufficient information on seed vigour, further vigour testing methods have been elaborated, the most widely used of which are the following:

- accelerated ageing test (Delouche and Baskin 1973) is designed to investigate seed storability and field emergence. The seeds are exposed to high temperature (41–45°C) and humidity for a period of 48–96 h to simulate the deterioration that occurs during long-term storage.
- tetrazolium test (Moore 1962) gives a rapid estimation of warm germination results. It can also be used to determine the viability of ungerminated seed at the end of the germination period.
- analysis of seedling growth rate (Burris et al. 1969).
- analysis of germination rate (Kotowski 1926).
- mechanical injury test (Koehler 1957, Simak and Kamra 1963).
- electrical conductivity test (Mathews and Bradnock 1968) indirectly determines the integrity of seed membrane systems.
- complex stressing vigour test, elaborated by Szirtes and Barla-Szabó (1981) and Szirtes et al. (1982), consists basically of a combination of accelerated ageing and cold tests. According to Barla-Szabó (1985) this method provides the greatest amount of information on seed vigour.

In the opinion of Delouche (1976) the results of cold tests are strongly influenced by four external factors: (1) seed dressing, (2) the duration of the cold test, (3) the temperature applied and (4) the soil in which the seeds are germinated. The cold test results therefore have no absolute value, but can be used for a comparison of seed lots.

Despite these difficulties, breeding for improved chilling tolerance has led to significant advances. It is sufficient to mention the northerly shift in maize-growing areas. The results achieved in breeding can be attributed to the great variability occurring within the maize species and to the genetic control of chilling tolerance.

5 Germplasm evaluation to improve cold tolerance through breeding

As early as 1914, Leath (see Neal 1950) set up a successful breeding programme aimed at improving the chilling tolerance of the Golden Glow cultivar. Mock and Skrada (1978) examined the emergence percentage and emergence index of 144 populations and the weight of 42-day-old plants at a temperature of 10°C, and found substantial variability for the traits tested. Pollmer (1967) studied the chilling tolerance of 600 cultivars in the laboratory and in staggered sowing trials in the field. Although great differences were found between the cultivars for chilling tolerance, the year had a considerable modifying effect. When McConnell and Gardner (1979) carried out combined laboratory (germination at 7.2°C) and field selection aimed at improving the chilling tolerance of two populations (Iowa Stiff Stalk Synthetic and Pioneer Cold Tolerant Synthetic), improvements of 8.8% and 9.9% per cycle, respectively, were recorded in the laboratory after four cycles of selection. This advantage could not be demonstrated in the field, probably due to the mild spring. Mock and Eberhardt (1972) evaluated the chilling tolerance of two populations (BSSS 2 and BSSS 13) and found sufficient variability in both populations to serve as a basis for the improvement of the trait.

Mock and Bakri (1976) used recurrent selection to improve the chilling tolerance of the same two populations in the field and laboratory. Selection resulted in a substantial improvement in the emergence percentage of the two populations and reduced the vegetation period. Using five cycles of recurrent selection, Hoard and Crosbie (1985) successfully improved the emergence percentage, emergence rate and germ dry mass of two populations (BSSS 2 and BSSS 13). Greater improvement in the emergence percentage was achieved in a colder environment, while – surprisingly – the germ dry weight improved to a greater extent when evaluated under warmer conditions. Selection for chilling tolerance also altered other, non-selected, agronomic properties of the populations, such as grain yield and lodging (Hoard and Crosbie 1985, 1986). Selection improved the ecological stability of chilling tolerance traits but caused no change in the environmental sensitivity of other agronomic properties. According to observations made by Hooker (1955) and Kovács (1961), variability for chilling tolerance could be detected between old, morphologically uniform inbred lines, making it possible to develop new chilling-tolerant sublines. Kovács (1989) introduced a new method of selection for chilling tolerance in pollen populations. Genetic analysis performed for breeding purposes demonstrated that the inheritance of chilling tolerance is complex (Pesev 1970).

Rodriguez et al. (2010) evaluated the European germplasms from EU Maize Landraces Core Collection to identify new cold-tolerant populations useful for improving adaptation to early sowing. After a preliminary screening of 95 populations the 11 populations with best germination and early growth under cold conditions were multiplied and evaluated in a cold chamber and in early field sowings. Some of these populations had similar emergence, higher vigour and, in some cases, higher yields than the control cold-tolerant hybrids.

All the sources cited suggested that chilling tolerance can be improved by breeding. Numerous references can be found in the literature to factors that are of importance in the development of chilling tolerance and exert a significant effect.

5.1 Genetic origin

Miedema (1979) studied the effect of low temperature on the emergence date of 21 maize varieties in early and late-sown plots. In general, types developed at higher altitudes emerged four days earlier than those designed for lower altitudes. There was a significant negative correlation between the emergence date and the shoot dry mass in early sown plots. The shoot meristem was located close to the surface in high-altitude varieties, but much lower in low-altitude variants. This difference could be attributed to differences in the growth of the mesocotyl. Herczegh (1983) evaluated breeding stock with various genetic backgrounds on the basis of the number of days to emergence. Inbred lines with Lancaster ancestry emerged almost three days later than Reid Yellow Dent lines. Genotypes of European origin emerged around the same time as the Reid Yellow Dent lines.

5.2 Special genes

Gupta and Kovács (1975) compared the emergence percentages and emergence dates of normal and *opaque* analogues of inbred lines and their hybrids. The normal variants emerged more rapidly than their opaque equivalents and with higher emergence percentages. These authors attributed the poorer chilling tolerance of the opaque analogues to three causes: the unfavourable pleiotropic effect of the *o2* gene, the thicker pericarp of the opaque analogues and the higher grain moisture content at harvest. Haskell (1952) found sweet corn mutants to have poorer chilling tolerance than the normal analogues. Haskell further suggested that the thin pericarp, which makes the kernels more susceptible to damage, was due not to the effect of the *su2* gene but to continual selection.

5.3 Cytoplasmic male sterility

Herczegh (1970) also compared the chilling tolerance of fertile and sterile analogues of inbred lines, and in most cases found the former to be better. Herczegh et al. (1970), however, reported that the sterile analogue was not necessarily less chilling-tolerant than the fertile variant, as it was found that the chilling tolerance of the inbred line might remain unchanged or even improve, depending on the sterility donor.

5.4 Hybridisation

It was demonstrated by Wortman (1950) and Pesev (1970) that F_1 hybrids were more chilling-tolerant than the parental lines. According to Pesev (1970) the better chilling tolerance of the hybrids was the consequence of complementary gene effects. Neptune (1953) and Rinke (1954) found a drop in the chilling tolerance of the lines as inbreeding progressed. Pinnell (1949) studied the emergence of double crosses, single crosses, inbreds and reciprocal crosses of maize at low temperatures and detected significant differences between the inbreds, which appeared to be heritable in crosses. The maternal parent was of primary importance in determining seedling stands in both single and double crosses. However, the level of pericarp damage could not account for a large part of the maternal differences observed. Szundy and Kovács (1981a,b) investigated the emergence date and emergence percentage of systematic heterozygous series and the hybrids developed on them and reported that genotypes with a 0 or 25% level of heterozygosity emerged

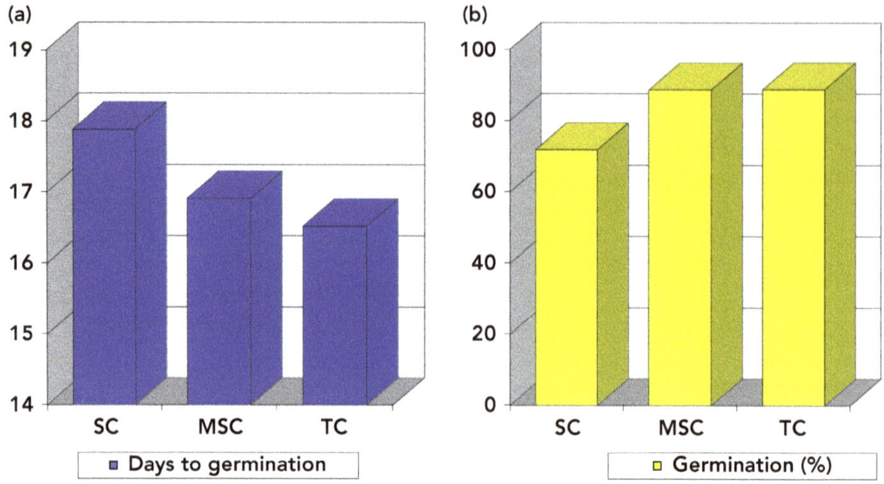

Figure 4 (a) Cold tolerance of different hybrid types (days from planting to germination). SC = single cross, MSC = modified single cross, TC = three-way cross. (b) Cold tolerance of different hybrid types (percent of germination). SC = single cross, MSC = modified single cross, TC = three-way cross.

significantly later than forms with 50 or 100% heterozygosity. As regards the emergence percentage, the 25 and 50% heterozygous forms proved to be better than those with 0 and 100% heterozygosity. There was a negative linear correlation between level of heterozygosity of the maternal parent and the emergence date of the hybrid. In the case of the hybrids, the emergence percentage rose and the time required for emergence declined with the level of heterozygosity of the female parent. This explains the better cold tolerance of three-way and double-cross hybrids (Fig. 4a,b).

Pesev (1970) found that single-cross hybrids in which both parents were tolerant to low temperatures were more tolerant than those whose pollen parent was susceptible.

In an experiment performed by Herczegh (1978), involving cold treatment at 8°C for 10 days, followed by emergence at 13.5°C, parent–progeny regression gave a significant positive correlation for emergence date. The heterosis effect reduced the emergence time by 10–12%. The emergence date of F_1 hybrids was decisively determined by the mean of the parental lines. Maryam and Jones (1983a) germinated maize seed *in vitro* at a constant temperature of 8°C. The mean germination time was between 9.7 and 19.3 days, but some lines did not germinate at all at this temperature. Hybrids developed on rapidly germinating maternal plants germinated rapidly, irrespective of the germination rate of the pollinator. In the case of hybrids developed on maternal plants with moderate or slow germination, the germination rate of the pollinator was manifested in the progeny. Rapid germination ability is reliably inherited from both the maternal and paternal side. When germination was carried out at 6°C the majority of the lines and hybrids failed to germinate even after 40 days. Maryam and Jones (1983b) demonstrated that the early growth of seedlings was strongly dependent on the maternal contribution.

5.5 Maternal effect

As shown in the literature, both heterosis and the level of heterozygosity of the maternal parent exert a significant effect on the development of chilling tolerance. However, the mother plant influences the chilling tolerance of the progeny not only through the level of heterozygosity but also via the pericarp, which is of maternal origin, and the endosperm, which has two-third maternal origin (Grogan 1970, Pesev 1970, Pinnell 1949). The decisive role of the maternal effect was also reported by Ivakhnyenko (1979) and Gonchar (1979). The diverse chilling tolerance of direct and reciprocal crosses was also mentioned by Rinke (1954) and Andrew (1954). The genetic effects involved in the inheritance of chilling tolerance were studied by Eagles and Hardacre (1979). Heterosis and the maternal effect were found to have a significant influence on the inheritance of germination percentage. McConnell and Gardner (1979) germinated the seed of maize inbred lines, hybrids and the F_2 and BC_1 generations in cold (7.2°C) soil and demonstrated additive, dominant and epistatic gene effects in the inheritance of emergence percentage.

6 Inheritance of cold tolerance

Ferenec (1980) tested the chilling tolerance of all possible combinations of 10 inbred lines, including reciprocals, at 6°C and revealed the role of dominance and overdominance, as well as additive effects, in the inheritance of chilling tolerance. Maryam and Jones (1983b) investigated the inheritance of chilling tolerance in maize using systematic genetic series. The scaling test could be applied for most of these series, indicating that inheritance was basically controlled by additive–dominant gene effects. Eagles (1982) evaluated the emergence date of diallel crosses in cold soil and found significant additive and dominant gene effects in the inheritance of this trait. Marton (1997) studied the chilling tolerance at emergence of six inbred lines and all possible hybrid progeny in two germinating media and detected significant GCA effects in both media, while the SCA effect was only significant in heat-sterilised soil.

Gupta and Kovács (1976) evaluated the chilling tolerance of opaque inbred lines and their hybrids at both emergence and in the seedling stage in a 'cold spell' experiment. In the homozygous state, the opaque gene had an unfavourable influence on seedling chilling tolerance. Substantial differences could also be detected in the level of chilling tolerance recorded in cold spell experiments on normal maize hybrids in the seedling stage (Gupta and Kovács 1974).

Stamp (1980) studied the effect of the genotype and the temperature on the characteristics of the shoots and roots of maize seedlings and found that the shoot dry mass was closely correlated with the root surface area and dry mass and with the number of lateral roots at both low and high temperatures. The heritability index of root traits exhibited greater temperature dependence than that of shoot traits.

Marton (1990) established the fact that the genetic control of chilling tolerance differed at emergence and in the seedling stage, and that the latter exhibited two forms. The first, which made the plants capable of withstanding short cold spells lasting 1–2 weeks, was characterised by relatively intensive growth during the cold spell, while the second, which helped the plants to acclimatise to slow warming in spring and to survive long cold periods of up to 3–5 weeks, involved relatively slow growth. It was suggested that these could be distinguished by evaluating changes in plant height as a function of time and by determining the percentage of shoot dry matter after the cold period.

The inheritance of seedling chilling tolerance was evaluated by Marton (2000) on a systematic genetic series at 10 different temperatures. The joint scaling test could be applied to the data, but the model systematically overestimated the values of certain genotypes (F_1, P_1, P_2), while underestimating those of the remaining genotypes (BC_1, BC_2, F_2). Marton and Kuti (2002) re-analysed the data, including a new genetic parameter, the level of heterozygosity of the maternal parent, in the analysis. The modified scaling test revealed that the level of heterozygosity of the female parent influenced initial development. It was also found that the influence of this parameter declined compared to the mean as the temperature rose.

When studying the correlation between the chilling tolerance of lines and their hybrids, Hodges et al. (1997) found that the chilling tolerance of the hybrids could not be predicted reliably from that of the parents. A similar conclusion was reached by Aidun et al. (1991). Dhillon et al. (1988) also examined the parent–progeny relationship and reported that the chilling tolerance of hybrid progeny could be improved by field selection.

The chilling tolerance of maize has been widely investigated, as shown by the large number of references in the literature, but not all authors define chilling tolerance in the same way; various methodologies are used and chilling tolerance is not always studied in the same phenophase. Nevertheless, it is clear that, in addition to significant heterosis effects and reciprocal differences, additive effects also play a role in the inheritance of chilling tolerance. This makes it possible to perform successful selection for chilling-tolerant genotypes even in the early stages of selection.

Practical breeding programmes try to concentrate on chilling tolerance in the wider sense, that is, as a function of the climatic conditions in the given location, in order to achieve chilling tolerance only from sowing to emergence, or to extend tolerance to the seedling stage. The latter is considerably more difficult, not only due to methodological problems, but also because of the diverse genetic control of chilling tolerance during emergence and in the seedling stage (Gupta and Kovács 1976, Zametra and Cuany 1982, Marton 1990). These difficulties are not insurmountable, however, as shown by the results achieved in practice. This was confirmed by Tollenaar and Kannenberg (1989) who expressed the opinion that the rise in maize yield averages achieved in previous years could be attributed largely to an increase in stress tolerance.

7 Physiological background of chilling tolerance

The most important abiotic stress factors affecting maize are cold and drought, which greatly limit the ability of the plants to obtain the necessary resources and thus to realise their genetic yield potential.

Stress is the response given by living organisms to non-specific stimuli in a non-specific form, known as the general adaptation syndrome. Three phases can be distinguished in this response: (1) the alarm reaction, (2) the stage of resistance and (3) the stage of exhaustion. The alarm reaction is the immediate response to the stress, and serves to mobilise the energy reserves of the organism and prepare it for 'fight or flight'. In the resistance stage the organism releases further energy reserves, but if these are not replaced and begin to be exhausted, after a time the resistance of the organism declines.

In severe cases, this third stage may result in the death of the organism (Selye 1936, 1978). Observations have shown that the organism exhibits numerous reactions that produce similar symptoms regardless of the type of stimulus causing the stress. This is the basis of the general adaptation syndrome theory. Selye primarily examined humans and animals, but the general aspects of the theory can also be applied to plants (Larcher 1995).

During the last 50 years, maize has become a major crop in northern regions with a cool climate, which often impairs the early development of the canopy and the photosynthetic productivity of the seedlings (Stirling et al. 1991, Fryer et al. 1998, Leipner et al. 1999). Maize leaves developing at low temperature are characterised by a change in the composition of the leaf pigments (Haldimann et al. 1995), changes in antioxidative defences (Leipner et al. 1997, Fryer et al. 1998, Kingston-Smith et al. 1999) and modifications in the protein composition of the thylakoid membrane (Nie and Baker 1991, Robertson et al. 1993). These changes probably reflect adaptation processes, which might influence the tolerance of the photosynthetic apparatus to low temperature.

7.1 Photosynthesis

Comparisons of genotypes indicate that there is wide genetic variation in the *Zea mays* species for the cold tolerance of the photosynthetic apparatus (Fracheboud et al. 1999, Haldimann 1999). Some authors have reported that changes in the fatty acids (Uemura et al. 1995, Hughes and Dunn 1996, Chinnusamy et al. 2007), increased antioxidant activity (Baek and Skinner 2003, Farooq et al. 2008, Zhou et al. 2012) and the expression of specific genes are important defence mechanisms in plant responses to the stress caused by low temperatures. The photosynthetic apparatus is one of the most sensitive parts of the plant with respect to the cold effect. In response to chilling stress, chlorotic spots appear on the leaves due to inhibition in the biosynthesis of chlorophyll molecules and in the development of etioplasts and chloroplasts at low temperature (Berry and Björkmann 1980, Yoshida et al. 1996, Böddi et al. 1997).

Stamp (1981) investigated the carboxylase activity and pigment content of leaves of some inbred lines after growing at optimal and suboptimal temperature. The phosphoenolpyruvate carboxylase activity (PEPCA) and the carotenoid content proved to be relatively thermostable, while the ribulose-1, 5-biphosphate carboxylase activity (RuBPCA) and the chlorophyll content was lower at suboptimal temperature. At low temperature, photosynthesis may be limited by the reduced RuBPCA activity, as there is a close correlation between the RuBPCA activity and the chlorophyll content in this temperature range. In order to improve photosynthesis, it is therefore necessary to find genotypes capable of incorporating chlorophyll into the developing leaves even at low temperature. Crookston (1983) studied the formation of anthocyanin in maize in early spring. In response to four days at a temperature of 2–3°C the leaves of sensitive genotypes turned purple. Anthocyanin formation was promoted not only by low temperature but also by the low uptake of phosphorus.

7.2 Photoinhibition

It has long been known that if the plant is exposed to more light than it can use for photosynthesis or neutralise via some type of regulating mechanism, there will be a drop in photosynthesising ability. This phenomenon is known as light inhibition or photoinhibition. It is now clear that other stress effects, such as cold stress, may strengthen the extent of photoinhibition. Looking at this from the opposite point of view, it may happen that a stress that the plant is able to withstand in the dark may become fatal in the light. As reported by Vass et al. (1992), the phenomenon of light inhibition, which damages the photosynthesising ability of plants exposed to high light intensity, may be manifested at relatively low light intensity at low temperatures. The photoinhibition induced by low temperature may be manifested not only in photosynthetic processes, but also in other stress markers, such as free amino acids and polyamines (Szalai et al. 1997). In the course of photoinhibition, reactive oxygen species (ROS) are formed at a greater rate, disrupting the equilibrium between absorbed and utilised energy and resulting in the over-excitation of the electron transport chain. Electrons may be transferred from the photosynthetic electron transport chain to O_2 molecules, resulting in the formation of superoxide radicals. Various mechanisms exist to eliminate the damaging effects of photoinhibition (Aroca et al. 2001), involving, on the one hand, the prevention of ROS formation, and, on the other hand, the dissipation of surplus energy in the form of heat via the de-epoxidation of violaxanthin. The water–water cycle taking place in the chloroplasts also serves to dissipate surplus energy. Antioxidant enzymes such as superoxide dismutase (SOD), ascorbate peroxidase (APX) and glutathione reductase (GR) are actively involved in the functioning of this cycle. The damaging effect of photoinhibition is also reduced by the neutralisation of ROS with the help of antioxidant compounds and enzymes (Aroca et al. 2001).

7.3 Antioxidant system

The response of the antioxidant system is especially important in the defence against chilling stress, because one of the main reasons for the damaging effect of low temperature is the accumulation of ROS (Prasad et al. 1994). Some authors consider that differences in the functioning of the antioxidant system may be responsible for the diverse levels of chilling susceptibility in maize genotypes (Prasad et al. 1994). A rise in the activity of most antioxidant enzymes, including SOD, APX, guaiacol peroxidase (POD) and GR, was observed in maize leaves after exposure to chilling, while there was a reduction in the catalase activity (Lee and Lee 2000). The differing chilling sensitivity of the two types of cells in maize leaves, mentioned above, may also be partially due to differences in the oxidative effect. Low hydrogen peroxide concentrations were recorded in the parenchyma sheath cells, while the mesophyll cells had higher H_2O_2 contents (Doulis et al. 1997). Differences were also demonstrated between the two types of tissue with respect to the distribution of antioxidants. NADPH synthesis, and thus the presence of GR, was characteristic of mesophyll cells, while APX was chiefly found in parenchyma sheath cells. Catalase could be detected in both cell types, but mesophyll cells mostly contained catalase1 and catalase3, whereas catalase2 was more abundant in parenchyma sheath cells. In response to low temperature the differences in the distribution of antioxidant enzymes became more pronounced (Pastori et al. 2000).

7.4 Carbon exchange rate

The maintenance of a high carbon exchange rate (CER) is among the most limiting factors for cold tolerance (Lee et al. 2002). At optimal temperatures, when CER variation between genotypes is small, other factors, such as the rate of development, might be of greater importance, leading to earlier canopy closure and thus to increased light interception (Richards 2000). Long periods of low temperature during early phases of development may therefore result in poor photosynthetic performance (Stirling et al. 1991). This may be connected to lower yields.

A negative relationship was found between specific leaf area (ratio of leaf area to leaf weight) and CER for three tropical and three temperate maize inbred lines sown early in the field (Verheul et al. 1996), indicating that SLA is negatively correlated with CER, even if the variation in CER is large. In contrast to the negative relationship between SLA and photosynthetic performance, high SLA, and thus a high rate of CO_2 assimilation per unit leaf mass, generally seems to be important for achieving fast growth. Poorter and Van der Werf (1998) concluded that the leaf area ratio (LAR, ratio of leaf area to plant weight), and more specifically SLA, are the most important factors in the variation of the relative growth rate. The introduction of exotic maize germplasm with temperature-stable, strong photosynthetic performance into European breeding programmes is considerably limited by its linkage to small SLA (Leipner et al. 1999b, Soldati et al. 1999, Stehli et al. 1999).

At constant light intensity, there is a strong linear correlation between the quantum efficiency of photosystem II (QEPSII) and CER in maize seedlings grown at suboptimal temperatures (Fracheboud et al. 2004). As QEPSII measurements can be performed much more rapidly than CER measurements, taking only a few seconds in comparison to several minutes, QEPSII is a useful tool for the routine screening of the photosynthetic performance of maize at suboptimal growth temperatures. However, as maize is not constantly exposed to suboptimal temperatures during its life cycle, constitutive high photosynthetic capacity, if achieved in combination with small SLA, would be an undesirable attribute (Leipner et al. 1999a).

7.5 Membrane structure and functions

Baker et al. (1983) and Long et al. (1983) studied the effect of cold treatment on the photosynthetic CO_2 assimilation (henceforth FC) and the photochemical function of thylakoid membranes. During cold treatment at 5°C the FC value was found to be only 7% of the value recorded at 20°C. When the cold-treated leaves were warmed to 20°C the original FC value was not regained. This means that photosynthesis not only was greatly inhibited by cold treatment, but was still unable to fix CO_2 efficiently when the temperature returned to a favourable level. The value of FC was directly proportional to the temperature of cold treatment and inversely proportional to the duration. Characteristic changes were observed in the functioning of photosystems 1 and 2. The phase transition occurring in membrane lipids in response to cold increased the activation energy of internal membrane reactions, thus slowing down these vital biochemical processes.

Numerous processes involved in the plant metabolism are linked to membranes. Many authors therefore hypothesise that the processes taking place in membranes during cold treatment may have a key role both in the development of cold damage and in acclimatisation to low temperatures (Lyons and Raison 1970, Lyons 1973, Lyons et al. 1979, Murata et al. 1983). For a long time, the membrane-lipid phase transition hypothesis provided some kind of explanation for the damage caused by chilling (Lyons 1973). According to this theory, membrane lipids are probably in the liquid crystal state, allowing them to act as semi-permeable barriers between the cell compartments. According to Lyons (1973) the poor functioning of the membranes at low temperature could be one cause of cold damage. Creencia and Bramlage (1983) observed increased electrolyte leakage and the uncoupling of oxidative phosphorylation in leaves exposed to low temperature. The loss of electrolytes and lack of ATP may differ in the case of intense or weak damage to the membranes. The temperature associated with phase transition depends on whether the fatty acids making up the lipids are saturated or unsaturated and on the quality of the hydrophilic groups (Vigh et al. 1988). Varieties with better chilling resistance contain a larger proportion of unsaturated bonds in the fatty acids than more sensitive varieties (Gubbels 1974).

According to Quinn (1988), however, no phase transition occurs in the vast majority of membrane lipids when plants sensitive to low temperature suffer cold damage, although smaller lipid fractions may suffer localised phase transition at low temperature. Differences in membrane composition may be partially responsible for the differences observed in chilling sensitivity. One indirect proof of this is that the introduction of a gene responsible for the desaturation of the lipids (desA) into both the cold-sensitive blue-green alga *Anacystis nidulans* and into certain higher plants resulted in a significant increase in cold tolerance (Wada et al. 1990, Ishizaki-Nishizawa et al. 1996).

7.6 Phytohormones

The development of acclimatisation to low temperature is subject to hormonal regulation. It is a well-known fact that the quantity of abscisic acid increases during low-temperature treatment (Janowiak and Dörffling 1996). A close correlation was demonstrated in various maize genotypes between chilling tolerance and the accumulation of abscisic acid (Janowiak et al. 2002). However, the abscisic acid level rises not only in response to low temperature, but also in the case of drought, for example, so it could be that the rise inabscisic acid level in response to cold is in fact the consequence of the water deficiency accompanying low-temperature stress. Several results suggest overlapping between the defence processes against the two types of stress, cold and drought (Fig. 5). It has been demonstrated that a short period of drought stress may enhance the chilling tolerance of maize (Sanchez-Diaz et al. 1993).

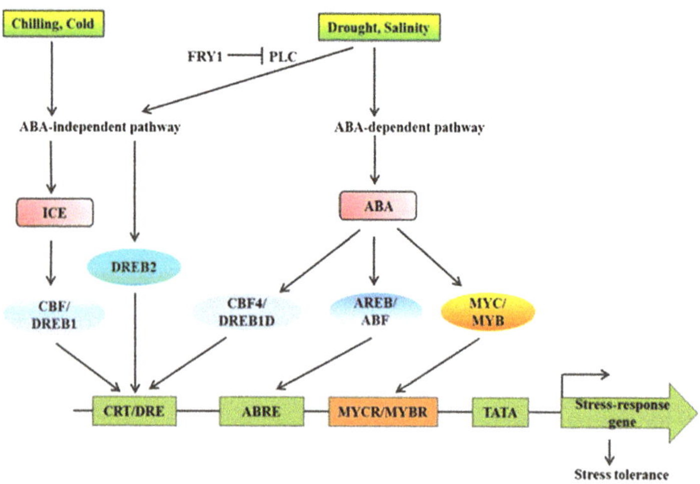

Figure 5 Overlapping between the defence processes against cold and drought (according to Rajesh Kumar 2014). FRY1, Inositol polyphosphate 1-phosphatase; PLC, phospholipase C; ABF, ABRE-binding factor; AREB, ABA-responsive element-binding protein; CRT, C-repeat element; DRE, dehydration-responsive element; CBF, C-repeat-binding factor; DREB, dehydration-responsive element binding; ABF, ABA-responsive element-binding factor; MYB, myoblastosis; ABA, abscisic acid.

Investigations aimed at finding compounds that could be used to improve the stress tolerance of crops form a separate field of research. Theoretically these compounds can be used in two ways: either by enhancing their synthesis within the plant or by using them to treat the plant externally (Janda 2007). Many results have been obtained with both methods. The damaging effect of low temperature, for example, could be mitigated by exposing plants to a short period of hydrogen peroxide treatment prior to cold stress (Prasad et al. 1994) or by applying preliminary treatment with salicylic acid (Janda et al. 1997, 1999). In young maize plants, not only salicylic acid but several related compounds, including possible precursors, such as benzoic acid or *ortho*-hydroxycinnamic acid, were able to reduce the damaging effects of low-temperature stress (Janda 2007).

Silva et al. (2015) characterised the tolerance of three maize inbred lines to cold temperatures during the germination process on the basis of physiological tests and by analysing the fatty acid content (palmitic, stearic, oleic, linoleic, linolenic and eicosenoic acids). The line showing the greatest susceptibility to low germination temperature had a high content of saturated fatty acids. The expression of α-amylase in seeds soaked for 72 h at a temperature of 10°C was lower than in seeds soaked at 25°C for the same length of time. The genes *AOX* and *Zm-AN13* seemed promising for use in identifying maize materials that are tolerant to low germination temperatures.

8 Genetic markers for cold tolerance

Over the past ten years, molecular markers have been extensively used to identify quantitative trait loci (QTLs) involved in the expression of traits of agronomic importance (Lee 1995), and for the genetic dissection of tolerance to abiotic stress (Ribaut and Poland 2000). Information on the number and characteristics of genomic regions responsible for cold tolerance will be useful for future germplasm improvement. To some extent, QTL analysis also allows the investigation of the causal relationships between different traits (Lebreton et al. 1995). This could provide useful information about the physiological responses of maize to low temperature.

Fracheboud et al. (2004) studied the effects of low growth temperatures (15°C) on the photosynthetic apparatus of maize. A set of 233 recombinant inbred lines (RILs) were investigated by means of chlorophyll fluorescence, gas exchange measurements and the analysis of photosynthetic pigments. The QTL analysis of five traits related to the functioning of the photosynthetic apparatus revealed a total of eight genomic regions that were significantly involved in the expression of the target traits. Four of these QTLs, located on chromosomes 1 (around 146 cM), 2 (around 138 cM), 3 (around 70 cM) and 9 (around 62 cM), were identified across several traits and the phenotypic correlation observed between these traits was confirmed at the genetic level. The two QTLs on chromosomes 1 and 9 were also expressed in leaves developed at near-optimal temperature (25°C), whilst the two QTLs on chromosomes 2 and 3 were specific to leaves developed at suboptimal temperature. QTL analysis conducted on traits related to the pigment composition of leaves developing at 15°C detected the QTL at around 70 cM on chromosome 3 for seven of the 11 traits analysed. This QTL accounted for up to 28% of the phenotypic variance of the quantum yield of electron transport at PSII in the fourth leaf after about 3 weeks at a suboptimal temperature. These results suggest that key gene(s) involved in the development of functional chloroplasts of maize at low temperature are located on chromosome 3, close to the centromere.

Biradar (2008) identified several QTLs for the chilling tolerance of photosynthesis in a Swiss dent maize mapping population derived from a cross between the chilling-sensitive line ETH-DL3 and the chilling-tolerant line ETH-DH7. Physiological and genetic analyses were conducted in backcross populations, in near-isogenic lines derived from the ETH-DL3 × ETH-DH7 population and in an alternative F_2 population in order to conduct an in-depth study of a major QTL for the chilling tolerance of photosynthesis, found in the telomeric region of the long arm of chromosome 6. A further study using two polymorphic SSR markers revealed an additional QTL in the telomeric region of the long arm of chromosome 6, which was also associated with the chilling tolerance of photosynthesis.

Yousef and Juvik (2002) described successful marker-assisted selection (MAS) for the enhancement of seedling emergence in sweet corn. The seedling emergence of three elite commercial sweet corn inbreds was improved by the introgression of three alleles using MAS.

Hu et al. (2016) used 243 lines of the intermated B73 × Mo17 (IBM) Syn4 RIL population for the QTL analysis of germination ability at low temperature. There were significant differences between the parental lines for germination-related traits both at low temperature (12°C/16 h, 18°C/8 h) and optimum temperature (28°C/24 h). Only three QTLs controlling the germination rate at optimum temperature were identified. Six QTLs controlling the low-temperature germination rate were detected on chromosomes 4, 5, 6, 7 and 9, and the contributions of individual QTLs explained between 3.39% and 11.29%. In addition, six QTLs controlling the low-temperature primary root length were detected in chromosomes 4, 5, 6, 7 and 9, with contribution rates of 3.96%~8.41% for individual QTLs. Five candidate genes putatively affecting seed germination and seedling growth at low temperature were selected for analysis.

Engelen et al. (2003) found a common QTL for cold tolerance and early vigour in their breeding material. Based on the investigation of numerous DH-derived lines and their crosses with SSR markers, Presterl et al. (2007) found QTLs for three additional chilling-related traits: leaf chlorosis, leaf purpling and frost damage. Some of these were co-localised with QTL for fresh matter yield.

These physiological and molecular breeding results will serve as a basis for further fine mapping, molecular marker-assisted breeding and functional studies on cold tolerance, and will provide useful tools for plant breeders attempting to improve the abiotic stress tolerance of maize breeding materials.

9 Summary

If maize production is to be sustainable, it is essential to carry out detailed studies on the defence mechanisms plants employ against environmental stress factors. Due to the northward shift in cultivation areas, the extreme weather events experienced during the growing season and the increased demand for hybrids with a shorter vegetation season, research on chilling tolerance is becoming increasingly important.

The physiological effect of cold (suboptimum temperature) is manifested in the reduced germination vigour of the seeds, weaker initial development and the yellowing of the leaves. In response to cold, photosynthesis is inhibited, the biosynthesis of chlorophyll molecules is retarded, the pigment composition of the leaves changes, the functioning of the antioxidant system is inhibited, ROS are accumulated and the protein composition and permeability of the membranes is altered. As a consequence, the plants develop more slowly, flowering and ripening are protracted and the yield decreases. Longer periods of cold weather or temperatures below freezing point may kill the plants.

Numerous methods have been elaborated for the evaluation of chilling tolerance, and experiments performed in the field or in controlled-temperature climatic chambers (phytotron) make up an important part of these. Chilling tolerance is greatly influenced by the moisture content of the seed and the soil, the duration of cold treatment, the susceptibility of the seed to pathogens and so on.

Many tools are available to breeders for the investigation and improvement of chilling tolerance. Using traditional methods, chilling-tolerant hybrids can be developed by selecting for inbred lines with good early or seedling chilling tolerance. Due to heterosis for earliness, these may have an even shorter growing period than the parents, combined with a much higher yield level. The increasingly widespread use of molecular breeding methods, MAS and the discovery of QTLs linked to chilling tolerance greatly assist and often shorten the breeding process.

The results achieved by breeders have led to the development of early flowering genotypes with short growing seasons and good tolerance of low temperatures in the seedling stage, which can be successfully grown in more northerly areas, thus increasing the added value of crop production and livestock farming. In practice this means that grain and silage maize hybrids can now be grown on areas where the unfavourable environmental conditions previously prevented their cultivation.

The advantages of cold-tolerant genotypes are manifested in many ways:

- earlier sowing is possible, even sowing in cooler northern regions or highlands;
- faster germination and emergence, better early vigour of the seedlings;
- earlier canopy closure provides good weed suppression, inhibits the evaporation of the soil;
- the faster development of the maximum leaf area provides better utilisation of the maximum sunlight (on the northern hemisphere it is on June 22);
- thus facilitates binding light energy by the greater assimilation capacity of photosynthesis;
- which can result in greater yields.

10 Future trends in research

It is in the interests of both plant breeders and growers to work with plants that tolerate environmental changes with as little damage as possible. If such crops are to be developed, it is essential to obtain knowledge on the defence and regulation processes of the plants.

Low temperature is the primary factor that affects the sowing of maize in early spring. Due to the tropical or subtropical origin of the species, the plants may suffer considerable damage even at temperatures of 10–12°C. It is therefore vital for maize breeding programmes to improve tolerance to low temperatures at the seed germination stage.

Factors affecting the response to cold are complex and the mechanisms of response to cold are not well understood (Yang et al. 2011). Thus, hybrids that are tolerant of low temperatures, especially during the germination phase, should be identified and characterised, together with the mechanisms that confer this tolerance.

New molecular marker technologies and comprehensive gene expression profiling methods provide opportunities to breed improved genotypes with stable grain yield under suboptimum environmental conditions (Bruce et al. 2002). In this context physiological genomics may improve our ability to manipulate the genomes of crop plants in order to improve their adaptation to stress (O'Sullivan et al. 1999, Cushmann and Bohnert 2000).

QTL analysis has improved our knowledge of the genetic background of a number of morpho-physiological traits involved in the response to various abiotic stress factors in maize. A number of candidate genes mapped near to QTLs regulating important morpho-physiological traits or grain yield have already been identified in maize. An updated compilation of mapped QTLs and major genes associated with abiotic stress tolerance in maize and other plants is available at http://www.plantstress.com/biotech/index.asp?Flag=1. Some of the most important traits include yield, flowering and phenological parameters, plant height, ear number, photosynthesis, chlorophyll fluorescence, leaf ABA concentration and root traits such as adventitious root formation, which is important for waterlogging and chilling tolerance, and root hair traits related to phosphorus deficiency and aluminium tolerance. Khavkin and Coe (1997) hypothesised that many apparent QTLs with major effects in maize are in fact clusters of genes (e.g. homeotic genes and other genes encoding for transcription factors), which are responsible for regulating development, and that many plant reactions to abiotic stresses rely on such gene clusters.

The identification and cloning of genes at target QTLs may further broaden our understanding of the genetic, physiological and functional bases of drought tolerance (Araus et al. 2008).

However, little is known about the maize QTLs involved in low-temperature germination ability. QTL mapping of low-temperature germination ability has been performed in rice, wheat and soya bean (Miura et al. 2001, Teng et al. 2001, Fujino et al. 2004, Jiang et al. 2006, Bága et al. 2007, Ji et al. 2009; Iwata and Fujino 2010, Zhang et al. 2012). In rice, dozens of QTLs for low-temperature germination ability were identified on ten chromosomes using different mapping populations, and it has been demonstrated that two QTLs are useful for enhancing low-temperature germination ability in breeding programmes. In maize, traits such as chlorophyll fluorescence parameters, leaf area, shoot dry weight and shoot nitrogen content in the seedling stage were investigated in order to map QTLs associated with chilling tolerance (Jompuk et al. 2005). Germination rate and primary root length are the most important traits involved in the response to cold stress at the germination stage. However, little attention has been paid to their genetic analysis, including QTL mapping. Therefore, it is imperative to carry out gene mapping on the low-temperature germination ability in maize, as this is the first step towards incorporating these traits in breeding programmes.

The newest molecular techniques go much further. When identifying a new locus affecting cold tolerance of a genotype, it can be engineered precisely in the way breeders want.

In the last years the CRISPR/Cas system became the major tool for targeted mutagenesis in plants (Schiml and Puchta, 2016). The gene-specific mutagenesis provides an accurate regulation of the gene expression and can be used for almost all species. By precision genome editing it is possible to modify the genome without using genetically modified organism techniques. Pyott et al. (2016) demonstrated a virus resistance engineered this way, which proved to be heritable and did not require the presence of a transgene. This technique is promising to be used to improve abiotic stress resistance of the plants as well. The QTLs and major genes for cold tolerance are now becoming well known and provide a good basis to start with.

These approaches have achieved rapid advances in plant biotechnology, but the regulatory status of crop varieties developed with this technology needs to be clarified (Stoger et al. 2016).

11 Where to look for further information

If you would like to work in this field, you are recommended to make enquiries in the department of plant physiology or breeding of your local university. Almost all major universities have a research team that deals with the physiological processes of maize. Significant research is underway in the United States in the Seed Tech Laboratory at Iowa State University (Ames) and in Europe at Wageningen University in the Netherlands, the University of Hohenheim in Germany, at ETH Zurich in Switzerland, in the northern research stations of INRA in France and, last but not least, in the MTA Centre for Agricultural Research in Martonvásár, Hungary, where the effects of changes in environmental factors (light, heat, CO_2, etc.) have been investigated in controlled climate chambers for many decades.

These research units are equipped with all the technical facilities, instruments and tools required to study this topic even at the molecular level. The importance of field experiments, however, must also be emphasised. It is essential to see how the various biochemical and physiological processes induced by stress influence plant survival and yields in practice.

Relevant information can be obtained from various organisations that deal with maize breeding, such as IFS, EUCARPIA, ESA, ASTA, NCGA, AGPM, DMK and CIMMYT.

It is also worth reading the articles on this subject published in main scientific journals , such as *Crop Science*; *Maydica*; *Plant Physiology*; *Seed Science and Technology*; *Plant, Cell & Environment*; *Genome* and so on, or searching scientific databases using the following key words: maize cold tolerance, chilling, freezing, early development and germination at low temperature. The annual meetings of EUCARPIA Maize and Sorghum Section are also worth to attend, in order to build scientific relationships for a possible cooperation.

12 References

Aidun, V. L., Migus, W. N. and Hamilton, R. I. (1991), Use of inbred seedling cold tolerance to predict cold tolerance in maize (*Zea mays* L.). *Can. J. Plant Sci.* 71, 663–7.

Andrew, R. H. (1954), Breeding maize for cold resistance. *Euphytica* 3, 108–16.

Araus, J. L., Slafer, G. A., Royo, C. and Serret, M. D. (2008), Breeding for yield potential and stress adaptation in cereals. *Crit. Rev. Plant Sci.* 27(6), 377–412.

Arnold, C. Y. (1975), Heat unit systems in corn production. *Ill. Res.* 17(2), 10–11.

Aroca, R., Irigoyen, J. J. and Sánchez-Díaz, M. (2001), Photosynthetic characteristics and protective mechanisms against oxidative stress during chilling and subsequent recovery in two maize varieties differing in chilling sensitivity. *Plant Sci.* 161, 719–26.

Baker, N. R., East, T. M. and Long, S. P. (1983), Chilling damage to photosynthesis in young *Zea mays*. II. Photochemical function of thylakoids in vivo. *J. Exp Bot.* 34, 189–97.

Baek, K. H. and Skinner, D. Z. (2003), Alteration of antioxidant enzyme gene expression during cold acclimation of near-isogenic wheat lines. *Plant Sci.* 165, 1221–7.

Bága M, Chodaparambil, S. V., Limin, A. E., Pecar, M., Fowler, D. B. and Chibbar, R. N. (2007), Identification of quantitative trait loci and associated candidate genes for low-temperature tolerance in cold-hardy winter wheat. *Funct. Integr. Genomic.* 7, 53–68.

Barla-Szabó, G. (1985), *Az ökológiai és genetikai tényezők hatása a búza- és kukoricavetőmag életerejére.* Kandidátusi értekezés, Martonvásár, p. 134.

Benz, B. F. (2005), Archaeological evidence of teosinte domestication from Guilá Naquitz, Oaxaca. *Proc. Nat. Acad. Sci. USA* 98(4), 2104–6.

Berry, J. and Björkmann, O. (1980), Photosynthetic response and adaptation to temperature in higher plants. annu. *Rev. Plant Physiol.* 31, 491–543.

Biradar, S. K. (2008), Molecular and physiological analyses of chilling tolerance in maize (*Zea mays* L.) seedlings. PhD Thesis. ETH Zurich, p. 68.

Black, M, Bewley, J. D. and Halmer, P. (2006), *The Encyclopedia of Seeds: Science Technology and Uses.* Plenum Press, NY.

Blacklow, W. M. (1972/a), Mathematical description of the influences of temperature and seed quality on inhibition by seeds of corn (*Zea mays* L.). *Crop Sci.* 12, 643–6.

Blacklow, W. M.(1972/b), Influense of temperature on germination and elongation of the radicle and shoot of corn (*Zea mays* L.). *Crop Sci.* 12, 647–50.

Bloc, D. and Gouet, J. P. (1976), *Influence des sommes de temperature sur la maturité du mais.* A.G.P.M. Paris, pp. 1–60.

Bochicchio, A. (1985), *Zea mays* L. and chilling conditions at sowing time: a review. *Maydica.* 30, 241–56.

Boswell, V. R. (1929), Factors influencing yield and quality of peas. *Maryland Agr. Exp. Sta. Bull.* 306–82.

Böddi, B., Keresztes, Á., Csapó, B., Kovács, J., Páldi, E. and Láng, F. (1997), Differences in the etioplast ultrastructure and chlorophyll biosynthesis time course of cold tolerant and cold sensitive maize lines under cold treatment. *Maydica.* 42, 305–11.

Brown, W. N. (1948), Germination of sweet corn seed in relation to the amount of seed injury sustained during harvesting an processing. M.Sc. Thesis. University of Illinois, p. 72.

Bruce, W. B., Edmeades, G. O. and Barker, T. C. (2002), Molecular and physiological approaches to maize improvement for drought tolerance. *J. Exp. Bot.* 53(366), 13–25.

Buican, D. (1969), Researches sur la transmission de la resistance aux basses temperatures· chez les hybrides de *Zea mays. Annales de Physiologie vegetale* 11, 431–40.

Bunting, E. S. (1976), Accumulated temperature and maize development in England. *J. Agric. Sci. Camb.* 87, 577–83.

Burris, J. S. (1989), Seed for inproving maize production. *Georgicon for Agric.* 2(Suppl.), 139–50.

Burris, J. S., Edje, O. T. and Wahab, A. H. (1969), Evaluation of various indices of seed and seedling vigor in soybeans. *Proc. Assoc. Off. Seed Anal.* 59, 73–81.

Burris, J. S. and Knittle, K. H. (1985), Freeze damage and seed quality in hybrid maize. *Proc. Annual Seed Technol.* 51–74.

Chinnusamy, V., Zhu, J. and Zhu, J. K. (2007), Cold stress regulation of gene expression in plants. *Trends Plant Sci.* 12, 444–51.

Christeller, J. I. (1984), Seedling growth of *Zea mays* at 13°C. Comparison of a corn Belt Dent hybrid and a hybrid selected for rapid plumule emergence at cool temperatures. *J. Exp. Bot.* 35, 955–64.

Clark, B. E. (1954), Factors affecting the germination of sweet corn in low temperature laboratory tests. Bulletin No. 769. New York State Agr. Exp. Sta. Geneva, New York.

Creencia, R. P. and Bramlage, W. D. (1983), Reversibility of chilling injuries to corn seedling. *Plant Physiol.* 47, 389–92.

Crookston, R. K. (1983), Purple corn: conspicuous color in spring not always bad. *Crops Soils Magazine.* 35(7), 16–18.

Cross, H. Z. and Zuber, M. S. (1972), Prediction of flowering dates in maize based on different methods of estimating thermal units. *Agron. J.* 64, 351–5.

Cushman, J. C. and Bohnert, H. J. (2000), Genomic approaches to plant stress tolerance. *Curr. Opin. Plant Biol.* 3(2), 117–24.

Delouche, J. C. (1976): Standardization of vigour tests. *J. Seed Technol.* 1, 75–85.

Delouche, J. C. and Baskin, C. C. (1973), Accelerated aging techniques for predicting the relativ storability of seed lots. *Seed Sci. Tech.* 1, 427–52.

Derieux, M. and Bonhomme, R. C. (1982), Heat unit requirements for maize in Europe. *I. Sowing-silking period Zea* 1, 11–20.

DeVries ML (2006), Methods for identifying frost injury in immature maize seed. PhD Diss. Dept. of Agronomy, Iowa State University, Ames, IA.

Dhillon, B. S., Sharma, R. K., Malhotra, V. V., A. S. Khehra, A. S. (1988), Evaluation of maize germplasm for tolerance to low temperature stress under field and laboratory conditions. *J. Agron. Crop Sci.*, 160, 89–93.

Dickson, J. G. (1923), Influence of soil temperature and moisture on the development of the seedling blight of wheat and corn caused by Gibberella saubinetii. *J. Agr. Res.* 23, 837–70.

Doulis, A. G., Debian, M. N., Kingston-Smith, A. H. and Foyer, C. H. (1997), Differential localization of antioxidants in maize leaves. *Plant Physiol.* 114, 1031–7.

Eagles, H. A. (1982), Inheritance of emergence time and seedling growth at low temperatures in four lines of maize. *Theor. Appl. Genet.* 62, 81–7.

Eagles, H. A. and Hardacre, A. K. (1979), Genetic variation in maize (*Zea mays* L.) germination and emergence at 10°C. *Euphytica*. 28, 287–95.

Engelen S, Reheul, D. and Gheysen, G. (2003), Molecular markers for cold tolerance and early vigour in maize (*Zea mays* L.). *Commun. Agric. Appl. Biol. Sci.* 68(2 Pt B), 367–74.

Farooq, M., Aziz, T., Cheema, Z. A., Hussain, M., et al. (2008), Activation of antioxidant system by KCl improves the chilling tolerance in hybrid maize. *J. Agron. Crop Sci.* 194, 438–48.

Ferenec, P. (1980), Prispevok k studin dedicnosti odolnosti kukurice proti chladu. *Vedecke Prace Vyskumneho Ustavu Kukurice v Trnave* 9, 61–70.

Fracheboud, Y., Haldimann, P., Leipner, J. and Stamp, P. (1999), Chlorophyll fluorescence as a selection tool for cold tolerance of photosynthesis in maize (*Zea mays* L.). *J. Exp. Bot.* 50, 1533–40.

Fracheboud, Y., Jompuk, C., Ribaut J-M., Stam, P. and Leipner, J. (2004), Genetic analysis of cold-tolerance of the photosynthesis in maize. *Plant Mol. Biol.* 56, 241–53.

Fryer, M. J., Anderws, J. R., Oxborough, K., Blowers, D. A. and Baker, N. R. (1998), Relationship between CO_2 assimilation, photosynthetic electron transport, and active O_2 metabolism in leaves of maize in the field during periods of low temperature. *Plant Physiol.* 116, 571–80.

Fujino, K., Sekiguchi, H., Sato, T., Kiuchi, H., Nonoue, Y., Takeuchi, Y., et al. (2004), Mapping of quantitative trait loci controlling low-temperature germinability in rice (Oryza sativa L.). *Theor. Appl. Genet.* PMID: 14624339. 108:794–9.

Gilmore, E. C. and Rogers, J. SD. (1958), Heat unit as a method of measuring maturity in corn. *Agron. J.* 50, 611–15.

Gonchar, I. O. (1979), Corn breeding for earliness and cold resistance in Central-chernozem Zone of Russia. In. *Proc. of the IX.th Meeting of the Maize and Sorghum Section of Eucarpia*. 291–7.

Grabe, D. F. (1976), Measurement of seed vigor. *J. Seed Techn.* 1(2), 18–32.

Grogan, C. O. (1970), Genetic variability in maize (*Zea mays* L.) for germination and seedling vigor at low temperatures. *Proc. 25th Corn and Res. Conf. ASTA*. 25, 90–8.

Gubbels, G. H. (1974), Growth of corn seedlins under low temperatures as affected by genotype, seed size, total fatty acid content of the seed. *Can J. Plant Sci.* 54, 425–6.

Gupta, D. and Kovács, I. (1974), Cold wave tolerance of maize seedlings. *Z. Acker- und Pflanzenbau.* 140, 306–11.

Gupta, D. and Kovács, I. (1975), Cold tolerance of parents, single-, three-way- and double crosses of opaque-2 maize inbreds and their normal analogues. *Euphytica*. 24, 245–50.

Gupta, D. and Kovács, I. (1976), Cold wave tolerance of the opaque-2 maize seedlings and their normal analogues, and relationships of the cold wave tolerance with cold test data. *Z. für Acker- und Pflanzenbau.* 143, 196–203.

Haldimann, P. (1999), How do changes in temperature during growth affect leaf pigment composition and photosynthesis in *Zea mays* genotypes differing in sensitivity to low temperature? *J. Exp. Bot.* 50, 43–550.

Haldimann, P., Fracheboud, Y. and Stamp, P. (1995), Carotenoid composition in *Zea mays* developed at sub-optimal temperature and different light intensities. *Physilogia Plantarum.* 95, 409–14.

Hall, H. (2009), Evaluation for fost damage in seed corn (*Zea mays*): with special emphasis on seed composition and moisture content. Graduate Thesis: Paper 11122, Iowa State University, USA.

Haskell, G. (1952), Genetics of cold tolerance in maize and sweet corn seed. *Heredity* 6, 377–85.

Herczeg, M. (1970), Some problems of cold tolerance. In Kovács, I. (Ed.), *Some methodological Achievements of the Hungarian Hybrid Maize Breeding*, 282–6. Akadémiai Kiadó, Budapest.

Herczeg, M. (1978), A kukorica hidegtűrésének javítása nemesítéssel. *Kandidátusi értekezés, Martonvásár*, p. 139.

Herczeg, M. (1983), Some experimental results on the cold tolerance of maize. In Bojanowski, J. and Mlyniec, N. (Eds), *Prof. XII. Meeting of Maize and Sorghum Section of Eucarpia*, pp. 14–18.

Herczeg, M. and Marton, L. C. (1986), Cold stress of maize in a temperature gradient chamber. In Dolstra, O. and Miedema, P. (Eds), *Breeding of Silage Maize*, 56–60. Pudoc, Wageningen.

Herczeg, M., Csetneki, A. and Oraby, F. T. (1970), Some data on cytoplasmic male sterility and "cold test" reaction in maize. In Kovács, I. (Ed.), *Some Methodological Achievements of the Hungarian Hybrid Maize Breeding*, 282–6. Akadémiai Kiadó, Budapest.

Hoard, K. G. and Crosbie, T. M. (1985), S1 line recurrent selection for cold tolerance in two maize populations. *Crop Sci.* 25, 1041–5.

Hoard, K. G. and Crosbie, T. M. (1986), Correlated changes in agronomic traits from S1 line recurrent selection for cold tolerance in two maize populations. *Science* 26, 519–22.

Hodges, D. M., Andrews, C. J., Johnson, D. A. and Hamilton, R. I. (1997), Sensitivity of maize hybrids to chilling and their combining abilities at two developmental stages. *Crop Sci.* 37, 850–6.

Hooker, A. L. (1955), Intra-inbred line variation in resistance to a Pythium seedling disease of corn. *Agron. J.* 47, 580–2.

Hooker, A. L. and Dickson, J. G. (1952), Resistance to Pythium manifest by excised corn embryos at low temperatures. *Agron. J.* 44, 443–7.

Hoppe, P. E. (1951), A new technique for incubating seed corn in cold soil for disease test. *Phytopathology.* XLI, 747–51.

Hoppe, P. E. (1955), Cold testing seed corn by the Rolled Towel Method. Bulletin 507. Agr. Exp. Sta. Madison. 1–5.

Hoppe, P. E. (1956), Cold testing. *Proc. Elevent Corn Ind. Res. Conf. ASTA*. 11: 68–74.

Hoppe, P. E. (1958), Correlation between laboratory cold test and field stands of corn. *Plant Dis. Rept.* 42, 367–72.

Hu, S., Lübberstedt, T., Zhao, G. and Lee, M. (2016), QTL Mapping of Low-Temperature Germination Ability in the Maize IBM Syn4 RIL Population. *PLoS ONE* 11(3), e0152795. doi:10.1371/journal.pone.0152795.

Hughes, M. A. and Dunn, M. A. (1996), The molecular biology of plant acclimation to low temperature. *J. Experim. Bot.* 47, 291–305.

Isely, D. (1950), The cold test of corn. *Proc. Int. Seed Test. Assoc.* 16, 299–311.

Isely, D. (1957), Vigor tests. *Proc. Ass. of Seed Anal. N. Am.* 47, 176–82.

Ishizaki-Nishizawa, O., Fujii, T., Azuma, M., Sekiguchi, K., Murata, N., Ohtani, T. and Toguri, T. (1996), Low-temperature resistance of higher plants is significantly enhanced by a nonspecific cyanobacterial desaturase. *Nature Biotech.* 14, 1003–6.

Ivakhnenko, A. N. (1979), Breeding maize for earliness and cold resistance. In *Proc. of IX.th Meeting of Maize and Sorghum Section of Eucarpia*, pp. 268–75.

Iwata, N. and Fujino, K. (2010), Genetic effects of major QTL controlling low-temperature germinability in different genetic backgrounds in rice (Oryza sativa L.). *Genome.* 53, 763–8. doi: 10.1139/g10-060 PMID: 20962882.

Janda, T. (2007), Termesztett növények abiotikus stresszfolyamatai és egyes védekező mechanizmusai, különös tekintettel az antioxidáns rendszerekre. *DSc tézisek, Budapest*, p. 130.
Janda, T., Szalai, G., Tari, I. and Páldi, E. (1997), Exogenous salicylic acid has an effect on chilling symptoms in maize (*Zea mays* L.) plants. In Sowinski, P., Zagdanska, B., Aniol, A. and Pithan, K. (Eds), *Crop Development for Cool and Wet Europian Climate*, 179–87. ECSP-EEC-EAEC, Brussels, Belgium.
Janda, T., Szalai, G., Tari, I. and Páldi, E. (1999), Hydroponic treatment with salicylic acid decreases the effect of chilling injury in maize (*Zea mays* L.) plants. *Planta*. 2008, 175–80.
Janowiak, F. and Dörrfling, K. (1996), Chilling of maize seedlings: Changes in water status and abscisic acid content in ten genotypes differing in chilling tolerance. *J. Plant Physiol.* 147, 582–8.
Janowiak, F., Maas, B. and Dörffing, K. (2002), Importance of abscisic acid for chilling tolerance of maize seedlings. *J. Plant Physiol.* 159, 635–43.
Ji, S. L., Jiang, L., Wang, Y. H., Zhang, W. W., Liu, X., Liu, S. J., et al. (2009), Quantitative trait loci mapping and stability for low temperature germination ability of rice. *Plant Breed.* 128, 387–92.
Jiang, L., Liu, S. J., Hou, M. Y., Tang, J. Y., Chen, L. M., Zhai, H. Q., et al. (2006), Analysis of QTL for seed low temperature germinability and anoxia germinability in rice (Oryza sativa L.). *Feld Crop Res.* 98, 68–75.
Jompuk, C., Fracheboud, Y., Stamp, P. and Leipner, J. (2005), Mapping of quantitative trait loci associated with chilling tolerance in maize (*Zea mays* L.) seedling grown under field conditions. *J. Exp. Bot.* 56, 1153–63. PMID: 15723825.
Kaerwer, H. E. (1953), Maturity in relation to seedling vigor. *Proc. 8th Corn Res. Conf. ASTA*. 8, 59–68.
Khavkin, E. and Coe E (1997), Mapped genomic locations for developmental functions and QTLs reflect concerted groups in maize (Zeamays L.). *Theor. Appl. Genetics*. 95(3), 343–52.
Kingston-Smith, A. H., Harbinson, J. and Foyer, C. H. (1999), Acclimation of photosynthesis, H_2O_2 content and antioxidants in maize (*Zea mays*) grown at sub-optimal temperatures. *Plant. Cell Environ.* 22, 1071–83.
Kisselbach, T. A. (1939), Effect of artificial dríing upon the germination seed corn. *Agron. J.* 31, 489–96.
Koehler, B. (1957), Pericarp injuries in seed corn: Prevalence in dent corn and relation to seedling blights. *Ill. Agri. Exp. Sta. Bull.* 617.
Kotowski, F. (1926), Temperature relations to germination of vegetable seeds. *Proc. Amer. Soc. Hart. Sci.* 23, 176–84.
Kovács, G. (1989), Hidegtűrésre történő szelekció kukorica pollenpopulációkban. *Növénytermelés*. 38, 9–13.
Kovács, I. (1958), *A kukorica hidegtűrőképességének fokozása különös tekintettel a koraiságra a termés nagyságára és biztonságára*. Kandidátusi disszertáció, Martonvásár.
Kovács, I. (1961), *A hidegpróba alkalmazása a kukorica hidegtűrő képességének fokozására irányuló nemesítésben*. Növénytermelés. 10, 27–42.
Kumar, R. (2014), Role of MicroRNAs in biotic and abiotic stress responses in crop plants. *Appl. Biochem. Biotechnol.* 174, 93–115. doi: 10.1007/s12010-014-0914-2
Lana, E. P. and Haber, E. J. (1952), Seasonal variability as indicated by cumulative degree hours with sweet corn. *Proc. Amer. Soc. Hort. Sci.* 59, 389–92.
Larcher, W. (1995), *Physiological Plant Ecology*. 3 rd edition, Springer-Verlag, Berlin, p. 506.
Lebreton, C., Lazicjanic, V., Steed, A., Pekic, S. and Quarrie, S. A. (1995), Identification of QTL for drought responses in maize and their use in testing causal relationships between traits. *J. Exp. Bot.* 46, 854–65.
Lee, D. H. and Lee, C. B. (2000), Chilling stress-induced changes of antioxidant enzymes in the leaves of cucumber: in gel enzyme activity assays. *Plant Sci.* 159, 75–85.
Lee, E. A., Stabler, M. A. and Tollenaar, M. (2002), Genetic variation and physiological discrimination for cold tolerance in maize (*Zea Mays* L.) during an early autotrophic phase of development. *Crop Sci.* 42, 1919–29.
Lee, M. (1995), DNA markers and plant breeding programs. *Adv. Agron.* 55, 265–344.

Lee, S. S. and Estes, G. O. (1982), Corn physiology in short season and low temperatures invironments. *Agron. J.* 74, 325–31.
Leipner, J., Fracheboud, Y. and Stamp, P. (1999a), Effect of growing season on the photosynthetic apparatus and leaf antioxidative defences in two maize genotypes of different chilling tolerance. *Environ. Exp. Bot.* 42, 129–39.
Leipner, J., Frachebound, Y. and Stamp, P. (1997), Acclimation by damage in maize leaves. *Plant. Cell Environ.* 20, 366–72.
Leipner, J., Stehli, A. and Soldati, A. (1999b), Photosynthetic performance of exotic maize (*Zea mays* L.) germplasm from tropical highlands at low and high temperature. *J. Appl. Bot.* 73, 20–4.
Livingston, J. E. (1951), Effect of low temperature on the germination of artificially dried seed corn. *Nebraska Agr. Exp. Sta. Res. Bull.* 169, 1–14.
Long, S. P., East, T. M. and Baker, N. R. (1983), Chilling damage to photosynthesis in young *Zea mays*. I. Effect of light and temperature variation on photosynthetic CO_2 assimilation. *J. Exp. Bot.* 34, 177–88.
Lunsford, J. N., Futrell, M. C. and Scott, G. E. (1975), Maternal influence on response of corn to Fusarium moniliforme. *Phytopathology.* 65, 223–5.
Lyons, J. M. (1973), Chilling injury in plants. *Annu. Rev. Plant Physiol.* 24, 445–66.
Lyons, J. M. and Raison, J. K. (1970), Oxidative activity of mitochondria isolated from plant tissues sensitive and resistant to chilling injury. *Plant Physiol.* 45, 386–9.
Lyons, J. M., Raison, J. K. and Steponkus, P. L. (1979), The plant membrane in response to low temperature: An overview. In Lyons, J. M., Graham, D. and Raison, J. K. (Eds), *Low Temerature Stress in Crop Plants: the Role of the Membrane*, 1–24. New York, Academic Pres.
Mangelsdorf, P. C.(1986), Origin of maize. *Science Am.* 10, 60–7.
Martinson, C. A. (1971), Low-temperature imhibition injury in seeds and susceptibility to seedling diseases. *Proc. 26th Corn Ind. Res. Conf. ASTA.* 26, 105–10.
Marton, L. C. (2000), A kukorica (*Zea mays* L.) fiatalkori hidegtűrése öröklődésének vizsgálata hőmérsékleti gradiens kamrában szisztematikus genetikai soron. *Növénytermelés.* 49, 475–85.
Marton, L. C. (1990), Kukorica beltenyésztett törzsek kelése és kezdeti fejlődése hőmérsékleti gradiens kamrában. I. A beltenyésztett törzsek kelése. *Növénytermelés.* 39, 305–11.
Marton, L. C. (1997), Inheritance of cold test index of maize in sterilised and normal soil. In Bedő, Z. (Eds), *Proc. Of Int. Symp.* On Cereal Adaptation to low temperatures stress. Martonvásár. 281–4.
Marton, L. Cs. and Kuti, C. (2002), A kukorica (*Zea mays* L.) fiatalkori öröklődésének értékelése modosított "joint schaling test"-tel. *Növénytermelés.* 51, 367–73.
Maryam, B. and Jones, D. A. (1983a), The genetics of maize (*Zea mays* L.) growing at low temperatures. I. Germination of inbred lines and their F1-s. *Euphytica.* 32, 535–42.
Maryam, B. and Jones, D. A. (1983b), The genetics of maize (*Zea mays* L.) growing at low temperatures. II. Germination of inbred lines and further generations at fluctuating temperatures. *Euphytica.* 32, 791–8.
Mathews, S. and Bradnock, W. T. (1968), Relationship between seed exudation and field emergence in peas and French beans. *Hort. Res.* 8, 89–93.
Matsuoka, Y., Vigouroux, Y., Goodman, M. M., et al. (2002), A single domestication for maize shown by multilocus microsatellite genotyping. *Proc. of the National Academy of Sciences USA.* 99(9), 6080–4.
McConnell, R. L. and Gardner, C. O. (1979), Inheritance of cold tolerance traits in corn. *Crop Sci.* 19, 847–52.
Mederski, H. J., Miller, M. E. and Weaver, C. R. (1973), Accumulated heat units for classifying corn hybrid mautrity. *Agron. J.* 65, 743–7.
Miedema, P. (1979), Potential use of CIMMYT gene pool material for improving low temperature andaptation in *Zea mays. Euphytica.* 28, 661–4.
Miedema, P. (1982), The effects of low temperature on *Zea mays. Adv. Agron.* 35, 93–128.
Miura, K., Lin, S. Y., Yano, M. and Nagamine, T. (2001), Mapping quantitative trait loci controlling low temperature germinability in rice (Oryza sativa L.). *Breeding Sci.* 51, 293–9.

Mock, J. J. and Bakri, A. A. (1976), Recurrent selection for cold tolerance in maize. *Crop Sci.* 16, 231–3.

Mock, J. J. and Skrada, W. H. (1978), Evaluation of maize plant introductions for cold tolerance. *Euphytica.* 27, 27–32.

Mock, J. J. and Eberhardt, S. A. (1972), Cold tolerance in adapted maize populations. *Crop Sci.* 12, 466–9.

Moore, R. P. (1962), Tetrazolium as a universally acceptable quality test of viable seed. *Proc. Int. Seed Test. Assoc.* 27, 795–805.

Murata, N., Wada, H., Omata, T. and Ono, T. (1983), Low temperature stress and membrane lipid phase int he blue-green algae. In Marcelle, R., Clijsters, H. and van Poucke M (Eds), *Effects of Stress on Photosynthesis*, 193–9.

Nagy, M., Szél, S. and Fotosné Kádár, K. (1984), A hidegkezelési idő hatása a különböző hidegtűrő kukorica vonalak kelésére. *Növénytermelés.* 33, 491–9.

Neal, N. P. (1950), Breeding corn for tolerance to cold. *Progress in corn production. ASTA.* 68–78.

Neal, N. P. and Davis, J. R. (1956), Seed viability of corn inbred lines as influenced by age and conditions of storage. *Agron. J.* 48, 383–4.

Neptune, G. (1953), Corn tolerance studies with hybrid seed corn. Ph.D. Thesis. Michigan State College, p. 43.

Nie Gy. and Baker, N. R. (1991), Modifications to thylakoid composition during development of maize leaves at low growth temperatures. *Plant Physiol.* 95, 184–91.

O'Sullivan, D. M., Edwards, D. and Edwards, K. J. (1999), Maize genomics. *AgBiotechNet.* 2, 1–6.

Pastori, G., Foyer, C. H. and Mullineuaux, P. (2000), Low temperature-induced changes in the distribution of H2O2 and antioxidants between the bundle sheath and mesophyll cells of maize leaves. *J. Exp. Bot.* 51, 107–13.

Pesev, N. V. (1970), Genetic factors affecting corn tolerance to low temperatures at emergence and germination. *Theoret. Appl. Gen.* 40, 351–6.

Presterl, T. L., Ouzunova, M., Schmidt, W., Möller, E. M., Röber, F. K., Knaak, C., Ernst, K., Westhoff, P. and Geiger, H. H. (2007), Quantitative trait loci for early plant vigour of maize grown in chilly environments. *Theor Appl Genet.* 114(6), 1059–70. Epub 6 Mar 2007.

Pinnell, E. L. (1949), Genetic and environmental factors affecting corn seed germination at low temperatures. *Agron. J.* 41, 562–8.

Pollmer, W. G. (1967), Correlations between characters of cold tolerance in maize. *Proc. of Fourth Meeting of the Maize and Sorghum Section of Eucarpia.* Montpellier, pp. 20–2.

Poorter, H. and Van der Werf, A. (1998), Is inherent variation in RGR determined by LAR at low irradiance and by NAR at high irradiance. A review of herbaceous species. In Lambers, H., Poorter, H. and Van Vuuren, M. M. I. (Eds), *Inherent Variation in Plant Growth Physiological Mechanisms and Ecological Consequences*, pp. 309–32. Backhuys Publishers, Leiden.

Prasad, T. K., Anderson, M. D. and Stewart, C. R. (1994), Acclimation, hydrogen peroxide and abscisic acid protect mitochondria against irreversible chilling injury in maize seedlins. *Plant Physiol.* 105, 619–27.

Pyott, D. E., Sheenan, E. and Molnár, A. (2016), Engineering of CRISPR/Cas9-mediated potyvirus resistance in transgene-free Arabidopsis plants. *Mol. Plant Pathol.* 17(8), 1276–88. doi: 10.1111/mpp.12417

Quinn, P. J. (1988), Effects of temperature on cell membranes. In Long, S. P., Woodward, F. I. (Eds), *Plants and Temperature.* Symp. Soc. Exp. Biol. Cambridge. The Company of Biologists Limited, 42, 237–58.

Reaumur, R. A. F. de (1735), Observations du thermometre, faites a Paris Pendent l'anée 1735. comparées avec ceelles qui ont été faites sous la ligne a l' Isle de France, a Alger et en quelque – une de nos Isle de l' Amerique Ne'm. Acad. des Sci. Paris, p. 545.

Ribaut J-M. and Poland, D. (2000), Molecular approaches for the genetic improvement of cereals for stable production in water-limited environments. A strategic planning workshop held at CIMMYT, El Batan, Mexico, 21–25 June 1999.

Richards, R. A. (2000), Selectable traits to increase crop pho-tosynthesis and yield of grain crops. *J. Exp. Bot.* 51, 447–58.
Rinke, E. H. (1954), Cold test germinations. *Proc. 8th Corn. Ind. Res. Conf. ASTA.* 8, 54–8.
Robertson, E. J., Baker, N. R. and Leech, R. M. (1993), Chloroplast thylakoid protein changes induced by low growth temperature in maize revealed by immunocytology. *Plant Cell Environ.* 16, 809–18.
Rodriguez, V. M., Romay, M. C., Ordas, A. and Revilla, P. (2010), Evaluation of European maize germplasm under cold conditions. *Genetic Res. Crop Evol.* 57(3), 329–35.
Rush, G. E. and Neal, N. P. (1951), The effect of maturity and other factors on stands of corn at low temperatures. *Agron. J.* 43, 112–16.
Sánchez-Díaj, M., Pérez de Juan, J. and Irigoyen, J. J. (1993), The effect of chilling on stomatal behaviour, transpiration, water absorption and carbon dioxide exchange rate in drought-hardened and non hardened maize (*Zea mays* L.). In Wilson, D., Thomas, H. and Pithan, K. (Eds), *Crop Adaptation To Cool, Wet, Climates*, 245–50. E GUYOT SA, Brussels.
Schiml, S. and Puchta H (2016), Revolutionizing plant biology: multiple ways of genome engineering by CRISPR/Cas. Plant Methods 12:8 Open Access. doi: 10.1186/s13007-016-0103-0
Schulz, F. A. and Bateman, D. F. (1969), Temperature response of seeds during the early phases of germination and its relation to injury by Rhizoctonia solani. *Phytopathology.* 59, 352–5.
Selye, H. (1936), A syndome produced by diverse noxious agents. *Nature.* 138, 32–45.
Selye, J. (1978), *Életünk és a stressz.* Akadémiai Kiadó, Budapest.
Silva-Neta, I. C., Pinho, E. V., Veiga, A. D., Pinho, R. G., Gumaraes, R. M., Caixeta, F., Santo, H. O. H. and Marques, T. L. (2015), Expression of genes related to tolerance to low temperatures for maize seed germination. *Genet. Mol. Res.* 14 (1), 2674–690.
Simak, M. and Kamra, M. (1963), Comparative studies on scots pine seed germinability with tetrazolium and x-ray contrast methods. *Proc. Int. Seed Test. Assoc.* 28, 3–18.
Soldati, A. Stehli, A. and Stamp, P. (1999), Temperature adaptation of tropical highland maize (*Zea mays* L.) during early growth and in controlled conditions. *European Journal of Agronomy.* 10(2), 111–17.
Stamp, P. (1981), Activities of PEP and RUBP carboxylase and contents in leaves of maize seedling in relation to genotype and rowing temperature. *Z. Pfanzenzücht.* 86, 20–32.
Stamp, P. (1984), Chilling tolerance of young plants demonstrated on the example of maize. *Adv. Agric. Crop Sci. Suppe. J. Agric. Crop Sci.* 7, 9–81.
Stehli, A., Soldati, A. and Stamp, P. (1999), Vegetative performance of tropical highland mazie (*Zea mays* L.) in the field. *Journal of Agronomy and Crop Science-Zeitschrift fur Acker und Pflanzenbau.* 183(3), 193–8.
Stirling, C. M., Nie Gy, Aguilera, C., Nugawela, A., Long, S. P. and Baker, N. R. (1991), Photosynthetic productivity of an immature maize crop: changes in quantum yield of CO2 assimilation, conversion efficiency and thylakoid proteins. *Plant, Cell Environ.* 14, 947–54.
Stoger, E., Kapusi, E. and Hilscher, J. (2016), Applying CRISPR/Cas to barley: Our experience with the technology, its acceptance and the national regulatory landscape. Conference on New Breeding Techniques. 26–27 September 2016. MTA ATK Martonvásár.
Svien, T. (1951), Factors affecting the germination of corn in the cold test. PhD Thesis. Iowa State College, p. 63.
Szalai, G., Janda, T., Bartók, T. and Páldi, E. (1997), Role of light in changes in free amino acid and polyamine contents at chilling temperature in maize (*Zea mays* L.). *Physiol. Plant.* 101, 434–8.
Szirtes, J. and Barla-Szabó, G. (1981), Módszer az őszi búza vetőmagvak vigorának meghatározására. *Növénytermelés.* 30, 493–500.
Szirtes, J., Barla-Szabó, G. and Papp, D. (1982), Módszer a kukorica vetőmagvak élettani vigorának meghatározására. A kelés előrejelzése különböző stresszviszonyok között. *Vetőmaggazdálkodás.* 9(1), 59–68.
Szundy, T. (1981), Eltérő heterozigóta szintű szülőkön előállított kukorica hibridek néhány tulajdonsága. *Kandidátusi értekezés, Martonvásár*, p. 154.

Szundy, T. and Kovács, I. (1981a), Különböző heterozigóta szintű kukorica genotípusok és hibridjeik hidegűrésének vizsgálata. I. Heterozigóta genotípusok kelési ideje. *Növénytermelés*. 30, 301–8.
Szundy, T. and Kovács, I. (1981b), Különböző heterozigóta szintű kukorica genotípusok és hibridjeik hidegtűrésének vizsgálata. II. Heterozigóta kukoria genotípusok kelési százaléka. *Növénytermelés*. 30, 385–90.
Tatum, L. A. (1942), The effect of genetic constitution and processing method on the ability of maize seed to germinate in cold soil. PhD Thesis. Iowa State College, Ames, p. 80.
Tatum, L. A. (1954), Seed permeability and cold test reaction in *Zea mays*. *Agron. J.* 46, 8–10.
Tatum, L. A. and Zuber, M. S. (1943), Germination of maize under adverse conditions. *J. Am. Soc. Agr.* 35, 48–59.
Teng, S., Zeng, D. L., Qian, Q., Kunihifo, Y., Huang, D. N. and Zhu, L. H. (2001), QTL analysis of rice low temperature germinability. *Chinese Sci. Bull.* 46, 1800–4.
Tollenaar, M. and Kannenberg, L. W. (1989), A physiological approach to maize breeding. In Workshop on maize breeding and maize production, Euromaize'88, Belgrad, pp. 119–28.
Uemura, M., Joseph, R. A. and Steponkus, P. L. (1995), Cold acclimation of Arabidopsis thaliana (Effect on plasma membrane lipid composition and freeze-induced lesions). *Plant Physiol.* 109, 15–30.
Vass, I., Styring, S., Hundall, T., Koivuniemi, A., Aro E-M. and Andersson, B. (1992), Reversible and irreversible intermediates during photoinhibition of photosystem II: Stable reduced QA species promote chlorophyll triplet formation. *Proc. Nat. Acad. Sci. USA*. 89, 1408–12.
Verheul, M. J., Picatto, C. and Stamp, P. (1996), Growth and development of maize seedlings under chilling conditions in the field. *Europ. J. Agron.* 5, 31–43.
Vígh, L., Horváth, I. and Farkas, T. (1988), A növények fagytűrő képessége és membránlipidjeik fázisállapotának kapcsolata. *A biológia aktuális problémái*. 26, 169–201.
Wada, H., Gombos, Z. and Murata, N. (1990), Enhancement of chilling tolerance of a cyanobacterium by genetic manipulation of fatty acid desaturation. *Nature*. 347, 200–3.
Wallace, H. A. and Bressmann, E. N. (1949), *Corn And Corn Growing*. John Willey and Sons, New York, London, p. 424.
Wortman, L. S. (1950), The inheritance of cold test reaction in *Zea mays* L. Ph.D. Thesis. University of Minnesota, p. 85.
Wortman, L. S. and Rinke, E. H. (1951), Seed corn injury at various stages of processing and its affect upon cold test performance. *Agron. J.* 43, 299–305.
Yang, G., Zou, H., Wu, Y., Liu, H., et al. (2011), Identification and characterization of candidate genes involved in chilling responses in maize (*Zea mays* L.). *Plant Cell Tiss. Org.* 106, 127–41.
Yoshida, R., Kanno, A., Sato, T. and Kameya, T. (1996), Cool-temperature-induced chlorosis in rice plants. *Plant Physiol.* 110, 997–1005.
Yousef, G. G. and Juvik, J. A. (2002), Enhancement of seedling emergence in sweet corn by marker-assisted backcrossing of beneficial QTL. *Crop Sci.* 42(1), 96–104.
Zametra, R. S. and Cuany, R. L. (1982), Research on cold tolerance in corn. Progress Rpt. for 1980 and 1981. Dept. of Agr. Colorado State Univ. Fort Collins, p. 17.
Zhang, W. B., Jiang, H. W., Qiu, P. C., Liu, C. Y., Chen, F. L., Xin, D. W., et al. (2012), Genetic overlap of QTL associated with low-temperature tolerance at germination and seedling stage using BILs in soybean. *Canada J. Plant Sci.* 92, 1381–8.
Zhou, J., Wang, J., Shi, K., Xia Xia, X. J., et al (2012), Hydrogen peroxide is involved in the cold acclimation-induced chilling tolerance of tomato plants. *Plant Physiol. Biochem.* 60, 141–9.

Part 2

Understanding and improving maize nutritional and sensory quality

Chapter 7

Nutritional and nutraceutical/functional properties of maize

Dharam Paul Chaudhary, Indian Institute of Maize Research (IIMR), India

1 Introduction

The nutritional well-being and health of its people is the major prerequisite for the development and prosperity of a society. Nutritional well-being depends primarily upon the quality of food. A large segment of the global human population still depend upon staple foods for all their dietary requirements. Maize (*Zea mays* L.) is one of the most important cereal crops used as a staple food worldwide, particularly in Asia and sub-Saharan Africa. It is consumed in different forms in different parts of the world, depending upon the taste and local custom. In the Indian subcontinent, maize flour is used for making *chapattis*. Maize is also processed and fermented in varied ways in different parts of the world. It is a quick growing plant; 3–4 months after sowing, a single maize kernel produces more than 600 kernels, thus making it one of nature's fastest multipliers of energy. This energy is stored as starch, oil and protein in the maize kernel.

Maize kernels consist of four major physical structures: the endosperm, the germ or embryo, the outer seed coat or pericarp, and the tip cap. The endosperm is the largest component (80–85%) and is mainly composed of starch with small concentrations of protein. The pericarp is the outer covering (5–6%) of the kernel and consists of hemicellulose, cellulose and lignin. The germ (9–10% of the kernel weight) is rich in fat (33%) and has relatively high levels of protein (18–19%) and minerals (Lawton and Wilson 1987, Prasanna et al. 2001). Maize grain has two types of endosperm – floury and horny. Floury endosperm contains loosely packed starch granules surrounding the central fissure, while horny endosperm has tightly packed, smaller starch granules towards the periphery.

http://dx.doi.org/10.19103/AS.2016.0001.12

Maize contains various bio-active constituents, such as carotenoids, anthocyanins, tocopherols and phenolic compounds that have many health-promoting and disease-preventing properties. As maize is widely used as a staple food, its nutritional quality needs to be understood and discussed in detail. This chapter enlarges upon the potential and limitations of the various nutritional components of maize, such as proteins, carbohydrates, oil and carotenoid composition.

2 Protein quality

2.1 Protein in maize

Although starch is the major component of maize, the quality of its protein is considered to be of greater importance as proteins are essential in the human diet for growth and development in children and the repair of wear and tear in adult cells. Proteins consist of 20 different amino acids, among which nine are considered essential as they cannot be synthesised and must therefore be supplied by diet. Protein quality refers to the presence of all essential amino acids in the appropriate concentrations in a food. A food protein containing essential amino acids in the correct concentration is termed a complete protein. Protein quality therefore is defined as a measure of the usefulness of a dietary protein for growth and maintenance of tissues. The best quality protein will provide an essential amino acid pattern very close to that of the tissue proteins. Usually, milk protein (casein) serves as the reference protein because of its superior quality. The recommended essential amino acid composition (mg/g protein) of reference protein required for children between two and five years of age is lysine 58 and tryptophan 11 (FAO/WHO Expert Consultation, 1991). The daily requirement of tryptophan is 4 mg/kg body weight/day for adults and 4.8 mg/kg body weight/day for children. The requirement of lysine is 30 mg/kg body weight/day and 35 mg/kg body weight/day for adults and children, respectively (WHO/FAO/UNU, 2007).

As observed above, much of the human population, particularly in the developing countries of Asia and sub-Saharan Africa, still derive all their protein requirements from cereals. As maize is a staple food for such a large population, its protein quality is of paramount importance. The protein quality of maize is, however, poor due to deficiencies of two main essential amino acids, lysine and tryptophan, and an excess of leucine in the endosperm protein of its kernel (Osborne and Mendel 1914). Maize protein contains 1.5–2.0% lysine, which is less than half of the recommended dose specified for human nutrition (Young et al. 1998). The poor quality of maize protein is attributed to the higher concentration of zeins in its endosperm protein (Vassal 2000). On the basis of solubility, maize endosperm proteins are classified into various fractions such as albumin (3%), globulin (3%), zein/prolamine (60%) and glutelin (34%), whereas the embryo protein largely consists of albumins (60%). Zeins have been classified into α (22 and 19 kDa), the most abundant being β (14 kDa), γ (27 and 16 kDa) and δ (10 kDa) (Wilson et al. 1981). The α-zein fraction is rich in cystein while the β and γ fractions are rich in methionine. In standard maize, the zein fraction generally contains higher proportion of leucine, phenylalanine, isoleucine, valine and tyrosine but low amounts of other essential amino acids such as threonine, histidine, cysteine, methionine and lysine and is almost devoid of tryptophan. The non-zein protein fraction is balanced and rich in lysine and tryptophan. Protein supplementation to correct such deficiencies is costly and energy inefficient for animal nutrition and is therefore not feasible for most developing countries which rely on cereals for their energy requirements. Therefore, increasing the amounts of protein-bound essential amino acids is the most important and economically viable means of dealing with nutritional deficiencies. Improving nutritional quality in cereal crops is particularly important as the benefits can be spread to hundreds of millions of people in a rapid and effective manner without changing traditional food habits.

2.2 Improving protein quality

Although it was demonstrated several years ago that maize protein is nutritionally deficient due to the limiting quantities of the two essential amino acids lysine and tryptophan (Osborne and Mendel 1914), the turning point in maize quality improvement came only in the early 1960s when Purdue researchers discovered high lysine mutants such as *opaque-2* (Mertz et al. 1964) and *floury-2* (Nelson et al. 1965). Introduction of these mutant genes into standard maize brings a two-fold increase in levels of lysine and tryptophan. Lysine-deficient zein (prolamine) is suppressed drastically while other fractions such as albumins, globulins and glutelins that are rich in lysine show a marked increase. This was a breakthrough discovery in maize quality improvement programme and soon the *opaque-2* gene was incorporated into the elite maize germplasm worldwide, producing a number of *opaque-2* varieties and hybrids. The maize thus produced is called *opaque-2* maize and possesses almost double the quantities of lysine and tryptophan when compared to standard maize. The nutritional superiority of the *opaque-2* maize over standard maize has been well established in feeding trials using albino rats (Lodha et al. 1976). In a nitrogen-balance study, the biological value of *opaque-2* maize protein was found to be 90% of the milk protein casein, as compared to the standard maize which was less than 70% of casein. The protein efficiency ratio (PER) value for *opaque-2* maize has been found to be 62–110% better than that of ordinary maize: the PER for *opaque-2* maize being 2.3–2.8 and that for standard maize 1.3–1.6. Furthermore, 1 g of dietary *opaque-2* maize resulted in 0.226 g weight gain in albino rats as against 0.131 g with standard maize (Lodha et al. 1976).

However, the excitement so created by the development of *opaque-2* maize dissipated soon as it was found to be inferior to standard maize in terms of yield and agronomic performance due to its soft and chalky endosperm, lower grain yield, poor kernel appearance, and susceptibility to insect and pest attack (Vassal, 2014). As a result, the newly developed maize became unpopular and was discarded by farmers. This was a major setback for researchers working on maize quality improvement across the world. However, the disappointment was soon dispelled by the identification of some partially hard endosperm kernels observed in the existing *opaque-2* stocks. These kernels were termed 'vitreous' or 'modified' as their endosperm had been modified from soft to hard.

A team led by the distinguished scientist Dr S. K. Vassal at CIMMYT, Mexico, played a pivotal role and identified some endosperm-modifier genes that could favourably alter the grain characteristics. This overcame a significant obstacle in popularising high-lysine *opaque-2* maize. The mechanism by which the endosperm modifiers changed the grain structure from chalky to vitreous in modified *opaque-2* is not yet clearly understood. The nutritional quality analysis showed that the modified *opaque-2* kernel possessed the desired concentrations of essential amino acids; lysine and tryptophan. This modified *opaque-2* maize with a hard endosperm is known as quality protein maize (QPM). This term therefore refers to maize homozygous for the *opaque-2* allele, with increased lysine and tryptophan content but without the negative secondary effects of a soft and chalky endosperm. The QPM developed in this way has about twice the concentration of lysine and tryptophan of standard maize and has a reduced leucine/isoleucine ratio.

As already stated, the nutritional quality of QPM is far better than that of standard maize. Based on feeding trials in small animals, it was found that the protein quality of QPM is almost 90% that of milk (Graham et al. 1980). The decreased leucine/isoleucine ratio in QPM liberates more tryptophan for niacin biosynthesis, and thus helps to combat pellagra (Vasal, 1999). QPM is an advantage in the poultry and piggery sectors as it helps these animals to increase their body weight faster when compared to standard maize (Burgoon et al. 1992; Osei et al. 1999).

A number of QPM hybrids are now available for commercial cultivation across the world, particularly in Africa, Asia and Latin America (Vasal 1999; Prasanna et al. 2001; Nuss and Tanumihardjo 2011). A QPM variety named 'Obatanpa' was introduced in Ghana in 1992, and in just 13 years, came to represent around 90% of maize seed sales in that country. During 2008, Ghana grew QPM on 350 000 hectares of land, which is one of the largest areas in the world put to this use (www.cimmyt.org).

3 Carbohydrate profile

Maize kernels contain simple as well as complex carbohydrates such as sucrose, starch, cellulose and hemicellulose. The various parts of the maize kernel differ in terms of their chemical composition. The pericarp is composed of high concentrations of crude fibre, consisting mainly of hemicellulose, cellulose and a very small content of lignin. The endosperm contains high levels of starch and some protein. Starch is the major carbohydrate in maize kernels, contributing around 70% of the kernel weight. Maize is considered as one of the fastest natural multipliers of starch as within 3–4 months it multiplies the starch content by approximately 600 times. Starch is the main product of the maize wet milling industry, which is further utilised in a number of other industries in its natural, modified, pregeletanised and dextrinised forms. These forms provide viscosity, texture and other desired properties to all types of food and paper products. Practically every industry in existence uses starch or its derivatives in one form or another.

The starch molecule is a homopolymer of repeating anhydroglucose units joined by α-glycosidic linkages, the aldehyde group of one unit being chemically bound to a hydroxyl group on the next unit through hemiacetal linkages. The 1,4-linkages produce a straight-chain starch molecule called amylose, while the 1,6-linkages serve as the branching point in branched chain starch molecules called amylopectin. In standard maize, amylose makes up 25–30% of the starch and amylopectin up to 70–75%. The starch composition in maize is genetically controlled and wide genetic variability has been observed in maize starch profiles. Maize is broadly categorised into three classes on the basis of starch composition: (1) waxy maize (high amylopectin content), (2) high amylose maize and (3) sugary maize (higher level of sugars) (Nelson and Pan 1995). The variation in starch composition is exploited in a variety of industrial applications.

Starches with a high percentage of amylose are difficult to gelatinise because of the extra energy needed to hydrate and disintegrate their firmly bonded, crystalline aggregates. After gelatinisation, such starches form firm gels and when properly prepared, produce strong, tough films. The major industrial uses of starch, such as paper coating and sizing, textile sizing, corrugated board manufacture and all adhesive applications, utilise this property, which is largely due to high amylose content. At the opposite end are the waxy starches, which are nearly 100% amylopectin. These gelatinise easily and yield nearly transparent, viscous pastes which retrograde slowly to weak gels. Therefore, due to ease of gelatinisation, waxy maize is used in the food industry.

Dietary starches also show variability in terms of digestibility in the humans. The recent recognition of incomplete digestion and absorption of starch in the small intestine as a normal phenomenon has raised interest in non-digestible starch fractions (Cummings and Englyst 1991; Englyst et al. 1992). Maize possesses wide genetic variability in respect of the digestibility of its starch molecule. Depending upon the rate of release and absorption of glucose in the gastrointestinal tract, starch is classified into three groups: rapidly digestible starch (RDS), slowly digestible starch (SDS) and resistant starch (RS). RDS can be rapidly hydrolysed by digestive enzymes while SDS is digested at a relatively slow rate (Englyst et al. 1992). However, a part is not digested in the small intestine and reaches the large intestine (colon), where it is fermented by the gut microflora, producing short-chain fatty acids, which are known to promote the optimal function of the viscera (Topping and Clifton 2001). This indigestible portion of starch is known as resistant starch. Waxy maize starch is more rapidly digested than high amylose starch. This is attributed to a larger surface area per molecule of the amylopectin than is the case in amylose. Many health benefits such as improved cholesterol metabolism and reduced risk of type II diabetes and colon cancer have been associated with the consumption of RS (Hoebler et al. 1999).

In addition to starch, crude fibres consisting of complex carbohydrates are important components of kernel carbohydrates. Maize is categorised as a coarse cereal, primarily due to its high fibre content when compared to rice and wheat which are fine cereals. The fibre components are mainly present in the pericarp, germ cell wall, endosperm cell wall and tip cap. Maize bran is composed of around 75% hemicellulose, 25% cellulose and 0.1% lignin on a dry weight basis (Sandstead et al. 1978, Van Soest et al. 1979). Although indigestible, the fibre components play an important role in peristaltic movement, thus normalising the digestive process. Simple sugars such as sucrose, glucose and fructose are also present in smaller quantities ranging from 1 to 5% of the kernel weight on a dry weight basis. These simple sugars are present in high concentrations in the maturing maize kernels. Sweet corn in the immature stage is widely used for human consumption throughout the world. Immature kernels contain relatively high levels of sugars and low levels of the starch which accumulates during development (Boyer and Shannon 1982). Sweet corn is genetically improved by selecting mutants producing high sugar levels in the kernel endosperm. Several genes involved in sweet corn improvement by increasing sugar and reducing starch content have been identified (Creech 1965, Boyer and Shannon 1982). The four most useful mutants are shrunken2 (*sh2*), brittle1 (*bt*), sugary1 (*su1*) and sugary enhancer1 (*se*). In general, these genes slow the conversion of sugar to starch in immature kernels.

4 Corn oil

Due to its quality and stability, corn oil is gaining importance both among the mix of products manufactured from the corn wet milling process and among food oils. Maize oil first emerged in the United States and large-scale production began in the 1910s. Since the 1950s, there has been a large consumption of maize oil in developed countries.

It plays an important role in human nutrition, being a concentrated source of energy which is very digestible, a rich source of vitamin E, a provider of essential fatty acids and characterised by high levels of unsaturated fatty acids and low levels of saturated fatty acids (Zai and Gao 2001). The term 'essential fatty acid' refers to fatty acids essential to biological processes. Only two fatty acids are known to be essential for humans: α-linolenic acid (an omega-3 fatty acid) and linoleic acid (an omega-6 fatty acid).

The Food and Agriculture Organisation and the World Health Organisation recommend that about 2–4% of energy should be acquired in the form of essential fatty acids, with an additional 3% energy for pregnant or breastfeeding women. A tablespoon of corn oil is sufficient to satisfy the daily essential fatty acid requirement of a healthy child or adult. Corn oil is rich in vitamin E (a natural antioxidant) which protects the body from the deleterious effects of free radicals, thus deferring the senescence of biological organisms. It belongs to a class of strong antioxidants which protect polyunsaturated fatty acids in membranes against degradation by reactive oxygen species such as ozone, singlet oxygen, peroxides and hyperoxides (Dormann 2003). The high concentration of polyunsaturated fatty acids is beneficial to human health due to its stabilising effect on blood cholesterol and regulation of blood pressure (Hauman 1985, Dupont et al. 1990, Sanjiv et al. 2014).

The ratio of saturated to unsaturated fatty acids is balanced in maize oil, which contains a higher relative amount of linoleic to oleic acids (Alexander 1988). Maize oil contains about 60% linoleic acid, 20% oleic acid, 18% palmitic acid, 2% stearic acid, small amounts of linolenic acid and trace amounts of other fatty acids. It has been reported that the two most important unsaturated fatty acids of maize oil (oleic and linoleic) had almost identical effects in reducing total plasma cholesterol and low-density lipid cholesterol (Mattson and Grundy 1985). Maize oil is found to be superior to sunflower oil in lowering body cholesterol levels (Iacono and Dougherty 1993). They further reported that no other vegetable oil is better than maize oil in lowering blood cholesterol levels and elevated blood pressure. Therefore, due to its high stability, pleasant taste and multifunctional usage, maize oil became the standard against which other oils were compared in assessing their cholesterol-lowering abilities (Hegsted et al. 1993). Maize oil is mainly used in cooking and has a high smoke point, which makes it valuable for frying purposes (Katragadda et al. 2010). It is also a key ingredient in some margarines.

The use of maize oil in non-food areas represents a negligible percentage of the total consumption. Small amounts are used in the manufacture of resins, plastics, lubricants and similar oils. A small quantity of highly refined oil is used by the pharmaceutical industry in certain dosage forms and for other purposes. The residues and by-products of maize oil refining are used in the form of soap stock or in animal feeds (CRA 2006).

Standard maize varieties produce around 2–5% of oil on a dry weight basis, whereas the genotype possessing an oil content of more than 6% is regarded as high-oil maize (Wang et al. 2009, Lambart 2001). Maize genotypes with higher levels of oil and lower proportions of carbohydrate are obviously higher in energy because the calorie content of oil is approximately 2.5 times higher than that of carbohydrates (Alexander 1988). Due to its higher calorific value, high-oil maize is of interest to many livestock feeders (Lambart 2001). Feeding high-oil maize not only improves the food energy utilisation rate but also reduces the amount of supplemental fats needed in livestock feed (Goldman et al. 1994). High-oil maize silage and grain improves dry matter intake and the milk yield of dairy cows during early lactation. Feeding these high-oil hybrids in animal diets reduces the need for soya bean supplementation (Brewbaker 2003). Adams and Jensen (1988) observed faster daily weight gain in pigs fed on high-oil maize as compared to those fed on standard maize grains. Moreover, the commercial cost of oil is almost three times higher than that of starch, making it the preferred product for the maize wet milling industry.

High-oil varieties of maize were developed at the University of Illinois (Dudley and Lambart 2004). Although these genotypes possessed a higher energy content, they displayed poor agronomic characteristics. Oil is mainly present in the germ of the maize kernel, which provides around 85% of the total kernel oil. High-oil maize differs in kernel composition as its germ size is larger than that of standard maize. As most of the oil is contributed by the germ, increasing the oil content in maize is directly proportional to the germ size. Increased germ size, however, is associated with reduction in starch as both the oil and starch are negatively correlated.

5 Micronutrients in maize

Corn has a wide range of kernel colours such as white, yellow, orange, purple and black. Apart from providing energy, coloured maize is also a rich source of phytochemicals and secondary metabolites such as phenolic compounds, carotenoids and flavonoids. These constituents are regarded as an important source of antioxidants (Montilla et al. 2011). Anthocyanin, carotenoid and phenolic contents in maize vary with colour. Yellow- or orange-coloured corn has been found to be a major source of carotenoids (Yang and Zhai 2010). The highest concentration of anthocyanin pigments in maize is present in the pericarp portion, whereas the aleurone layer contains small amounts (Moreno et al. 2005).

Carotenoids have several important biological functions in the human body. The most widely studied and well-understood nutritional role is their provitamin A activity. Vitamin A deficiency is a major cause of premature death in developing nations, particularly among children. Vitamin A, which has many vital systemic functions in humans, can be produced within the body from certain carotenoids, notably β-carotene. In addition to its nutritional importance, maize represents one of the most important sources of carotenoids (Vallabhaneni and Wurtzel 2009). Maize exhibits considerable natural variation for kernel carotenoids with some genotypes accumulating levels as high as 66.0 μg/g (Harjes et al. 2008). The predominant carotenoids in maize kernels, in decreasing order of concentration, are lutein, zeaxanthin, β-carotene, β-cryptoxanthin and α-carotene. Yellow maize has more carotenoids than floury maize. Generally, provitamin A carotenoids constitute only 10–20% of the total carotenoids in maize, whereas 30–50% are represented by zeaxanthin and lutein. In typical maize, concentrations of provitamin A carotenoids, that is, α-carotene, β-carotene and β-cryptoxanthin, range from 0 to 1.3, 0.13 to 2.7 and 0.13 to 1.9 nmol/g, respectively (Kurilich and Juvik 1999). Several publications have also reported phytochemical contents in different types of maize, including carotenoids (Ibrahim and Juvik 2009), anthocyanins (Cui et al. 2012) and phenolic compounds (Xu et al. 2010, Montilla et al. 2011). It has been reported that waxy corn is a good source of carotenoids, anthocyanins, phenolics and antioxidants (Hu and Xu 2011).

Ferulic acid is an important phytochemical in maize and its concentration varies in different maize types. High-carotenoid maize contains higher amount of total ferulic acid compared to white, yellow, red and blue maize. Most of the ferulic acid in maize is present in bound form (Adom and Liu 2002). The bulk of phenolics (phenolic acids, flavonoids and conjugated amines) are concentrated in the pericarp and aleurone layers as well as in the germ, with traces in the endosperm (Sen et al. 1994).

However, maize is not a good source of minerals and is generally considered to be deficient in iron and zinc, although the mineral composition of cereal grains may be affected by a number of factors such as soil type and fertility, soil moisture, the environment, crop genotype and interactions among nutrients (Feila et al. 2005). The human body requires iron (Fe) for basic cellular functions including the proper functioning of muscle, brain and red blood cells, and is an integral part of various enzymes (Hallberg 1982). Zinc (Zn) is an essential mineral for many biological functions and is a component of more than 300 enzymes responsible for the synthesis and degradation of various biomolecules, *viz.*, carbohydrates, proteins and lipids (Sandstorm 1997).

Over 60% and 30% of the world's populations are deficient in iron and zinc, respectively (White and Broadley 2009). These deficiencies affect people of all ages but are more prominent in children and pregnant women. The estimated average requirement for humans is 1460 μg/day of Fe and 1860 μg/day for Zn. The target levels of Fe and Zn in maize have been fixed as 60 μg/g and 38 μg/g, respectively (Bouis and Welch 2010). Sufficient genetic variation for kernel iron and zinc in maize has been reported (Banziger and Long 2000, Pixley et al. 2011). Menkir (2008) evaluated a vast set of tropical-adapted inbred lines for mineral concentrations. The best inbred lines identified from each trial had 32–78% more iron and 14–180% more zinc than their trial average. Ullah and Farooqi (2010) also analysed grains of ten maize varieties for mineral composition and reported the range of Na (540.30–620.41 ppm), K (2915–3471 ppm), Ca (410–590 ppm), Fe (38.02–56.14 ppm), Zn (37.05–52.4 ppm), Mg (985.2–1125.3 ppm) and Cu (11.02–14.25 ppm). Wide genetic variability has been reported from India in kernel iron and zinc across a diverse set of standard maize and QPM inbreds (Gupta et al. 2015). An anti-nutritional component, phytic acid/phytate plays a major role in reducing the bioavailability of iron and zinc. Nearly 80% of the total phosphorus in the maize grain is present as phytic acid (Raboy et al. 2000). Development of low phytate maize is, therefore, another important aspect in enhancing the availability of micronutrients in maize.

6 Future trends

An increasing amount of maize is now being grown and the crop will gain importance due to its diverse applications. Changing climatic conditions, depletion of groundwater and global warming are some of the factors responsible for projecting maize as a future crop. In India, the production and productivity of maize has increased over the last few years. In addition to its major utility as animal feed, maize could serve as an important food source for the growing human population. However, research interventions are still needed to improve the nutritional quality and to popularise such nutritionally improved maize.

Consistent efforts by maize scientists across the world have led to the successful development of QPM cultivars. The nutritional superiority of QPM over standard maize is well established. QPM could be of great value to vulnerable groups, particularly the rural poor in developing countries. QPM tastes sweeter than standard maize and therefore can be consumed without any change in traditional food habits.

Scientists have already developed provitamin A-enriched inbreds of maize (VQL1, VQL2, V335, V345, HKI1105, HKI323 and HKI161) containing kernel β-carotene ranging from 8.6 to 17.5 μg/g (Muthusamy et al. 2014). The hybrids developed from these improved inbreds also showed higher kernel β-carotene when compared to the original hybrid (Muthusamy et al. 2014). Efforts should now be directed to the development of biofortified maize genotypes with a combination of various micronutrients such as lysine, tryptophan, provitamin A components, antioxidants, iron and zinc.

Biofortification is the development of micronutrient-dense staple crops by using the best traditional breeding practices combined with modern biotechnology. As the third most important cereal staple food crop worldwide (FAPRI 2009) and a major cereal staple for African consumers (FAOSTAT 2010), maize qualifies as a suitable crop for biofortification.

7 Where to look for further information

The international wheat and maize improvement center (CIMMYT), Mexico is an International organization involved in the development of nutritionally improved maize. The Indian Institute of Maize, Ludhiana, Punjab, India is a premier institute involved in the development of nutritionally superior maize. The institute played a pivotal role in developing many quality protein maize genotypes which are cultivated widely across India.

8 References

Adams, K. L. and Jensen, A. H. (1988). High-fat maize in diets for pigs and sows. *Anim. Feed. Sci. Technol.* 17: 201–12.

Adom, K. K. and Liu, R. H. (2002). Antioxidant activity of grains. *J. Agric. Food Chem.* 50: 6182–7.

Alexander, D. E. (1988). Breeding special nutritional and industrial types. In Sprague, G. F. and Dudley, J. W. (eds), *Corn And Corn Improvement*, 3rd ed., Agronomy Monograph No. 18. ASA, CSSA, and SSSA, Madison, Wisconsin, USA, pp. 869–80.

Banziger, M. and Long, J. (2000). The potential for increasing the iron and zinc density of maize through plantbreeding. *Food Nutr. Bull.* 21: 397–400.

Bouis, H. E. and Welch, R. M. (2010). Biofortification – a sustainable agricultural strategy for reducing micronutrient malnutrition in the global South. *Crop Sci.* 50: S20–32.

Boyer, C. D. and Shannon, J. C. (1982). The use of endosperm genes for sweet corn improvement. In Janick, J. (ed.), *Plant Breeding Reviews*, AVI Publishing Co, Westport, CT, Vol. 1, pp. 139–54.

Brewbaker, J. L. (2003). *Corn Production in the Tropics: The Hawaii Experience*. College of Tropical Agriculture and Human Resources, University of Hawaii at Manoa, USA.

Burgoon, K. G., Hansen, J. A., Knabe, D. A. and Bockholt, A. J. (1992). Nutritional value of quality protein maize for starter and finisher swine. *J. Anim. Sci.* 70: 811–17.

CRA (Corn Refiners Association) (2006). *Corn Oil*, 5th ed., Corn Refiners Association, Washington, DC.

Creech, R. G. (1965). Genetic control of carbohydrate synthesis in maize. *Genetics* 52: 1175–86.

Cui, L., Gao, R., Dong, S., Zhang, J., Liu, P., Zhang, H., Meng, J. and Shi, D. (2012). Effects of ear shading on the anthocyanin contents and quality of kernels in various genotypes of maize. *Aus. J. Crop Sci.* 6: 704–10.

Cummings, J. H. and Englyst, H. N. (1991). Measurement of starch fermentation in the human large intestine. *Can. J. Physiol. Pharmacol.* 69: 121–9.

Dormann, P. (2003). Corn with enhanced anti-oxidant potential. *Nat. Biotechnol.* 21: 1015–16.

Dudley, J. W. and Lambert, R. J. (2004). 100 generations of selection for oil and protein in corn. *Plant Breed. Rev.* 24: 79–110.

Dupont, J., White, P. J., Carpenter, M. P., Schaefer, E. J., Meydani, S. N., Elson, C. E., Woods, M. and Gorbach, S. L. (1990). Food uses and health effects of corn oil. *J. American Oil Nutr.* 9: 438–70.

Englyst, H. N., Kingman, S. M. and Cummings, J. H. (1992). Classification and measurement of nutritionally important starch fractions. *Eur. J. Clin. Nutr.* 46: S33–50.

Feila, S., Mosera, B., Jampatongb, S. and Stampa, P. (2005). Mineral composition of the grains of tropical maize varieties as affected by preanthesis drought and rate of nitrogen fertilization. *Crop Sci.* 45: 516–23.

Goldman, I. L., Rocheford, T. R. and Dudley, J. W. (1994). Molecular marker associated with maize kernel oil concentration in an Illinois high protein 9 Illinois low protein cross. *Crop Sci.* 34: 908–15.

Graham, G. G., Placko, R. P. and Maclean, W. C. (1980). Nutritional value of normal, *opaque2* and, *sugary 2* and *opaque 2* maize hybrids for children and infants 2. Plasma free amino acids. *J. Nutr.* 110: 1070–4.

Gupta, H. S., Hossain, F. and Muthusamy, V. (2015). Biofortification of maize: An Indian perspective. *Ind. J. Genet.* 75: 1–22.

Hallberg, L. (1982). Iron absorption and iron deficiency. *Hum. Nutr. Clin. Nutr.* 36: 259–78.

Harjes, C. E., Rocheford, T. R., Bai, L., Brutnell, T. P., Kandianis, C. B., Sowinski, S. G., Stapleton, A. E., Vallabhaneni, R., Williams, M., Wurtzel, E. T., Yan, J. B. and Buckler, E. S. (2008). Natural genetic variation in lycopene epsilon cyclase tapped for maize biofortification. *Science* 319: 330–3.

Hauman, B. F. (1985). Corn oil. *J. American Oil Chem. Soc.* 62: 1524–31.

Hegsted, D. M., Ausman, L. M., Johnson, J. A. and Dallal, G. E. (1993). Dietary fat and serum lipids; an evaluation of the experimental data. *Am. J. Clin. Nutr.* 57: 875–83.

Hoebler, C., Karinthi, A., Chiron, H., Champ, M. and Barry, J. L. (1999). Bioavailability of starch in bread rich in amylose: metabolic responses in healthy subjects and starch structure. *Eur. J. Clin. Nutr.* 53: 360–6.

Hu, Q. P. and Xu, J. G. (2011). Profiles of carotenoids, anthocyanins, phenolics, and antioxidant activity of selected color waxy corn grains during maturation. *Agri. Food Chem.* 59: 2026–33.

Iacono, J. M. and Dougherty, R. M. (1993). Effects of polyunsaturated fats on blood pressure. *Ann. Rev. Nutr.* 13: 243–60.

Ibrahim, K. E. and Juvik, J. A. (2009). Feasibility for improving phytonutrient content in vegetable crops using conventional breeding strategies: case study with carotenoids and tocopherols in sweet corn and broccoli. *Agri. Food Chem.* 57: 4636–44.

Katragadda, H. R., Fullana, A. S., Sidhu, S. and Carbonell-Barrachina, Á. A. (2010). Emissions of volatile aldehydes from heated cooking oils. *Food Chem.* 120: 59–65.

Kurilich, A. C. and Juvik, J. A. (1999). Quantification of carotenoid and tocopherol antioxidants in *Zea mays. J. Agric. Food Chem.* 47: 1948–55.

Lambert, R. J. (2001). High-oil corn hybrids. In Hallauer, A. R. (ed.), *Specialty Corns*, CRC Press, Boca Raton, 2nd ed., pp. 131–54.

Lawton, J. W. and Wilson, C. M. (1987). Proteins of the kernel. In White, P. J. and Johnson, L. A. (eds), *Corn Chemistry and Technology*, Am. Assoc. of Cereal Chemists, St. Paul, MN, pp. 313–54.

Lodha, M. L., Gupta, H. O., Ram, P. C. and Singh, J. (1976). Some biochemical characteristics of modified phenotype strains of *opaque-2*. *Curr. Sci.* 45: 285–6.

Mattson, F. H. and Grundy, S. M. (1985). Comparison of effects of dietary saturated, monounsaturated, and polyunsaturated fatty acids on plasma lipids and lipoproteins in man. *J. Lipid Res.* 26: 194–202.

Menkir, A., Liu, W., White, W. S., Maziya-Dixon, B. and Rocheford, T. (2008). Carotenoid diversity in tropicaladapted yellow maize inbred lines. *Food Chem.* 109: 521–9.

Mertz, E. T., Bates, L. S. and Nelson, O. E. (1964). Mutant gene that changes protein composition and increases lysine content of maize endosperm. *Science* 145: 279–80.

Montilla, E. C., Hillebrand, S., Antezana, A. and Winterhalter, P. (2011). Soluble and bound phenolic compounds in different bolivian purple corn (*Zea mays* L.) cultivars. *Agric. Food Chem.* 59: 7068–74.

Moreno, Y. S., Hernandez, D. R. and Velazquez, A. D. (2005). Extraction and use of pigments from maize grains (*Zea mays* L.) as colorants in yogurt. *Arch. Latinoam Nutr.* 55: 293–8.

Muthusamy, V., Hossain, F., Nepolean, T., Choudhary, M., Saha, S., Bhat, J. S., Prasanna, B. M. and Gupta, H. S. (2014). Development of β-carotene rich maize hybrids through marker-assisted introgression of *β-carotene hydroxylase* allele. *PLOS ONE* 9(12): e113583.
Nelson, O. and Pan, D. (1995). Starch synthesis in maize endosperm. *Ann. Rev. Plant Physiol. Plant Mol. Biol.* 46: 475–96.
Nelson, O. E., Mertz, E. T. and Bates, L. S. (1965). Second mutant gene affecting the amino acid pattern of maize endosperm proteins. *Science* 150: 1469–72.
Nuss, E. T. and Tanumihardjo, S. A. (2011). Quality protein maize for Africa: closing the protein inadequacy gap in vulnerable populations. *Adv. Nutr.* 2: 217–24.
Osborne, T. B. and Mendel, L. B. (1914). Nutritive properties of proteins of the maize kernel. *J. Biol. Chem.* 18: 1–16.
Osei, S. A., Dei, H. K. and Tuah, A. K. (1999). Evaluation of quality protein maize as a feed ingredient for layer pullet. *J. Anim. Feed Sci.* 8: 181–9.
Pixley, K. V., Palacios, N. and Glahn, R. P. (2011). The usefulness of iron bioavailability as a target trait for breeding maize (*Zea mays* L.) with enhanced nutritional value. *Field Crops Res.* 123: 153–60.
Prasanna, B. M., Vasal, S. K., Kassahun, B. and Singh, N. N. (2001). Quality protein maize. *Curr. Sci.* 81: 1308–19.
Raboy, V., Gerbasi, P. F., Young, K. A., Stoneberg, S. D., Pickett, S. G., Bauman, A. T., Murthy, P. P.N., Sheridan, W. F. and Ertl, D. S. (2000). Origin and seed phenotype of maize *low phytic acid 1-1* and *low phytic acid 2-1*. *Plant Physiol.* 124: 355–68.
Sandstead, H. H., Munoz, J. M., Jacob, R. A., Kelvay, L. M., Reck, S. J., Logan, G. M. Jr., Dintzis, F. R., Inglett, G. E. and Shvey, W. C. (1978). Influence of dietary fiber on trace element balance. *Am. J. Clin. Nutr.* 31: 5180–4.
Sandstorm, B. (1997). Bioavailability of zinc. *Eur. J. Clin. Nutr.* 51: S17–19.
Sanjeev, P., Chaudhary, D. P., Sreevastava, P., Saha, S., Rajenderan, A., Sekhar, J. C. and Chikkappa, G. K. (2014). Comparision of fatty acid profile of specialty maize to normal maize. *J. Am. Oil Chem. Soc.* 91(6): 1001–5.
Sen, A., Bergvinson, D., Miller, S., Atkinson, J., Gary, F. R. and Thor, A. J. (1994). Distribution and microchemical detection of phenolic acids, flavonoids, and phenolic acid amides in maize kernels. *J. Agric. Food Chem.* 42: 1879–83.
Topping, D. I. and Clifton, P. M. (2001). Short-chain fatty acids and human colonic function: roles of resistant starch and nonstarch polysaccharides. *Physiol. Rev.* 81: 1031–64.
Ullah, I., Ali, M. and Farooqi, A. (2010). Chemical and nutritional properties of some maize (*Zea mays* L.) varieties grown in NWFP, Pakistan. *Pak. J. Nutr.* 9: 1113–17.
Vallabhaneni, R. and Wurtzel, E. T. (2009). Timing and biosynthetic potential for carotenoid accumulation in genetically diverse germplasm of maize. *Plant Physiol.* 150: 562–72.
Van Soest, P. J., Fadel, J. and Sniffen, C. J. (1979). Discount factors for energy and protein in ruminant feeds. In *Proceedings, Cornell Nutrition Conference For Feed Manufacturers*, Ithaca, NY, Cornell University, pp. 63–75.
Vasal, S. K. (1999). The quality protein maize story. In *Improving Human Nutrition Through Agriculture: The Role of International Agricultural Research*, October 5–7, IRRI, LasBanos, Phillippines, pp. 1–16.
Vasal, S. K. (2000). The quality protein maize story. *Food Nutr. Bull.* 21: 445–50.
Vassal (2014). Breeding challenges and perspectives in developing and promoting quality protein maize germplasm. In Chaudhary, D. P., Kumar, S. and Langyan, S. (eds), *Maize: Nutrition Dynamics and Novel Uses*, Springer, pp. 161. doi: 10.1007/978-81-322-1623-0.
Wang, H. W., Wu, B. Y., Song, T. M. and Chen, S. J. (2009). Effects of long-term selection for kernel oil concentration in KYHO, a high-oil maize population. *Crop Sci.* 49: 459–66.
White, P. J. and Broadley, M. R. (2009). Biofortification of crops with seven mineral elements often lacking in human diets- iron, zinc, copper, calcium, magnesium, selenium and iodine. *New Phytol.* 182: 49–84.

WHO/FAO/UNU (2007). Protein and amino acid requirements in human nutrition. 1764 Report of a Joint WHO/FAO/UNU Expert Consultation, WHO Technical Report Series, No 935, Geneva.
Wilson, C. M., Shewry, P. R. and Miflin, B. J. (1981). Maize endosperm proteins compared by sodium dodecyl gel electrophoresis and isoelectric focussing. *Cereal Chem.* 58: 275–81.
Xu, J. G., Hu, Q. P., Wang, X. D., Luo, J. Y., Liu, Y. and Tian, C. R. (2010). Changes in the main nutrients, phytochemicals, and antioxidant activity in yellow corn grain during maturation. *Agric. Food Chem.* 58: 5751–6.
Yang, Z. and Zhai, W. (2010). Identification and antioxidant activity of anthocyanins extracted from the seed and cob of purple corn (*Zea mays* L.). *Inn. Food Sci. Emer. Technol.* 11: 169–76.
Young, V. R., Scrimshaw, N. S. and Pellet, P. L. (1998). Significance of dietary protein source in human nutrition: animal and/or plant proteins? In Waterlow, J. C., Armstrong, D. G., Fowden, L. and Riley, R. (eds), *Feeding a World Population of More than Eight Billion People*, Oxford University Press in associate with Rank Prize Funds, New York, pp. 205–21.
Zai, S. W. and Gao, J. (2001). Nutritional value and developing prospect of high-oil corn. *Cereal Feed Ind.* 6: 41–2.

Chapter 8

Biofortification of maize

Eliab Simpungwe, HarvestPlus, Zambia

1 Introduction

While a precise definition of 'biofortification' has not been coined yet, generally, it is a term referring to the process of enhancing micronutrient density of crops through plant breeding, genetic engineering or agronomic management, such as fertilizer application (e.g. zinc foliar application), without compromising their yield potential and particular context-specific culinary traits. The aim is to increase the bioavailable concentrations of essential elements in edible portions of crop plants to avail consumers of food crops vitamins and minerals in quantities sufficient to improve nutrition. Biofortified crops can deliver various vitamins and mineral elements the body requires for growth and optimal function. The proponents of biofortification contend that micronutrient deficiencies are caused by diets characterized by high intake of staple foods but low intake of vegetables, fruits, animal and fish products, which are rich sources of vitamins and minerals (Philip and Martin 2005; Bouis and Haddad 1990; Grusak and Cakmak 2004; Graham et al. 2001 and Bouis et al. 2009).

The coining of the term 'biofortification' is credited to Steve Beebe, head of the HarvestPlus bean programme in early 2000 (Morgan 2013). HarvestPlus and teams of leading research scientists collaborating with the HarvestPlus Program have pioneered and championed 'biofortification' of staple crops targeting iron, zinc and vitamin A, the three micronutrients listed among others by the World Health Organization as most limiting in terms of global public health. The HarvestPlus Program is part of the Consultative Group on International Agricultural Research Program on Agriculture for Nutrition and Health and

http://dx.doi.org/10.19103/AS.2016.0001.13

is coordinated by the International Center for Tropical Agriculture and the International Food Policy Research Institute. There are other schools of researchers working on biofortification of food crops with various other nutrients such as protein biofortification of maize, sorghum, cassava and rice, folic biofortification of tomato and so on (GKC 2007). This chapter presents biofortification of maize with provitamin A carotenoids, which is currently the largest micronutrient biofortification effort in maize that has achieved widespread application.

Notwithstanding, measures to enhance zinc content in maize are also being pursued. However, although the breeding approach for zinc maize is something that is sound and sustainable, it has limitations. Cakmak (2007) explains that maize biofortification with zinc is a long-term process requiring a variety of breeding activities and huge resources. He cites four challenges: (1) identification of a useful genetic variation and the most promising parents, (2) long-term crossing and back-crossing activities, (3) stability of other target traits (e.g. high grain Zn concentrations) across the different environments that feature huge variation in soil and climatic conditions and (4) adaptation of the newly developed biofortified genotypes over a range of crop and soil management practices applied in the target regions or countries. The newly developed genotypes should be able to extract sufficiently large amounts of Zn from potentially Zn-deficient soils and accumulate it in whole grain at sufficient levels for human nutrition. The challenge is that soils that are widespread in major cereal-growing regions have several adverse soil chemical factors (i.e. high pH values and low soil moisture and organic matter) that could potentially diminish the expression of high grain Zn trait and limit the capacity of newly developed (biofortified) cultivars to absorb adequate amount of Zn from soils to contribute to daily Zn requirement of human beings.

Evidence presented in studies on the biofortification of maize with lysine indicates the superiority of quality protein maize (QPM) over other maize types in terms of the amount of absorbed nitrogen needed to provide the necessary amino acids for the different metabolic functions (FAO 1992; Galinat 1977) with proven human nutrition benefits. No doubt then that the production and consumption of QPM in maize-eating countries has the potential to provide nutrition and economic benefits.

The premise of biofortification is that nutrition of poor people can be improved if they consumed staple crops rich in micronutrients since their diets are dependent primarily on few staples they grow and eat (Bouis and Haddad 1990). Since nutrients are abundant in food crops, agriculture can be a primary supplier of these nutrients. Spatially, biofortification is considered to overlap briefly with fortification and supplementation programmes, while maintaining a unique niche in populations living from subsistence agriculture and residing outside the reach of fortified food and health care delivery systems. Biofortification is considered different from conventional fortification in that it aims to increase nutrient levels in crops during plant growth rather than during processing of the crops, as is done in fortification. Biofortified crops can be grown year after year by farmers everywhere, who may plant and replant their own seed, may buy fresh seed from their local outlets every year (in the case of hybrid crops), share seed and eat their own produce. In contrast, other interventions to improve nutrition attract recurrent costs and in many cases do not reach the very vulnerable households in remote rural areas consistently (Bouis 2003).

The biofortification technology emerged as a response to the widespread public health burden of micronutrient malnutrition referred to as 'hidden hunger', which has for a long time afflicted poor rural and urban populations. Hidden hunger occurs when intake and absorption of vitamins and minerals are too low to sustain good health and development (GHI 2014). Statistics provided by the United Nations indicate that approximately 1.5 billion people suffer from iron deficiency[1] and about a third of all preschool children (about 190 million) suffer from vitamin A deficiency.[2] The fraction of developing-country populations at risk of inadequate zinc intake is estimated to be 12–66% (average 17.3%).[3]

Biofortification is achieved largely through traditional ('selective') plant breeding, where parent lines with high vitamin or mineral levels are crossed over several generations to produce plants that have the desired nutrient and agronomic traits (Saltzman et al. 2012). However, transgenic approaches are advantageous when the nutrient does not naturally exist in a crop or when sufficient amounts of bioavailable micronutrients cannot be effectively bred into the crop. Furthermore, transgenic breeding can sometimes offer micronutrient gains beyond those available to conventional breeders. The difficulty, however, with transgenes is that many countries lack legal frameworks to allow release and commercialization of these varieties, and public understanding of this technology tends to be at a very early stage.

Philip and Martin (2005) contend that mineral elements most frequently lacking in human diets are iron (Fe), zinc (Zn) and iodine (I), although other elements, such as calcium (Cal), magnesium (Mg), copper (Cu) and selenium (Se), can be deficient in the diets of some populations. It is estimated that, of the world's 6 billion people, 60–80% are Fe deficient, >30% are Zn deficient, 30% are I deficient and about 15% are Se deficient (Philip and Martin 2005). Deficiency of vitamin A is considered one of the most prevalent nutritional maladies in developing countries resulting in impaired growth, vision, reproduction and immunity (Sommer 1995). Severe vitamin A deficiency is directly linked to xerophthalmia, which is a continuum from night blindness to permanent damage to the cornea, resulting in what would otherwise be preventable blindness, especially among children. The condition of suppressed immunity leading to greater susceptibility to infectious disease, greater mortality from these diseases and more profound and pronounced periods of illness can be linked to vitamin A deficiency (Rice et al. 2004; Emily and Tanumihardjo 2010). Dietary vitamin A deficiency causes eye disease in 40 million children every year and places 140–250 million at risk for health disorders. Children affected with vitamin A deficiencies get sick more frequently and are more intensely ill than those that are not deficient. The long-term effects ultimately carry forward to their adulthood and affect the productivity of entire societies (West 2002; Sommer et al. 1996; Harjes et al. 2008).

2 Justification for maize biofortification

Graham et al. (2001) postulate that hunger resulting from a lack of sufficient energy-giving food is regarded as the most serious problem in developing countries and as a result, the food security debate has for a long time primarily focused disproportionately on the provision of adequate calories. This is because the feeling of hunger is satiated largely by the intake of macronutrients that provide calories such as carbohydrates, protein and fat. Agriculture's primary role has, therefore, been seen to be that of delivering these macronutrients. This perception has created

1. Stevens, G. A., et al. (2013). Global, regional, and national trends in hemoglobin concentration and prevalence of total and severe anaemia in children and pregnant and non-pregnant women for 1995–2011: a systematic analysis of population-representative data. *Lancet Glob Health* 1: e16–25.
2. WHO (2009). *Global prevalence of vitamin A deficiency in populations at risk 1995-2005*. WHO Global Database on Vitamin A Deficiency. World Health Organization.
3. Wessells, K. R., Singh, G. M. and Brown, K. H. (2012). Estimating the global prevalence of inadequate zinc intake from National Food Balance Sheets: effects of methodological assumptions. *PLoS ONE* 7 (11): e50565. doi:10.1371/journal.pone.0050565.

an unfortunate gap between food and health. Agriculture has been seen to be responsible for providing food, while nutrition is widely considered to be a health issue.

Recent discourses on food security and nutrition postulate a paradigm shift for agriculture that links it directly with human health, calling for agriculture to address the gaps at the intersection of food, food and people, people and health (Graham et al. 2001). Since agriculture is the primary source of vitamins and minerals, Welch et al. (1997) argue that widespread micronutrient malnutrition can be linked to dysfunctional agricultural food systems. Agricultural technologies have the potential to deliver quality food crops that can enhance human nutrition and health. To affirm this assertion, Grusak and Cakmak (2004) state that 16 mineral elements are deemed essential for humans and plant foods can provide all of these elements. While limited dietary diversity is largely blamed for micronutrient deficiencies, the quality of the food consumed is a serious contributory factor too. In recent times, many resource-poor populations are dependent on staple crop monocultures, in contrast to the more traditional micronutrient-rich crops of pulses, fruits and a wide range of vegetables (Graham 2001).

To combat dietary micronutrient deficiencies, various interventions have been used and these include food fortification, supplementation and promotion of diet diversification. The persistence of hidden hunger points to limitations in all these interventions as currently implemented. For instance, supplementation and fortification methods come with huge costs and in many cases do not reach the vulnerable households in remote rural parts of developing countries where the problem is severe, while access to a diverse diet rich in animal sourced foods, fruits and vegetables is a function of many variables that include crop seasonality, food prices, access to markets and income (Harjes et al. 2008). These factors limit accessibility to sources of micronutrients by rural populations in developing countries.

In his pursuance to find a lasting solution to the challenge of hidden hunger, Dr. Howarth Bouis, one of the founders and the first director of HarvestPlus, was drawn to the study of food consumption patterns. Following a systematic inquiry of food consumption data sets, he established that dietary quality was the underlying cause for micronutrient malnutrition (Bouis and Haddad 1990). He also noted that rural households that were affected by hidden hunger were dependent on agricultural livelihoods that depended on staple food[4] crops to meet energy requirements. The missing piece was a sustainable and readily accessible source of adequate vitamins and minerals to improve dietary intake. In his analysis, Dr. Bouis observed that staple foods that the rural poor populations popularly cultivated and consumed could become the primary carriers of essential vitamins and minerals. These are mainly rice in South Asia, wheat in Central Asia and maize for most of sub-Saharan Africa. While these are rich sources of energy, they are typically not micronutrient dense (Pixley et al. 2013).

Dr. Bouis reckoned that micronutrients could be increased in the common grains, roots, tubers and legumes that were the basis of the everyday diets of the poor. Researchers were led to consider the enhancement of popularly consumed staple crops with essential micronutrients, a process that came to be popularly known as biofortification, as a means for reaching large populations of rural poor households through the food crops they themselves grew and ate in large quantities. Preliminary research into germplasm screening demonstrated that it was possible to produce staple crops that were simultaneously high in both micronutrients and yields by using conventional breeding.

4. 'A staple food is one that is eaten regularly and in such quantities as to constitute the dominant part of the diet and supply a major proportion of energy and nutrient needs' (FAO 1995 in Emily and Tanumihardjo 2010).

Biofortification provides the opportunity for agriculture to fortify staple crops with the much-needed micronutrients and make a direct linkage between human health and nutrition. The potential to increase the micronutrient density of staple foods by conventional breeding has become possible with the emergence of biofortification technology. Scientists have established that adequate genetic variation in concentrations of β-carotene, other functional carotenoids, iron, zinc and other minerals, exists among cultivars, making the selection of nutritionally appropriate breeding materials possible. Also, there are strong indications that micronutrient-density traits are stable across environments. Graham et al. (2001) state that in all crops studied, it is possible to combine the high micronutrient density trait with high yield. Therefore, it is possible to improve the content of several limiting micronutrients together, thus pushing populations towards nutritional balance.

Biofortification is possible through both conventional and transgenic pathways. However, because of regulatory and political restrictions on the use of transgenic approaches and because significant progress can be made through conventional breeding, a large percentage of resources are currently devoted to conventional breeding, even though transgenic approaches are in some cases potentially advantageous when compared with conventional breeding (Saltzman et al. 2013).

3 Suitability of maize for biofortification

Emily and Tanumihardjo (2010) observe that 22 of the 145 countries surveyed in 1984 had maize consumption of more than 100 g per person per day, suggesting a significant global dependency on maize as a major food source, as well as a delivery mechanism for essential nutrients. For this reason, maize and maize-based food products are targets for nutritional enhancement. Biofortified maize is enriched with β-carotene, which is converted in the body to vitamin A. Researchers have established that the β-carotene in the maize is converted to vitamin A at a higher rate than what was originally predicted for maize and higher than the rate for β-carotene in vegetables – including spinach and carrots, among others (Li et al. 2010). The initial HarvestPlus goal for provitamin A carotenoid enhancement efforts is 15 μg/g.

As a staple crop, maize exhibits considerable variation in kernel carotenoid composition, including some of the highest carotenoid concentrations for cereal crops, with some lines accumulating as much as 80 μg total carotenoids per gram, dry weight (DW). If maize had very little or no naturally occurring variation for provitamin A carotenoid (pVAC) concentrations, it would have been extremely difficult to improve the concentration of these carotenoids through conventional breeding approaches (Pixley et al. 2013).

Carotenoids serve a variety of functions in plants, including acting as antioxidants, photoprotectants and accessory pigments for light harvesting and substrates for production of volatile compounds in flowers, fruit and seed (Emily and Tanumihardjo 2010). Approximately, 700 naturally occurring carotenoids are known, typically yellow, orange and red pigments primarily produced by photosynthetic organisms and also by certain fungi and bacteria (Emily and Tanumihardjo 2010). However, the depth of colour is not correlated to provitamin A activity. Of the 50 known carotenoids that are consumed in human diets, only 3 are vitamin A precursors (β-carotene, β-cryptoxanthin and α-carotene), which may all be found in maize (Menkir et al. 2008). Provitamin A carotenoids can be converted by oxidative cleavage in the body to retinol, the active form of vitamin A, which is stored in the liver (Owens et al. 2014).

While white and most yellow maize grown and consumed throughout the world has only 0.5–1.5 μg/g of β-carotene, provitamin A maize lines averaged 23 μg/g for total carotenoids (range 5.5–66.0 μg/g) and a range of 0.06–13.6 μg/g of provitamin A carotenoids specifically (Harjes et al. 2008). Maize is also a good source of nonprovitamin A carotenoids, including lutein and zeaxanthin, which play beneficial roles in human health (Krinsky, Landrum and Bone 2003). For instance, in addition to benefits for dietary vitamin A, the consumption of carotenoid-rich foods is also associated with reduced risks of developing cancer (Agarwal and Rao 2000). The most important and best-defined function of carotenoids in animals is that it is a dietary source of provitamin A.

The delivery of biofortified maize to target populations commenced only in recent years. Therefore, relatively little is known about its economic impact and wider ramifications. Nonetheless, *ex ante* studies from India and other developing countries suggest that generally biofortified crops can reduce the problem of micronutrient malnutrition in a cost-effective way when they are targeted at specific situations (Qaim et al. 2007). Lividin and Keith (2015) assert that maize biofortification is a long-term investment that shows great promise in becoming a highly cost-effective addition to the public health interventions for combating micronutrient deficiencies. However, doing so requires strategic stakeholder engagements and collaborations, meaningful government support, rigorous marketing campaigns, targeted nutrition messaging and education in order to increase the likelihood that adopting farmers will prioritize production and home consumption.

Tunamihadjo et al. (2015) have demonstrated that orange maize biofortified with provitamin A has shown to be efficacious in maintaining vitamin A liver reserves in humans, preschool children to be precise, even within the adverse context of prevalent infection, as was the case in rural Zambia where the study was implemented.[5] Furthermore, negative feedback mechanisms of β-carotene absorption and bioconversion to vitamin A address concerns with the risk of hypervitaminosis A, which is only associated with preformed vitamin A.

4 Breeding of provitamin A biofortified maize

The emphasis on maize biofortification was to enhance β-carotene concentrations in the kernels. Generally, it is observed that tropical maize contains more β-cryptoxanthin and less β-carotene than temperate maize, and hence, most of the initial breeding sources of high provitamin A germplasm were selected from temperate regions. Subsequently, citing Harjes et al. (2008) and Yan et al. (2010), Pixley et al. (2013) explain that genetic association mapping studies using three diverse maize germplasm panels, selected to encompass a wide range of carotenoid contents and ratios, identified favourable alleles of genes encoding two key enzymes in the carotenoid biosynthetic pathway. These were lycopene epsilon cyclase and β-carotene hydroxylase 1 (crtRB1). They affect the accumulation of β-carotene in grain. On account of the use of molecular markers identified for these two enzymes, the screening of germplasm became faster and cheaper than selections based on quantification of carotenoid concentrations with high performance liquid chromatography. Germplasm carrying the favourable crtRB1 allele in homozygous form was identified with

5. *Am. J. Clin. Nutr.* 2014, 100: 1541–50.

β-carotene concentrations of up to 26 μg/g DW and total provitamin A as high as 30 μg/g DW. These newly identified sources of provitamin A are currently being used in breeding programmes, leading to rapid gains in provitamin A carotenoid concentrations in new germplasm (Pixley et al. 2013).

An essential first step in breeding maize for enhanced carotenoid concentrations involves an assessment of the carotenoid diversity of adapted maize inbred lines. Initially, trials were conducted (i) to explore the genetic variation in carotenoid concentrations among tropical-adapted yellow maize inbred lines, (ii) to assess the potential for concurrent improvement of different carotenoids and (iii) to determine the consistency of carotenoid concentrations in plants produced in different environments (Menkir et al. 2008). The first breeding crosses were made in 2004 and involved the best (elite) white-grained inbred lines with 0 μg/g of provitamin A, from the International Centre for Maize and Wheat Improvement (CIMMYT) programme in Harare, Zimbabwe and temperate yellow lines, with provitamin A concentrations ranging from 5 to 8 μg/g, concentrations that were obtained from the University of Illinois in the United States. Each F1 cross was back-crossed to its respective elite line prior to initiating new line development. Selection during the first two generations of inbreeding was primarily for deep yellow or orange grain colour, as well as for resistances to lodging, and plant and ear diseases. The breeding programme followed a cyclic process that developed better varieties competitive for all crucial traits, including producibility of the seed, grain yield, disease resistance, drought tolerance, food processing quality, taste and various characteristics that determine acceptability to farmers and consumers.

Generally, genotypic variation in different environments can greatly influence crop performance across different crop scenarios. However, an increasing body of evidence suggests that the expression of provitamin A is relatively stable under different growing conditions, making biofortified provitamin A maize suitable across many environments. For a detailed description of the processes involved in the breeding of provitamin A biofortified maize, and in particular, the development of the first wave of varieties released in Zambia, the reader may consult Pixley et al. (2013) and Pfeiffer and McClafferty (2007). It suffices here to state that the successful development of new biofortified maize varieties that are productive and desirable to farmers and consumers is a result of a breeding programme that integrates the knowledge from nutrition, social science and plant genetics (Pixley et al. 2013).

The research costs for plant breeding and crop development are likely the single largest cost component of maize biofortification, but are a one-time investment, incurred at the outset. This is because international and national agricultural research systems that develop modern varieties of maize already exist, and therefore, research costs for biofortification are essentially the incremental costs of enhancing micronutrient density. Once biofortified maize varieties are developed and in-country trials and local adaptation research costs are incurred, only routine maintenance breeding to ensure that the trait remains stable is required (Pixley et al. 2013).

5 Target setting for provitamin A carotenoids

The first question asked when developing biofortified crops is the extent to which the levels of micronutrient content in the crops should be increased to achieve a measurable improvement in the micronutrient status of consumers. To address this question, HarvestPlus seeks four main types of information (current findings in parentheses):

i. the amount of additional micronutrients that would allow the majority of a population to meet their daily nutrient requirement (i.e. >50% for vitamin A) and prevent deficiencies from occurring;
ii. the average amount of the staple food consumed daily by different age groups (i.e. ~300 g for women of child-bearing age);
iii. the losses of nutrients that occur with usual storage, processing and cooking practices (~70%); and
iv. the bioavailability of micronutrients in the context of the local diet and health status (~16–32%).

On the basis of available information on nutrient needs, nutrient retention after traditional storage and cooking of maize, the daily intake of maize and maize products and the bioavailability estimates, HarvestPlus (HarvestPlus 2014) set a preliminary target of 15 μg/g of provitamin A, predominantly β-carotene. The first wave of three varieties released in Zambia in 2012 carried β-carotene concentrations ranging from 7 to 8 μg/g, which translated to 50% of the full target level of the 15 μg/g originally set. Nevertheless, these varieties could provide 30% of the estimated average requirement of vitamin A to young children who consume a daily intake of 200 g DW (HarvestPlus 2001, 2013; Tanumihardjo 2008). Despite significant initial degradation of pVACs during ambient storage, relatively low losses during cooking and higher than previously assumed bioavailability result in a significant contribution of biofortified maize to daily vitamin A needs and thus support the feasibility of agronomic provitamin A enhancement efforts (Li et al. 2007; Mugode et al. 2014). The released varieties were characterized by an orange kernel phenotype. For comparison, typical yellow maize varieties have only 0.25–2.5 μg/g, while the baseline content of provitamin A carotenoids in white maize is virtually zero.

A bioavailability study (Li et al. 2010) conducted at Iowa State University indicated a 6:1 conversion rate of β-carotene to retinol, twice as efficient as the 12:1 rate assumed in setting the original target of 15 μg/g (new information which, all things being equal, would allow a reduction of the full target level to 7.5 μg). A more recent study by Tufts University (Muzhingi et al. 2011) showed an even more efficient bioconversion ratio of 3:1 among healthy men. However, after considering new evidence on provitamin A degradation during storage (50%) and cooking (25%), retention was later estimated to be 25% lower than originally assumed. Furthermore, average maize consumption in rural Zambia was measured at 287 g/day for women and 172 g/day for children 2–5 years of age, instead of the assumed 400 g/day and 200 g/day, respectively. This information led to maintaining the breeding target of 15 μg/g as originally set (HarvestPlus 2013).

To help clarify the impact of genotype-specific effects on retention, additional studies on retention during storage and processing, specifically testing pre-release varieties, were completed during the first half of 2012. The effect on the degradation of provitamin A varied among the pre-released genotypes evaluated. One maize genotype showed 60% retention of provitamin A after the maize was roasted for about 20 minutes, while virtually no degradation was observed among the other genotypes. The retention of provitamin A ranged from 61 to 85% after boiling maize (green) for 35 minutes at 95°C. During storage of maize grains, most of the degradation occurred in the first 15 days of storage (52–56% retention), which then stabilized retaining 30–33% of provitamin A after 6 months of storage. In conclusion, most of the provitamin A degradation in biofortified maize hybrids occurred during storage than during cooking and the magnitude of the effect varied among the different genotypes (Mugode et al. 2014).

6 Delivering biofortified maize

For proof of concept that biofortification can be successful, HarvestPlus pursues research to answer three critical questions (Saltzman 2012):

i. Can breeding increase the micronutrient density in food staples to reach target levels that will make a measurable and significant impact on nutritional status?
ii. When consumed under controlled conditions, will the extra nutrients that are bred into the food staples be bioavailable and absorbed at sufficient levels to improve micronutrient status?
iii. Will farmers grow the biofortified varieties and will consumers buy/eat them in sufficient quantities?

Much of the evidence to address the three questions has already been generated by HarvestPlus, working in collaboration with its partner research institutions that include CIMMYT, IITA, NARS and several universities. With reference to biofortified maize, the responses to the question above have been affirmed as follows:

i. Several provitamin A maize varieties have since been released, carrying sufficient provitamin A carotenoids to improve nutrition and without compromising yields. This has confirmed that breeding can indeed increase the micronutrient density in food staples to reach target levels that would make a measurable and significant impact on nutritional status. The first wave of three provitamin A orange maize varieties was released in Zambia, in September 2012, followed by other releases in Brazil, Ghana, Nigeria, Zimbabwe and Malawi. HarvestPlus has followed a systematic way for developing, releasing and introducing biofortified crops to communities for production and consumption. The process is outlined in Fig. 1 below.
ii. Despite significant combined losses of pVACs during the first month of storage (~70%) and further loss during cooking of traditional dishes (25%), at current average maize intakes, orange maize can still contribute significantly to daily vitamin A requirements. In support of this assertion, efficacy studies have proved that when consumed under controlled conditions, extra nutrients bred into the maize are absorbed and nourish the body at sufficient levels to improve the micronutrient status of consumers. In a randomized, placebo-controlled provitamin A biofortified maize efficacy trial conducted in 140 rural Zambian children, the findings indicated that the orange maize group continued to absorb and convert β-carotene to vitamin A, increasing their liver reserves (Gannon et al. 2014; Li et al. 2007). Furthermore, findings from a feeding trial involving 60 rural Zambia children with a mean age of 29 months indicated that feeding zinc-biofortified maize to vulnerable populations can meet zinc requirements and provide an efficient dietary alternative to local diets that are limited in nutrient content. This study was set to compare the quantity of zinc absorbed from hammer-milled zinc-biofortified maize with that from control maize even when the

flour was fortified to a higher level of zinc than biofortified maize. Results from this study showed that zinc absorbed from biofortified maize was double the amount absorbed from postharvest fortified maize flour (Chomba et al. 2015).

iii. Zambia has demonstrated that farmers are willing to adopt and consumers will buy and eat provitamin A biofortified maize even when it is presented in an unfamiliar orange colour. Within the first two years of launching the product, over 100000 farmers were growing the crop and orange maize products were being snatched off the shelves of retail shops in the capital city.

Three factors are critical to farmer adoption of biofortified maize, namely, acceptance of colour change, preference of taste and availability of well-crafted dissemination infrastructure. Provitamin A biofortified maize presents itself in yellow to deep orange colour, associated with high provitamin A concentrations. In communities accustomed to white maize consumption, adoption of this crop requires that both producers and consumers actively accept the colour and associated sensory changes in addition to acceptable yield levels.

Creed (1998) observes that the rise of consumer expectation and awareness leads to reluctance to accept certain foods presented for consumption. The food service industry, therefore, has been compelled to provide scientific proof of not only nutritional benefits but also sensory superiority of newly introduced foods. In many white maize-consuming communities, consumers present anecdotal sentiments linked to emotional and psychological factors that influence their choice of white over coloured maize types. In Zambia, for example, many consumers frown upon yellow maize merely because of its historical connotation that yellow was developed for stockfeed and is accepted for consumption only when there is a famine. Given this scenario, HarvestPlus scientists ensure that acceptance studies are conducted as part of the delivery processes of provitamin A biofortified maize.

The crop delivery strategy outlined in Fig. 1 was successfully applied in Zambia to develop and deliver provitamin A biofortified orange maize. Zambia was identified as a target population on the basis of the severity of vitamin A deficiency. During the discovery phase that lasted from 2004 to 2008, baseline maize consumption trends were used to validate nutrient breeding targets and plant breeders identified and screened crop germplasm that went into crop development. From 2009 to 2013, HarvestPlus engaged with the development stage involving all breeding activities. Initial product development activities were undertaken at CIMMYT and IITA research centres to develop varieties with improved nutrient content and high agronomic performance, as well as preferred consumer qualities. When promising high-yielding, high-nutrient lines emerged, they were tested by the Zambia Agricultural Research Institute (ZARI) and private seed companies. The best-performing lines were then selected for submission to the government's release committee. Parallel to crop improvement, nutrition research studies to measure retention of micronutrients in the maize under typical processing, storage and cooking practices were undertaken. To provide evidence of efficacy, randomized controlled efficacy trials demonstrating the impact of biofortified crops on micronutrient status and functional indicators of micronutrient status, including visual adaptation to darkness following consumption of provitamin A crops, were undertaken (Saltzman et al. 2013).

In general, the period after 2013 is recognized as the delivery phase of HarvestPlus biofortified crops. By 2015, HarvestPlus facilitated the release of over 130 crops in more than 25 countries including more than 10 provitamin A biofortified maize varieties. The

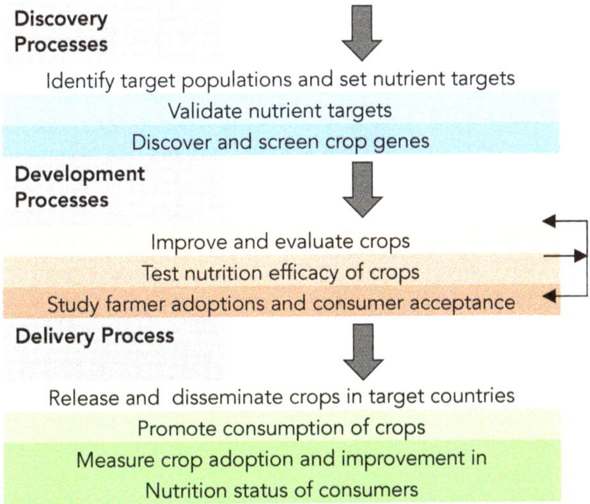

Figure 1 HarvestPlus impact pathway for delivery of biofortified crops (Saltzman et al. 2012).

dissemination strategy for maize has been to commercialize the seed through private sector companies as a way of keeping the seed systems sustainable. Advocacy, awareness creation, promotions and market creation are achieved through joint initiatives with the public sector, civil society organizations and the NGO community.

7 Case study: the Zambia maize biofortification programme

While the potential for biofortification is promising, it needs to be proven that it is a feasible concept that can work at all stages from breeding, seed production and distribution, farmer adoption, food production and consumer acceptance. This calls for an integrated delivery strategy that engages and links several aspects along the value chain of a product function. For provitamin A maize in Zambia, the crucial aspects included crafting supply and demand systems for seed and grain, influencing change in dietary habits and influencing policy and regulatory frameworks. Overall, securing government support was pivotal, without which acceptance of a new crop could be stifled, especially when the crop was presented in unfamiliar colour and raised suspicions of genetic modifications. The successful introduction of provitamin A biofortified orange maize in Zambia is an interesting case study that highlights the practicality, the methods and the challenges of undertaking the biofortification initiative.

7.1 The status quo

For the first time in the history of Zambia, in September 2012, the Zambia Variety Release Committee, under the auspices of the Seed Certification and Control Institute (SCCI) of

Zambia, approved the release of three provitamin A biofortified orange maize varieties. The three orange maize varieties were among 11 other white maize varieties released that year. Zambia releases an average of 16 maize varieties every year, resulting in no single variety dominating more than 10% market share on the seed market (Smale et al. 2014; De Groote et al. 2011). By 2015, one of the private seed companies that had received the licence to market one of the provitamin A maize varieties had gained slightly over 1% market share and enabled over 100000 farmers to grow provitamin A orange maize. Also in 2015, for the first time in the history of Zambia, four maize milling companies expressed interest in buying orange maize grain and introducing orange maize flour in retail shops. The sales were not disappointing. The number of farmers adopting provitamin A orange maize was projected to grow to over half a million in four years. More than four private seed companies were lined up to participate in producing over a 1000 metric tonnes of seed, and if these were bought by farmers, it would place the provitamin A maize seed market at 10% market share of the maize seed industry. This would be an unprecedented outcome considering that the Zambian population is not accustomed to the orange-coloured maize, preferring largely white maize and abhorring yellow maize, which was considered famine food best suited for livestock.

The early success of HarvestPlus in introducing provitamin A biofortified orange maize in Zambia was dependent on several factors, most importantly, strategic engagement with the whole maize value chain, facilitating efficient operations from seed supply, demand creation through promotions and advocacy, developing market chains that linked producers, processors, retailers and consumers and getting a buy-in from the government and NGOs.

7.2 The context

Zambia is a landlocked country in Southern Africa, measuring 743390 km2 (about 30% agricultural land) and with population count of 14.5 million by 2013. The Zambian economy has historically been based on copper mining. However, agriculture plays an important part, providing more jobs than the mining industry.

White maize is by far the most important crop consumed by the large majority of Zambians. HarvestPlus Baseline Study conducted in 2011 showed that almost all farming households grew white maize and up to 50% consumed their own home grown maize throughout the year. Most smallholder producers cultivated up to 5 ha with most area planted to maize. Average unit yields for smallholder producers stood at 2 tonnes/ha in most provinces. In Zambia, maize production is mostly done by smallholder producers, complemented by medium- and large-scale commercial farmers. It is estimated that 75% of maize growers are small-scale farmers (De Groote et al. 2012).

The maize crop is mostly rain-fed, which leads to substantial fluctuations in production volume from year to year. Where irrigation facilities are available, maize may be grown in the dry season (from May to October), but the area under irrigation is below 5% of the total maize cropped area. Zambia can boast to be self-sufficient in white maize production, even achieving exports in a number of years. For instance, in 2011/2012, white maize production surpassed 3 million metric tonnes, dramatically surpassing the estimated domestic annual consumption. In the 2013/2014 season, Zambia recorded the highest white maize harvest in history, amounting to 3.4 million metric tonnes against the national consumption requirements of 2.8 million metric tonnes (Fig. 2).

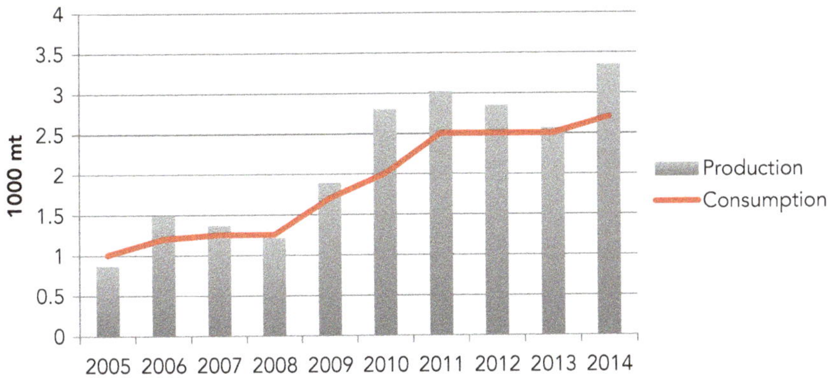

Figure 2 Annual production and consumption of white maize in Zambia.

Zambia's Food Reserve Agency, which was created to secure and administer strategic reserves of white maize, has managed to stabilize essential maize supplies for the country to the extent that it has decided to reduce its annual procurements of maize over the years from as high as 1.7 million metric tonnes in 2012 down to 500 000 metric tonnes in 2015. Zambia's crop production index, which is a value that shows agricultural production for each year compared to the base period (base period being 2004–2006 = 100) and includes all crops except fodder crops, has hovered above 100 for many years, sometimes rising to 172, indicating increasing quantities of agricultural commodities produced.

Paradoxically, in the midst of the increasing and diverse agricultural production, an insidious malnutrition problem looms in rural settings and urban compounds. Zambia is classified among countries with high levels of macro- and micronutrient deficiencies, especially among young children, pregnant and lactating women (Harris and Drimie 2012). Undernutrition is endemic in many parts of the country and poses a serious threat to the well-being of children and women. According to statistics in the Zambia Demographic and Health Surveys, undernutrition is responsible for 52% of all deaths occurring in children below the age of five. Stunting affects 40% of children under five years, 15% are underweight, a figure that has not changed over ten years, while wasting has increased from 5 to 6% over ten years. Anaemia affected 53% of children under five and 30% in women of child-bearing age (ZDHS 2015; CSO 2015). Vitamin A deficiency remains a serious public health problem with 54% of children under five and 13% among women of child-bearing age being affected (IZiNCG 2004). According to an indirect method of estimating rates of zinc deficiency by using food balance sheets, on the basis of the amount of zinc absorbable in foods and the population's theoretical zinc requirements, it is estimated that 38% of the population in Zambia was at risk of low zinc intake (IZiNCG 2004).

Household livelihood studies reveal that inadequate intakes of several micronutrients in Zambia are widespread because staple diets are predominantly maize based and intakes of plant-based and animal source foods are low (Mason and Jayne 2009; Chapoto et al. 2010). Food consumption patterns across the country show inadequate intakes of vitamins and minerals among large populations of rural households.

It is only against this background that HarvestPlus justified the introduction of provitamin A biofortified orange maize in Zambia to contribute towards reducing vitamin A deficiency levels. HarvestPlus prioritizes intervention countries for biofortification by using a Biofortification Prioritization Index[6]. Based on this index, Zambia ranks number 3 out of 107 countries suitable for investing in biofortified maize. Most efforts to address micronutrient deficiencies in Zambia have focused on providing high-dosage supplements, often through widespread periodic distribution of vitamin A capsules to children under five years and iron and folic acid capsules to women during antenatal. However, supplements maintain serum micronutrient levels for several months before needing to be repeated and may not reach all residents in remote rural areas. If micronutrient deficiencies could be addressed by biofortified staple crops, especially those grown in rural areas for subsistence, the complex supplementation programmes could be complemented in rural areas, resulting in significant positive benefits for mother and child health, growth, survival and cognitive functioning. Average per capita consumption of maize in Zambia is more than 100 kg per annum (287 g per day) for adult women, and maize provides more than half of the calories and proteins for most Zambians (Pixley et al. 2013), while providing only traces or no provitamin A at all (Nuss and Tanumihardjo 2010).

Recognising the importance of maize for Zambia, HarvestPlus initiated biofortification of maize with provitamin A, as a means to improve nutrition for a large majority of the Zambian population. The maize biofortification and dissemination initiative embarked upon in Zambia was the first of its kind anywhere in the developing world. It was officially launched in 2008 when HarvestPlus formalized its partnership with the ZARI, a national research organization, to expand local breeding and testing of provitamin A biofortified orange maize, to a full operational scale. Five competitive three-way provitamin A biofortified orange maize hybrids were identified and three of these were officially released by the release committee of the Zambian government in 2012. The varieties were named GV662A, GV664A and GV665A. The letters GV stood for 'Golden Valley', the research centre where most evaluation work was undertaken, and the letter 'A' signified vitamin 'A'. The figure 600 represents the medium-maturing series of maize varieties in Zambia.

The HarvestPlus mission in Zambia became one of creating and mainstreaming provitamin A orange maize value chain from the scratch. While Zambia had white and yellow maize varieties, provitamin A, orange maize-coloured varieties never existed. The orange-coloured maize was unfamiliar to the Zambian population. As a result, it was easily associated with the disliked yellow-coloured varieties, which were considered stockfeed. The yellow-coloured maize was linked to a history of famine that occurred in the 1990s, when yellow maize was distributed as famine relief food. The yellow maize had unfamiliar taste and was understood to have been grown for livestock in North America. To this day, Zambians hold a general dislike for, and frown upon, coloured maize varieties, considering them inferior to white maize. The Zambian population holds white maize in high esteem, preferring the highly refined white maize mealie meal for consumption. The maize biofortification project had to surmount this challenge by creating acceptance among

6. BPI combines three subindices consisting of country-level crop production and consumption data primarily from Food and Agriculture Organization (FAO) and micronutrient deficiency-level data primarily from the World Health Organization (WHO) (Asare-Marfo et al. 2013). BPI ranks countries according to their suitability for investment in biofortified crops.

Zambian maize consumers, to consume orange maize as *nshima*[7], the format in which 80% of maize is consumed.

7.3 Country crop delivery strategy

The Zambia maize biofortification project's crucial question was whether farmers would adopt the orange-coloured maize and whether consumers would buy/eat in sufficient quantities?

In moving from research and development of suitable varieties to the creation of adoption and consumption, a combination of strategies that worked together like cogs in an engine were launched (Fig. 3). Conceptualizing the maize value chain as a 'system' was effectual in successfully launching biofortified orange maize into the Zambia maize industry. In its broadest conception, a 'system' is described as a complex of interacting components together with relationships among them permitting functionality that achieves an outcome (Laszlo and Kripper 1998). Hence, HarvestPlus recognized and appreciated the complex interactions that existed between seed supply, seed and grain markets and the important role that the government played in each one of these. The Zambia maize industry is driven by a supply of mainly hybrid seed produced exclusively by a private sector of more than 14 companies, some of which are international, such as SeedCo, which operates in 13 African countries. At least 84% of maize growers use hybrid seed, with the rest of smallholder producers relying on recycled local open pollinated varieties, which also yielded less than half their compatriot hybrids (Smale et al. 2013, 2014; De Groote et al. 2012). This scenario justified the strategy of releasing hybrid varieties of orange maize, rather than open pollinated varieties and licencing seed to private seed companies for commercialization. Among the first three seed companies licenced to commercialize biofortified orange maize was SeedCo. The three companies that were licenced to commercialize biofortified maize seed held a combined seed maize market share of over 50%. The 'Seed Companies and Agrodealers cog' in Fig. 3 was crucial in ensuring successful seed supply and distribution. Agrodealers (seed outlets) were trained in the nutritional benefits of orange maize as well as seed handling and agronomic practices. They, therefore, became influential in assisting farmers to make a choice for orange maize at selling points.

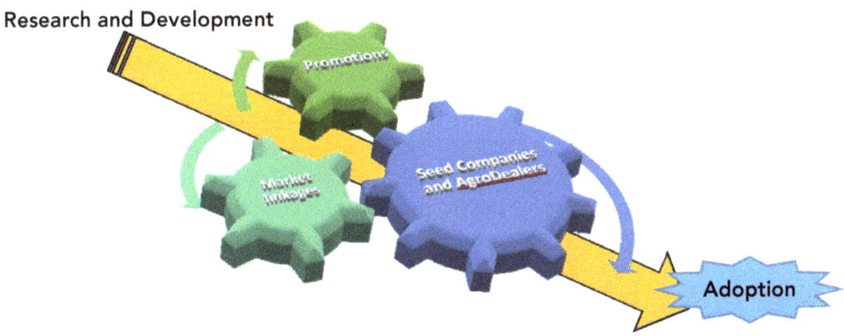

Figure 3 A systems approach in the delivery of provitamin A biofortified orange maize in Zambia.

7. *Nshima* is stiff porridge prepared from maize flour. It is the popular format in which maize is consumed in Zambia, besides roasting and boiling fresh cobs. At least 80% of the maize consumed in Zambia is consumed as *nshima*. Other ways of consuming maize include samp (maize rice) and as a drink (prepared as maize juice).

Even though over 50% of maize growers in Zambia grow and consume their own maize, availability of grain markets was crucial to trigger the adoption of orange maize. De Groote et al. (2011) established that more than 50% smallholder maize growers sold all or part of their maize. Therefore, as illustrated in Fig. 3, getting the 'markets' cog moving was crucial. This was achieved by creating market linkages between farmers and grain processors to catalyse grain markets.

To move further forward, a third cog; 'promotions', as indicated in Fig. 3, was crucial in creating adoption. This cog was driven largely by actors in the government as the following:

- Ministry of Health, by recognizing and endorsing orange maize as a feasible intervention for micronutrient malnutrition.
- Ministry of Agriculture, by providing enabling policy environment and committing extension services towards farmer advisory services on nutrition benefits and agronomic practices relating to orange maize. The most important enabling policy enacted by the Ministry of Agriculture was the inclusion of orange maize seed under the Farmer Input Support Programme (FISP). FISP was a government subsidy programme that allowed for widespread access to seed and fertilizer by rural farming households.
- Ministry of Education in collaboration with World Food Program to adopt orange maize for school feeding programmes.
- NGOs by incorporating orange maize into their crop promotional activities.

The systems approach illustrated in Fig. 3 was intended to establish a self-sustaining biofortified orange maize value chain, like the one that existed for white maize in the country.

7.4 Assessing success

The initial successes in the delivery of biofortified maize in Zambia can be attributed to the following factors:

i. Instituting a multi-stakeholder platform
ii. Creating government ownership
iii. Evidence-based delivery

i. Instituting a multi-stakeholder platform

The concept of multi-stakeholder platform or innovation platform has become popular as an institutional framework for handling, resolving and managing complex, sectoral cross-cutting issues. The idea is that when multiple stakeholders whose values, activities and interests overlap in delivering a public good or service, they undertake collective action and can evade dissonance and create impetus in achieving the required outcomes (Simpungwe 2002; Tenywa et al. 2011). In the Zambian case, the nutrition sector, agriculture, health, education, community development, NGOs and the scaling-up nutrition civil society organizations held a common interest in tackling malnutrition. Subsequently, HarvestPlus harnessed this common interest by creating a 'platform', a 'forum' or 'space' wherein biofortified maize initiative was introduced, discussed, embraced, supported and implemented as a complementary solution for meeting each of the stakeholder's interests of resolving malnutrition in Zambia.

Soon after the commencement of the maize biofortification programme in Zambia, a multi-stakeholder platform called National Provitamin A Maize Steering Committee (NPASC) was set up with members drawn from relevant government departments including Ministry of Agriculture and Ministry of Health, the nutrition community, private seed companies, NGOs and civil society organizations such as farmer's union, the consumer association and civil society for scaling up nutrition. The platform allowed participating organizations to appreciate the rationale behind the development of provitamin A biofortified maize for Zambia and to begin supporting delivery of the crop. Besides acting as an advisory body, the NPASC role was to provide direction for product development and contribute towards advocacy and promotional activities. The NPASC was influential in

- securing the interest of government to embrace biofortification as a feasible food-based intervention to address malnutrition;
- securing the interest of private sector seed companies to participate in provitamin A maize field trial evaluations and seed production and
- attracting the participation of food processors to use biofortified maize in food products and to take the new maize to retail markets.

ii. Creating government ownership

Creating government ownership was crucial for ensuring that biofortified maize was not perceived as a foreign, outsider intervention. Vitamin A deficiency was a public health concern that required government leadership in its resolution. Therefore, it was important to position biofortification as a government-led intervention. Hence, strategically, the NPASC was chaired by the Director of ZARI, a government institution in the Ministry of Agriculture and Livestock. The ZARI maize team also led the in-country evaluation of provitamin A biofortified maize varieties. The Director of Public Health and Environment in the Ministry of Health (MoH) was a member of the NPASC, whose influence was crucial in securing MoH endorsements of the crop and its products at marketing stage.

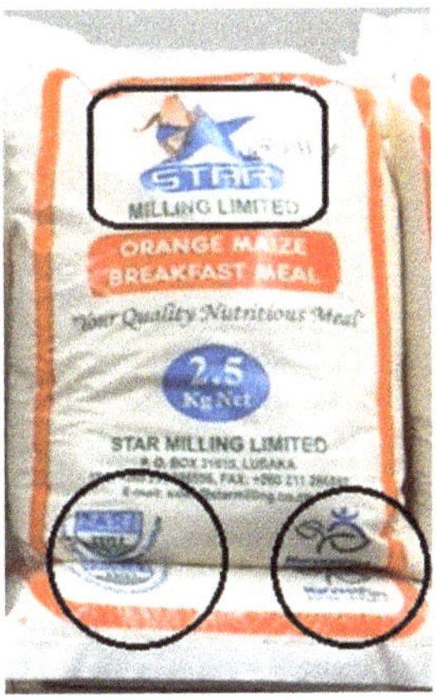

Figure 4 Provitamin A biofortified maize flour in-store package depicting co-branding: ZARI, HarvestPlus and Star Milling Branding.

To further reinforce ownership of provitamin A biofortified maize by the government, co-branding was used as a tool to create an alliance between HarvestPlus, the government and the private sector, a typical private–public partnership (PPP) (Fig. 4). Co-branding allowed farmers and consumers to associate the new maize with the government and Zambian companies, resulting in acceptance of the product as a proudly Zambian product, since the logos on the in-store packages immediately created a visible identity of the product in terms of ownership.

Once full production of seed and products was reached in 2015, the government Minister of Agriculture undertook an official flagging-off of the crop, again signifying full ownership of the project and sending a clear message of the support and ownership by government.

iii. Evidence-based delivery

Consumer acceptance of biofortified maize hinged on the value proposition of the nutritional superiority of the maize while maintaining acceptable yields. Ahead of the release of biofortified maize, the first acceptance study was instituted to understand consumer acceptance by eliciting willingness to pay and sensory evaluation data. The results revealed that (a) orange maize was not confused with yellow maize and has the potential to compete with white maize even in the absence of a nutrition campaign, (b) there is a premium for orange maize with nutrition information and (c) different modes of nutritional message dissemination have the same impact on consumer acceptance (Meenakshi et al. 2012). Another study was conducted around the farmer field days, which introduced farming communities to orange maize varieties planted on experimental farmers' field and allowed farmers an opportunity to taste various foods made with orange maize varieties. In this study, farmers were asked to evaluate the various production and consumption attributes of orange maize varieties and compare them to the white varieties in their own fields. After the farmers had a chance to attend the field days, observe the production characteristics of orange maize varieties in the trial fields and also consume food prepared with orange maize, they indicated a high valuation (measured in terms of willingness to pay) of the seed of the orange varieties, with an average premium of 40% compared to the conventional white varieties. The results of this study revealed a strong appreciation for both the production and the consumption qualities of orange maize varieties, compared with the conventional white varieties. Farmers appreciated the yield, cob size and cob-filling characteristics of the new orange varieties, as well as the taste and aroma of the orange maize preparations compared to their white counterparts (Chibwe et al. 2013). Other unpublished[8] studies that followed after the official release of orange maize in Zambia indicated that based on the consumption characteristics of both boiled fresh maize and *nshima*, orange maize is superior to white maize, particularly in taste and aroma.

In addition to the above three major drivers of early successes of maize biofortification in Zambia, several other initiatives added impetus:

- Country-wide promotions of biofortified maize was conducted by the government extension service using crop demonstration programmes, farm schools and agricultural shows. The crop demonstration plots were effectual in convincing farmers that the new crop was as good as their existing white maize and was superior in taste.

8. Some unpublished studies include the evaluation of acceptance of orange maize among school children undertaken by HarvestPlus Zambia Country Program in collaboration with World Food Program Zambia Office, which supports the Ministry of Education in the School feeding programme. The study tested the potential for substituting white with orange maize in school lunches. The findings indicated that over 80% of the children preferred orange maize over white maize for its superior taste and aroma. The only negative sentiments were those linked to its unfamiliar colour.

- The National Food and Nutrition Commission, an arm of the Ministry of Health, included biofortification in its National Food and Nutrition Strategic Plan, as an official food-based intervention to micronutrient malnutrition, earning biofortification government recognition and subsequently financial support.
- Advocacy and promotional activities led by HarvestPlus were rigorous and far reaching. They encompassed multiple partners and multiple channels, absorbing 35% of the in-country operational budget. Public media was used to send messages on the existence and benefits of the newly developed provitamin A orange maize. This included radio, television, newspapers and use of popular music. The goal was to initiate mindset change among a historically white maize-consuming population to begin to accept consumption of orange-coloured maize as a better alternative to white maize due to its superior taste and nutrition benefits.
- Enlisting of opinion leaders as champions for provitamin A orange maize. Opinion leaders included traditional leaders such as tribal chiefs and political leaders. For instance, the tribal leader from the eastern province of Zambia became the first adopter and largest grower of orange maize in Zambia. His influence among his subjects resulted in the eastern province taking the lead in biofortified maize production in Zambia during the first three years. Senior Chief Mzamane of the Nsenga people in the eastern province even attended the second global conference in biofortification held in Kigali in March 2013 as a panel discussion member.
- Undertaking of widespread food-tasting fairs where orange maize *nshima* was served to prove to consumers that it has superior taste. For example, HarvestPlus offered a free orange maize lunch to the parliament cafeteria to expose parliamentarians to the new product. Food-tasting events were effectual in impressing consumers that orange maize tasted as good, if not better, than the much preferred white maize.

8 Future trends

Biofortification can be considered a major field of research that is fast expanding at the nexus of plant breeding, human health and nutrition security. Several scientists are actively working to extend the frontiers of knowledge on basic genetics, molecular and cellular mechanisms that predict accumulation of micronutrients in crops and the efficacy of plant micronutrients. Advancing knowledge to increase the density of micronutrients in seeds without compromising yield and sensory qualities, as well as avoiding the presence of anti-nutrients which limit bioavailability of essential nutrients, will remain a priority. Specifically, HarvestPlus observes that priority areas of further research in maize biofortification include the following:

- Further description of genetic variability of provitamin A in maize germplasm with a view to developing higher carotenoid content varieties, with better productivity, resilience and eco-efficiency in the face of emerging climate change challenges.
- Integration of multiple micronutrients in single crop germplasm, such as a combination of provitamin A and zinc in maize.
- Genome mapping and subsequent use of molecular markers-assisted selection[9] to expedite the time taken in developing new crop varieties.

9. For details on the application of Market Assisted Selection (MAS), refer to Guimaraes et al. 2007.

- Innovative crop promotions that integrate the use of modernistic multimedia interfaces such as popular contemporary music, road shows, concerts, street theatre, cinema and cell phone technologies to hasten adoption.
- Enhancement of knowledge on how to counteract anti-nutrients while increasing nutrition promoters.

9 Where to look for further information

Role of agriculture in improving nutrition:

Emily, T. and Tanumihardjo, S. (2010). *Maize: A Paramount Staple Crop in the Context of Global Nutrition*. Comprehensive Review in Food Science.

Cakmak, I. (2002). Plant nutrition research: priorities to meet human needs for food in sustainable ways. *Plant Soil*, 247, 3–24.

Cakmak, I., Graham, R. and Welch, R. M. (2002). Agricultural and molecular genetic approaches to improving nutrition and preventing micronutrient malnutrition globally. In I. Cakmak and R. M. Welch (eds), *Encyclopedia of Life Support Systems*. UNESCO-EOLSS Publishers Co. Ltd., UK, pp. 1075–99.

Grusak, M. A. and DellaPenna, D. (1999). Improving the nutrient composition of plants to enhance human nutrition and health. *Annu. Rev. Plant Physiol. Plant Mol. Biol.* 50, 133–61.

HarvestPlus biofortification programme including Zambia case studies:

www.harvestplus.org

Saltzman, A., Birol, E., Bouis, E. H., Boy, E., DeMoura, F., Islam, Y. and Pfeiffer, W., (2012). *Biofortification: Progress Towards a More Nourishing Future*. Global Food Security, Elsevier.

HarvestPlus (2013). *Zambia Product Development and Delivery Plan*. HarvestPlus Mimeo, Washington, DC.

Biofortification:

Bouis, H. E. (1996). Enrichment of food staples through plant breeding: a new strategy for fighting micronutrient malnutrition. *Nutr. Rev.* 54, 131–7.

Frossard, E., Bucher, M., Machler, F., Mozafar, A. and Hurrell, R. (2000). Potential for increasing the content and bioavailability of Fe, Zn and Ca in plants for human nutrition. *J. Sci. Food Agric.* 80, 861–79.

Welch, R. M. and Graham, R. (2004). Breeding for micronutrients in staple food crops from a human nutrition perspective. *J. Exp. Bot.* 55, 353–64.

10 References

Agarwal, S. and Rao, A. V. (2000). Carotenoids and chronic diseases. *Drug Metab. Drug Interact.* 17, 189–210.

Asare-Marfo, D., Birol, E., Gonzalez, C., Moursi, M., Perez, S., Schwarz, J. and Zeller, M. (2013). *Prioritizing Countries for Biofortification Interventions Using Country-Level Data*. HarvestPlus Working Paper 11. International Food Policy Research Institute, Washington, DC.

Bouis, H. E. (2003). Micronutrient fortification of plants through plant breeding: can it improve nutrition in man at low cost? *Proc. Nutr. Soc.* 62, 403–11.
Bouis, H. E. and Haddad, L. (1990). *Agricultural Commercialization, Nutrition, and the Rural Poor: A Study of Philippine Farm Households*. Lynne Rienner Publishers, Boulder, Colorado.
Bouis, H. E., Hotz, C., McClafferty, B., Meenakshi, V. J. and Pfeiffer, W. (2009). *Biofortification: A New Tool to Reduce Micronutrient Malnutrition*. Paper prepared for presentation at the 19th International Congress of Nutrition held in Bangkok, Thailand.
Cakmak, I. (November 2007). *Enrichment of Cereals Grains with Zinc: Agronomic or Genetic Biofortification?* Article in *Plant and Soil*. Springer Science + Business Media, Berlin.
Central Statistical Office (CSO) (2015). *Zambia Demographic and Health Survey 2013–14*. Lusaka, Zambia.
Chapoto, A., Govereh, J., Haggblade, S. and Jayne, T. S. (2010). *Staple Food Prices in Zambia*. Prepared for the COMESA Policy Seminar on 'Variation in Staple Food Prices: causes, Consequence, and Policy Options', pp. 25–6. Maputo, Mozambique.
Chibwe, E., De Groote, H., Gitonga, Z. and Birol, E. *A Participatory Evaluation of Vitamin A Biofortified Maize in Zambia*, Paper presented at the 4th African Association of Agricultural Economists Conference, 22–25 September 2013, Hammamet, Tunisia.
Chomba, E., Westcoast, C., Westcoast, J., Mpabalwani, E., Krebs, N., Patinkin, Z., Palacious, N. and Hambidge, K. (2015). Zinc absorption from biofortified maize meets the requirements of young rural Zambian children. *J. Nutr.* 114, 204933v1.
Creed, P. G. (1998). *A Study of the Sensory Characteristics of Food Produced by the Sous Vide System: The Measure of Pleasure*. A Thesis submitted in Partial fulfilment of the Requirements of Bournemouth University, for the degree of Doctor of Philosophy. Bournemouth University, Poole.
De Groote, H., Gitonga, Z., Smale, M., Asare-Marfo, D., Kasuta, E., Birol, E. and Sonder, K. (2012). *Adoption and Diversification of Modern Maize Varieties in Zambia: A Baseline Report Prepared for HarvestPlus*. Mimeo, HarvestPlus, Washington DC.
Emily, T. and Tanumihardjo, S. (2010). Maize: a paramount staple crop in the context of global nutrition. *Comprehensive Review in Food Science*. Institute of Food Technologists Vol 9, 417–36. doi:10.1111/j.154-4337.2010.00117.
FAO (1992). *Maize in Human Nutrition. Food and Agricultural Organisation of the United Nations*. Rural Infrastructure and Agribusiness Division, Rome.
Galinat, W. (1977). The origin of corn. In G. F. Sprague (ed.), *Corn and Corn Improvement*, Am. Soc. Agron, pp. 1–47. Madison, WI, USA.
Gannon, B., Kaliwile, C., Arscott, S., Schmaelzle, S., Chileshe, J., Kalungwana, N., Mosonda, M., Pixley, K., Masi, C. and Tanumihardjo, S. (2014). Biofortified orange maize is as efficacious as a vitamin A supplement in Zambian children even in the presence of high liver reserves of vitamin A: a community-based, randomized placebo-controlled trial1–6. *Am. J. Clin. Nutr.* doi: 10.3945/ajcn.114.087379.
Global Hunger Index (GHI) (2014). *Addressing the Challenge of Hidden Hunger*. Chapter 3. International Food Policy Research Institute, Washington, DC.
Global Knowledge Centre on Crop Biotechnology (GKC) (2007). *Biotechnology and Biofortification*. Pocket K No. 27. International Services for the Acquisition of Agri-biotechnology Applications.
Graham, R. D. and Welch, R. M. (1996). *Breeding for Staple-Food Crops with High Micronutrient Density: Agricultural Strategies for Micronutrients*. Working Paper No 3. International Food Policy Research Institute, Washington, DC.
Graham, R. D., Welch, R. M. and Bouis, H. (2001). Addressing micronutrient malnutrition through the nutritional quality of staple foods: principles, perspectives, and knowledge gaps. *Adv Agron.* 70, 77–142.
Graham, R. D., Welch, R. M., Saunders, D. A., Ortiz-Monasterio, I., Bouis, E. H., Bnierbale, M., De Haaan, S., Burgos, G., Thiele, G., Liria, R., Meisner, C. A., Beebe, E. S., Michael, J. P., Kadian, M., Peter, R. H., Gupta, K. R. and Twomlow, S. (2001). Nutritious subsistence food systems. *Adv. Agron.* 92, Elsevier.

Grusak, A. M. and Cakmak, I. (2004). Methods to Improve the Crop-delivery of Minerals to Humans and Livestock. Plant Nutrition Genomics. SM/TechBooks. BY037-Broadley-v1.cls. Chapter 12.

Guimaraes, E., Ryuane, J., Schef, B., Sonnino, A. and Dargie, J. (eds) (2007). *Marker Assisted Selection: Current State and Future Perspectives in Crops, Livestock, Forestry and Fish*. Food and Agriculture Organisation of the United Nations (FAO), Italy.

Harjes, C., Rocheford, T. R., Bai, L., Brunel, P. T., Kandianis, B. C., Sowinski, G. S., Stapleton, E. A., Vallabhaneni, R., Williams, M., Wurtzel, T. E., Yan, J. and Buckler, S. E. (2008). Natural genetic variation in Lycopene epsilon cyclase tapped for maize biofortification. *Science* 319, 330–3.

Harris, J. and Drimie, S. (2012). Toward an Integrated Approach for Addressing Malnutrition in Zambia. A literature Review and Institutional Analysis. IFPRI Discussion Paper 01200. Poverty, Health and Nutrition Division.

HarvestPlus (2013). *Zambia Product Development and Delivery Plan*. Mimeo, HarvestPlus, Washington, DC.

HarvestPlus (2014). *Biofortification Progress Briefs*. Mimeo, HarvestPlus, Washington, DC. Available online at www.HarvestPlus.org.

Hotz, C. and Brown, K. (2004). Assessment of the risk of zinc deficiency in populations and options for its control. International Zinc Nutrition Consultative Group Technical Document 1.

International Zinc Nutrition Consultative Group (IZiNCG) (2004). Assessment of the risk of zinc deficiency in populations and options for its control. In C. Hotz and K. H. Brown (eds), *Food Nutr. Bull.* 25, S91–S204.

Krinsky, N. I., Landrum, J. T. and Bone, R. A. (2003). Biologic mechanism of the protective role of lutein and zeaxanthin in the eye. *Annu. Rev. Nutr.* 23, 171–201.

Laszlo, A. and Kripper, S. (1998). Systems theories: their origins, foundation and development. In J. S. Jordan (ed.), *Systems Theory and A Priori Aspects of Perception,* Chapter 3, pp. 47–74. Elsevier Science, Amsterdam.

Li, S., Tayie, F., Young, M., Rocheford, T. and White, W. (2007). Retention of Provitamin A carotenoids in high-carotene maize (Zea mays) during traditional African household processing. *J. Agric. Food Chem.* 55(26), 10744–50.

Li, S., Nugroho, A., Rocheford, T. and White, W. (2010). Vitamin A equivalence of the b-carotene in b-carotene–biofortified maize porridge consumed by women. *Am. J. Clin. Nutr.* dol: 10.3945/ajcn.2010.29802.

Lividini, K and Fiedler, J. (2015). Assessing the promise of biofortification: a case study of high Provitamin A maize in Zambia. *Food Policy*. 54, 65–77. Elsevier Publication.

Mason, N. and Jayne, T. S. (2009). *Staple Food Consumption Patterns in Urban Zambia: Results from the 2007/2008 Urban Consumption Survey*. Food Security Research Project Working Paper 42. Michigan State University, Lusaka.

Meenakshi, J. V., Banerji, A., Manyong, V., Tomlins, K., Hamukwala, P., Zulu, R. and Mungoma, C. (2010). *Consumer Acceptance of Provitamin A Orange Maize in Rural Zambia*. HarvestPlus Working Paper 4.

Meenakshi, J. V., Banerji, A., Manyong, V., Tomlins, K., Mittal, N. and Hamukwala, P. (2012). Using a discrete choice experiment to elicit the demand for a nutritious food: Willingness-to-pay for orange Maize in rural Zambia. *J. Health Econ.* 31, 62–71.

Menkir, A., Liu, W., White, W., Maziya-Dixon, B. and Rocheford, T. (2008). Carotenoid diversity in tropical-adapted yellow Maize inbred lines. *Food Chem.* 109, 521–9, Elsevier.

Morgan, J. (2013). CSA News. Biofortification, Lasting Solution to Micronutrient Malnutrition and World Hunger. *Science*. https://crops.org/files/publications/csa/biofortification.pdf.

Mugode, L., Ha, B., Kaunda, A., Sikombe, T., Phiri, S., Mutale, R., Davis, C., Tanumihardjo, S. and De Moura, F. (2014). Carotenoid retention of Biofortified Provitamin A Maize (*Zea mays* L.) after Zambian traditional methods of milling, cooking and storage. *J. Agric. Food Chem.* 62, 6317–25.

Muzhingi, T., Gadaga, T., Siwela, A., Grusak, M., Russell, R. and Tang, G. (2011). Yellow maize with high β-carotene is an effective source of Vitamin A in healthy Zimbabwean men. *Am. J. Clin. Nutr.* doi: 10.3945/ajcn.11006486.

Nuss, E. T. and Tanumihardjo, A. S. (2010). Maize: a paramount staple crop in the context of global nutrition. *Compr. Rev. Food Sci. Food Saf.* 9, 417–36.

Owens, B. F., Lipka, A. E., Magallanes-Lundback, M., Tiede, T., Diepenbrock, C. H., Kandianis, C. B., Kim, E., Cepela, J., Mateos-Hernandez, M., Buell, C. R., Buckler, E. S., DellaPenna, D., Gore, M. A. and Rocheford, T. (2014). A foundation for provitamin A biofortification of maize: genome-wide association and genomic prediction models of carotenoid levels. *Genet. Soc. Am.* 198(4), 1699–716.

Pfeiffer, H. and McClafferty, B. (2007). Biofortification: breeding micronutrient-dense crops. In S. Kang and P. M. Priyadarshan (eds), *Breeding Major Food Staples*. Blackwell Publishing, Hoboken.

Philip, J. and Martin, R. (2005). Biofortifying crops with essential mineral elements. *Trends Plant Sci.* 10(12), 586–93. Elsevier.

Pixley, K., Palacios, N., Babu, R., Mutale, R., Surles, R. and Simpungwe, E. (2002). Biofortification of maize with Provitamin A carotenoids. In S. A. Tanumihardjo (ed.), *Carotenoids and Human Health*. Humana Press, New York (2013), 271–92. ISBN 978-1-62703-202-5. doi: 10.1007/978-1-62703-203-2_17.

Qaim, M., Alexander, J. S. and Meenakshi, J. V. (2007). Economics of biofortification. *Agric. Econ.* 37, 119–33.

Rice, A., West, K. and Robert, E. B. (2004). Vitamin A deficiency. In M. Ezzati, A. D. Lopez, A. Rodgers and C. J. Murray (eds), *Quantification of Health Risks*. WHO, Geneva.

Saltzman, A., Birol, E., Bouis, H. E., Boy, E., De Moura, F., Islam, Y. and Wolfgang, H. P. (2013). Biofortification: Progress toward a more nourishing future. *Glob. Food Secur.* 2(1), 9–17.

Simpungwe, E. (2006). *Water, Stakeholders and Common Ground: Challenges for Multi-Stakeholder Platforms in Water Resource Management in South Africa*. PhD Dissertation. Wageningen University, Netherlands.

Smale, M., Moursi, M., Birol, E. and De Groote, H. (2013). *Hybrid Seed Use and Diversity of Diets Among Women in Smallholder Maize-Growing Households in Zambia*. HarvestPlus Working Paper No. 12. www.HarvestPlus.org.

Smale, M., Simpungwe, E., Birol, E., Kassie, G., De Groote, H. and Mutale, R. (2014). The changing structure of the Maize seed industry in Zambia: prospects for orange Maize. *Agribusiness*, 1–14. Published online in Wiley Online Library: wileyonlinelibrary.com/journal/agr.

Sommer, A. (1995). Vitamin A deficiency and its consequences: a field guide to detection and control. 3rd ed. WHO. Geneva. http://www.who.int/iris/10665/40535.

Sommer, A., Tarwotjo, I., Djunaedi, E., West, P. K., Loeden, A., Tilden, R. and Mele, L. (1986). Impact of Vitamin A supplementation on childhood mortality: a Randomized controlled community trial. *Lancet* 327(8491), 1169–73.

Tenywa, M. M., Rao, K. P. C., Tukahirwa, J. B., Buruchara, R., Adekunle, A. A., Mugabe, J., Wanjiku, C., Mutabazi, S., Fungo, B., Kashaija, N. I. M., Pali, P., Mapatano, S., Ngaboyisonga, C., Farrow, A., Njuki, J. and Abenakyo, A. (2011). Agricultural innovation platform as a tool for development oriented research: lessons and challenges in the formation and operationalization. *Learn. Publics J. Agric. Environ. Stud.* 2(1), 117–46.

Welch, R. M., Combs Jr., G. F. and Duxbury, J. M. (1997). Towards a greener revolution. Issues *Sci. Technol.* 14, 50–8.

West, P. K. (2002). Extent of vitamin A deficiency among preschool children and women of reproductive. *Age. J. Nutr.* 132, 2857S–66S.

Chapter 9

Assessing and improving the nutritional quality of maize

Elena Lima-Cabello and Paula Robles-Bolivar, Spanish National Research Council (CSIC), Spain; and Jose C. Jimenez-Lopez, Spanish National Research Council (CSIC), Spain and The University of Western Australia, Australia

1 Introduction: maize kernel composition

Maize (*Zea mays* L.) belongs to the grass family *Poaceae*, which is recognized as the most strategically and economically important plant family in the production of crops for human consumption. Maize itself is a fundamental crop in many countries, where it constitutes the main source of nutrition for both humans and animals (FAO, 2013).

The maize seed is called the kernel, whose total dry weight is made up of the endosperm (80–85%), the pericarp (5–6%) and the germ (9–10%), comprising 73% starch, 9% proteins, 4% oil and 14% other constituents such as fibre. The largest structure in the kernel is the endosperm, which is made up of 90% starch and 10% protein. The endosperm therefore contains the majority of the kernel's protein (Prasanna et al., 2001).

The starch from maize has multiple uses in human and animal diets, as well as in industry and manufacturing. One of the most important uses of maize is in the food industry, where it serves as a raw material for the extraction of oil and the production of starch; food sweeteners for soft drinks; thickening agents for the preparation of many foods such as soups, sauces and custard powder; and bonding agents for ice cream cones. Starch can

http://dx.doi.org/10.19103/AS.2016.0001.15

also be used in the production of edible pharmaceutical products, and it is increasingly used as a renewable source for other products such as high fructose corn syrup, polymer-based fibres, ethanol and other biofuels (James et al., 2009).

Maize is not just important for human nutrition because of its protein content. It is also the major component of many people's diets. With a growing world population, increasing maize grain yield is necessary to meet market demand.

An important goal for current and future breeding programmes is the ability to manipulate the quality and quantity of starch in maize kernels. However, knowledge concerning the functional mechanisms involved in starch biosynthesis and accumulation is still scarce. Improving the quality and quantity of proteins in maize is also a central task. For this, it is necessary to know which protein macromolecules are most beneficial for human consumption. This means discovering the properties indicative of maize protein quality, which are related to (i) an adequate food energy/protein relation, (ii) the food protein content, (iii) whether essential amino acids are supplied in patterns similar to those which exist in human tissues and finally (iv) the digestibility/bioavailability of these proteins (Henley et al., 2010). The WHO, in partnership with the FAO and UNU, recommends a daily protein intake of 45 mg/g body weight/day and 30 mg/kg body weight/day of the essential amino acid lysine (WHO/FAO/UNU Expert Consultation, 2007). The supply of essential amino acids, particularly lysine, used to be a significant problem in legume and cereal-based diets (Young et al., 1988). For example, maize contains 28.6 mg/g protein, and so a diet based on maize is likely to provide lower than the recommended levels of protein (Henley et al., 2010).

Maize endosperm starch is chemically composed of two homopolymers of α-D-glucose, amylose and amylopectin. Although starch metabolism is complex, four main classes of enzymes have been identified so far as playing critical roles in starch biosynthesis. These are adenosine diphosphate glucose pyrophosphorylases (AGPases), starch synthases (SSs), starch-branching enzymes (SBEs) and debranching enzymes (DBEs). Endosperm is the main storage organ of the maize seed, and it has been found to have the following composition: (i) non-zein fraction [albumins (3%), globulins (3%), glutelins (34%)] with balanced amino acids content and (ii) zeins/prolamins (60%). The zein/prolamin fraction is rich in leucine (18.7%), phenylalanine (5.2%), isoleucine (3.8%), valine (3.6%) and tyrosine (3.5%) but poor in essential amino acids such as threonine (3%), histidine and cysteine (1%), methionine (0.9%) and lysine (0.1%), and has no tryptophan (Vasal, 2000).

Zein proteins can be grouped based on their solubility; their ability to interact forming disulphide unions; and their nucleotide sequence of α- (22 and 19 kDa), β- (15 kDa), γ- (50, 27 and 16 kDa) and δ-zeins (10 kDa and 18 kDa) (Coleman and Larkins, 1999). Due to the large size of the zein fraction, the contribution of lysine and tryptophan from the non-zein fraction is diluted. This means that the overall quality of maize protein is low. Thus, a diet containing a high proportion of maize, and therefore a high zein/non-zein ratio may cause deficiencies in lysine and other essential amino acids such as tryptophan, threonine and methionine, which humans are unable to synthetize (Vasal, 2000).

Zein proteins are packaged inside membranous, enclosed bodies called protein bodies (PBs) that formed during endosperm development. They consist of initially small PBs that are filled with mainly γ- and β-zeins deposits. When the PBs grow, an increasing amount of α- and δ-zeins accumulate in their core. At a mature stage of seed development, γ- and β-zeins linked by disulphide bonds create a superficial mesh that encapsulates α- and δ-zein locules and promotes PB filling, so that they are finally filled to capacity (Woo et al., 2001). The vitreous nature of the mature kernel will directly depend on the PB's composition and the spatial organization of the zeins α, β, γ and δ (Mertz et al., 1964; Coleman et al., 1995, 1997; Segal et al., 2003; Kim et al., 2004, 2006; Holding et al., 2007).

2 Overview of methods for improving the protein and starch content of maize kernels

The evolution of plant breeding has contributed to the rapid progress of our civilization, and it continues to be important for improving the quality of our crops and the sustainability of our agriculture. Three factors have underpinned the rapid improvements in crop production: (i) improved varieties through the development and adoption of new breeding technologies, (ii) expansion of the area under irrigation and (iii) widespread use of fertilizers. Of these, only the first can be regarded as sustainable. The most promising scientific solution to ensuring food security is therefore the genetic improvement of crops.

Traditional plant breeding is based on phenotypic selection. Although very effective, this breeding method suffers from several limitations. Selection programmes have frequently focused on genes that have major effects which give a clear phenotype, such as race-specific disease resistances. For additive gene effects, large populations can be used to give a reasonable probability of combining desirable alleles at multiple loci into a single line. For complex traits, for which we possess little or no clear genetic understanding, breeders treat the germplasm in their programme as a population and endeavour to shift the mean population response in a particular direction. With intensively bred species, such as maize, breeding programmes often evaluate a million lines each year to maintain good rates of genetic gain.

Crop improvement is dependent upon genetic variation. Traditional breeding has focused primarily on improving maize nutritional quality through targeting the protein and starch content. However, this is limited by the initial genetic diversity of the crop.

The first attempts to enhance the nutritional qualities of maize followed two strategies: (i) altering the germ to endosperm ratio, in order to increase germ size and decrease endosperm size, thus improving protein quality; and (ii) pursuing traditional hybridization using populations selected for their natural high lysine content. However, maize genetic variation is relatively narrow, and so this second strategy did not obtain successful results (Bjarnason and Pollmer, 1972).

The rate of genetic gain in breeding programmes can be increased in two ways, either by increasing the amount or significance of the variation available for selection or by accelerating the selection process to produce new varieties more rapidly. Genomics can provide support for both approaches. Interesting inbred lines and collections of varieties can be found in structured crossover programmes, which produce hybrid maize populations with desirable characteristics. In these lines and collections, heterosis (the hybrid vigour phenomenon) shows that the hybrid organisms have the capacity to develop superior features compared to their genetically diverse parental lines. The underlying genetic mechanism which causes heterosis remains unknown, but it is being extensively studied in maize because of the opportunity provided by intensive hybrid breeding programmes (Lee and Tracy, 2009).

Furthermore, with the advent of modern genetic, genomic and molecular biology, breeders' options have been substantially broadened, with the possibility of the purposeful transfer of genetic information from related or diverse organisms into different plant species in order to attempt specific performance improvements.

Omics platforms are currently improving our ability to discover genes and pathways that control specific traits and provide screening and analysis platforms to support selection strategies. The most obvious application for genomic technologies is in the area of gene discovery, as part of the forward genetics or positional cloning approach. Positional cloning is greatly facilitated by access to a full genome sequence, but it is still feasible for species for which we have only limited sequence data. However, several other factors limit gene discovery by positional cloning. Perhaps the most important is the reliability of the phenotyping assay that is used to detect the gene or quantitative trait locus (QTL). QTL mapping in maize breeding is an interesting method for identifying loci for quantitative traits of interest (gene discovery) without prior genetic knowledge. A variety of QTLs for the starch content in maize kernels have been identified (Wang et al., 2015). As part of these studies, 33 single-nucleotide polymorphisms (SNPs) associated with starch content in maize kernels were identified using the single regression method, and 127 SNPs were identified using the subsampling method in a nested association mapping population (Cook et al., 2012).

For traits of low heritability or high genotype x environmental interaction, it can be difficult to unequivocally assign a phenotype. The situation can be even more complex for traits determined by multiple genes of varying effect. However, recently developed genomics and genotyping technologies are fast, low cost and high throughput, and this allows the screening of large populations and the use of SNP markers to increase marker density. This can help to overcome these problems. This workflow has been widely applied to construct genetic linkage maps and in the QTL mapping of wheat, rice, sorghum and maize (Zou et al., 2012; Chen et al., 2014; Li et al., 2014; Maccaferri et al., 2015).

Another alternative mechanism to generate SNPs, while at the same time increasing the variability of cultivars, is the RNA editing process. Through this process, transcripts in the organelles of flowering plants can be altered by the conversion of cytidines to uridines. This alteration occurs in RNAs co- or post-transcriptionally. The absence of editing in some mutants leads to the production of improper proteins that can result in seedling lethality (Bahve et al., 1990).

During the past decade, the possibility of using meganucleases (Dudley et al. 2004), zinc finger nucleases (ZFNs), transcription activator-like effector-based nucleases (TALEN), and recently, the CRISPR-Cas system in genome editing has provided new strategies for genetic manipulation in plants, assisting in the engineering of desired plant traits by modifying endogenous genes. For example, site-specific gene addition in major crop species can be used for 'trait stacking', whereby several desired traits are physically linked to ensure their co-segregation during the breeding process (James et al., 1995). Progress in this area has recently been reported in *Z. mays*, where disruption of the target locus was achieved by ZFN-induced double-strand breaks (DSBs) and the resulting non-homologous end-joining (NHEJ) or homologous recombination, leading to targeted mutations ('edits').

DSBs are central to improving the frequency (up to 1000-fold or more) of gene editing and site-specific gene insertion via homology-directed repair (HDR) (Puchta et al., 1993; Choulika et al., 1995; Smih et al., 1995; Puchta et al., 1996). The technologies which use them have three major applications in plant biotechnology: gene mutagenesis resulting in gene knockouts, gene editing allowing modification of a gene product or metabolic pathway in a desired way and site-specific trait insertion to allow gene stacking in plant breeding programmes. In this final use, the integration of trait genes into a specific location is particularly attractive, and the process can be completed through HDR.

The recently discovered Cas96 guide RNA system has revolutionized the field (Doudna and Charpentier, 2014; Hsu et al., 2014). This technology was developed based on a bacterial adaptive defence system which includes clustered, regularly interspaced, short palindromic repeats (CRISPR) and CRISPR-associated (Cas) endonuclease to protect the host from plasmid and viral infection (Barrangou et al., 2007).

The Cas9-gRNA system has been shown to work in many different organisms, including plants. To date, it has been used to generate targeted gene modifications in multiple plant species, including *Z. mays* (Liang et al., 2014). The majority of these genome modifications proceed via NHEJ-generated mutations and gene knockouts, which then allow gene function analysis. This has then led to ALS2 gene editing, resulting in plants resistant to chlorsulfuron, and site-specific gene integration in maize.

These combined approaches may further improve the precision and accuracy of genome modification in plants. This will mean that our abilities in targeted mutagenesis and the editing of specific genes or gene families to yield new allelic variants will no longer place limits on our ability to discover the functions of genes and establish their relationships to complex biological pathways. Adoption of robust genome modification methods to introduce genetic variation, aided by inexpensive sequencing and new plant transformation methods, will accelerate the development of new breeding techniques to address the challenges of modern agriculture.

3 Improving the protein content of maize using mutant lines

New technologies allowing the use of molecular markers provide opportunities for genotypic selection when the genetic position of key loci is known. As marker densities have increased, it has become feasible to use markers to monitor the genetic background of lines. The selection of mutants from maize germplasm is making an important contribution to improving the protein content of maize. One interesting strategy is the use of genetic mutants with high lysine content as a base from which the nutritional quality of maize can be improved. These mutants have not only been used to alter the transcription and translation of zein genes but also to affect PB formation, amino acid biosynthesis, plastid development and cytoskeletal functions (Holding, 2014). However, with the exception of the *opaque2* and *floury2* mutants, most of these mutants have never been applied in maize breeding, because of known endosperm modification disadvantages (Pajic et al., 2000):

Opaque grain maize emerged from a spontaneous mutation in Connecticut in the 1920s, called the opaque2 mutant (*o2*) (Singleton, 1939). Studies during the mid-1960s in Purdue University revealed that in homozygosis, the recessive *o2* mutant allele raises the lysine content of the grain endosperm by more than 69% compared to normal maize (Mertz et al., 1964). Other amino acids are also increased, such as tryptophan (by two to three times), histidine, arginine, aspartic acid and glycine (Bressani, 1991). The affected gene *O2* encodes a leucine-zipper class transcription factor that regulates the transcription of 22-kDa α-zein family members, with the result that the total α-zein fraction is indirectly reduced (Schmidt et al., 1990). The leucine-zipper class transcription factor also regulates transcription of lysine ketoglutarate reductase, promoting the accumulation of lysine-free endosperm amount (Lohmer et al., 1991) and pyruvate orthophosphate dikinase

(Maddaloni et al., 1996). Recessive mutations that produce opaque phenotypes affect the regulatory network, intervening in storage protein metabolism. In contrast, semi-dominant and dominant non-vitreous mutation phenotypes implicate altered genes, which directly encode storage proteins (Gibbon and Larkins, 2005).

During the early 1970s, attention was focused on the production of *o2* gene carrier inbred lines and maize populations. These were then made into crossovers, producing important commercial o2 hybrid lines. One of the best hybrids developed was -ZP 704*o2*, which could compete with the common commercial maize varieties in terms of yield. However, the initial interest was disrupted by the appearance of unwanted agronomical characteristics in endosperm o2 phenotypes (Pajic et al., 2000).

Later on, the development of the Quality Protein Maize (QPM) variety (Geevers and Lake, 1992) made it possible to avoid the unfortunate consequences of the o2 mutant and other early mutants, including their undesirable commercial and agronomic features such as reduced grain yield, chalky and dull kernel appearance, vulnerability to ear rots and susceptibility to stored grain pests due to soft endosperm.

The development of QPM involved manipulating three distinct genetic systems (Krivanek et al., 2007): (i) the simple recessive allele of the opaque2 gene, (ii) the modifiers/enhancers of the o2-containing endosperm to confer higher lysine and tryptophan levels and (iii) the genes that modify the o2-induced soft endosperm to hard endosperm.

The breeding programmes that developed QPM seeds (Mo2) were able to avoid the opaque2 starchy endosperm, instead creating a hard kernel with acceptable agronomic and commercial features, while at the same time preserving the high protein quality content of the original mutant. To do this, breeders had to select for distinct, genetically separable factors simultaneously, such as the recessive character, the variable penetrance and pleiotropic effects of o2 genes in parallel with endosperm hardness modifier loci or opaque 2 modifiers (*Mo2*s), amino acid composition genes and acceptable yield characteristics (Holding et al., 2008).

The underlying mechanisms of *Mo2*s that reverse the opaque endosperm phenotype remain unknown. However, 27-kDa γ-zein might be partially responsible for this reversal of phenotype, as QPM contains two to three times more of this than wild varieties or *o2*. There is also a direct correlation between the level of prolamins present and the modified vitreousness of the endosperm (Lopes and Larkins, 1995). It is thought that 27-kDa γ-zeins have an important role in PB formation initiation, as disulphide bonds between these molecules create a rigid and glassy matrix contributing to a vitreous endosperm texture and appearance (Wu et al., 2010). However, two copies of 27-kDa γ-zein are necessary to form a completed vitreous endosperm (Yuan et al., 2014).

In addition to opaque2, other recessive opaque mutants have been discovered, namely *o1*, *o5*, *o7*, *o9-o11* and *o13-o17*. Although zein content is normal in opaque1 (*o1*), the phenotype of this mutant's endosperm appears to be different because of its misshapen and slightly smaller PBs, as well as its dilated ER appearance. O1 encodes a member of the plant myosin XI family, a protein usually associated with ER and PBs (Wang et al., 2012). In opaque5 (*o5*), the opaque phenotype is produced by an altered gene encoding a monogalactosyl-diacylglycerol synthase (MGD1) that produces reduced galactolipid content and so alters starch production (Myers et al., 2011). Interestingly, the opaque7 (*o7*) mutant differs from wild varieties because of a 12-bp in-frame deletion in the O7 gene, which encodes an acyl-activating enzyme-like protein (AAE3-like) with an important role in membrane biogenesis and stability by contributing to acylation of 19-kDa α-zein post-translational precursors (Miclaus et al., 2011).

Floury mutants (*fl1*, *fl2* and *fl3*) exhibit variable expression for kernel opacity and protein quality. In the floury 1 (*fl1*) mutant, the zein level and shape and size of PBs are normal. However, the distribution of different zeins exhibits modification, with 22-kDa α-zein in the γ-zein typical area (Holding et al., 2007). Floury 2 (*fl2*) is associated with a soft, starchy endosperm phenotype with reduced zein fraction and doubled lysine levels. It is caused by a defective signal peptide of the 24-kDa polypeptide, a precursor of 22-kDa α-zein, that accumulates in the ER membrane producing small and irregular PBs (Coleman et al., 1997).

As an example of a mutant with dominance effects, we can consider the maize mutant *Mucronate* (*Mc*), which exhibits reduced deposition of zein during endosperm maturation, producing an opaque phenotype. Mc is made by a 38-bp deletion in 16-kDa γ-zein and shows synergistic action with opaque-2 in zein (Kim et al., 2006).

In contrast, the Defective endosperm (*De-B30*) mutation produces an aberrant 19-kDa α-zein signal peptide, triggering opaque and starchy grain and abnormal PBs through aggregation in ER membrane synthesis (Kim et al., 2004). In the *mto140* mutant, tyrosine and phenylalanine accumulate in the endosperm because of anomalous arogenate dehydrogenase (AroDH) activity, interrupting the amino acid biosynthesis pathways. Increased lysine levels and low levels of zein synthesis result, as in other opaque phenotypes (Holding et al., 2010).

Both spontaneous and induced mutants suffer from feedback complications and, as a result, have experienced limited success. Therefore, alternative strategies based on genetic engineering, biotechnological techniques and new methodologies might be required to ensure more specific alterations to gene expression, and thus to avoid undesirable epistasis, additives or synergistic interactions which affect the features of the final product because of the introduction of RNAi constructs against sequence encoding zeins.

RNAi lines have the advantage that transgenes act in a dominant way, avoiding the major breeding problem of maintaining the homozygous state of a recessive mutant. The silencing of 19-kDa (Huang et al., 2004) and 22-kDa (Segal et al., 2003) α-zeins generates seeds with reduced expression of zein mRNA, and a consequent 15–20% increase of lysine content. In addition, simultaneous 19-kDa and 22-kDa α-zein RNAi lines produced by adding a coupled GFP marker have allowed researchers to select RNAi and QPM hybrid plants. These exhibit a green, fluorescent and vitreous endosperm phenotype (Wu and Messing, 2011). As the RNAi silencing mechanism does not have a knockout effect, it allows the synthesis of the remaining zein protein to produce a vitreous endosperm without the necessity of *Mo2*s (Wu and Messing, 2012).

Another alternative to alter protein levels involves changing the free lysine level so as to modify the plant's natural metabolism. Lysine synthesis is controlled by negative feedback mechanisms. The two main enzymes which act as control steps are aspartate kinase (AK) and dihydrodipicolinate synthase (DHDPS), both of which are inhibited in the presence of lysine. Expression of insensitive AK or DHDPS allows higher lysine accumulation in the maize endosperm (Galili and Amir, 2013). Boarding lysine basal degradation decreases Lys-ketoglutarate reductase/saccharopine dehydrogenase (LRK/SDH) activity through a mediated RNAi suppression mechanism, and therefore provokes an increase in lysine content (Kirma et al., 2012).

Therefore, acylation and stabilization of lysine tRNA-expressing *Arabidopsis thaliana* lysyl tRNA synthetase (AtKRS) in maize endosperm also contributes to improving its content, recoding lysine into zeins (Wu et al., 2007). The level of methionine is also increased, stabilizing the mRNA transcript of the Dzs10 gene. This gene encodes a seed-specific methionine storage protein, and so it deletes the regulatory sequences that are dependent on enhancers (Lai and Messing, 2002).

Another modern approach to increasing the levels of essential amino acids in maize grain involves the introduction of heterologous genes which will contribute higher levels of essential amino acids in a seed-specific manner. Seed-specific expression of Sb40, a pollen-specific protein from *Solanum berthaultii* connected to 19-kDa α-zein promoter, results in significantly higher levels of lysine and total protein in maize seeds (Yu et al., 2004). Analogous results can also be obtained through the introduction of its potato homolog, SBgLR (Lang et al., 2004). Adding both SBgLR and TSRF1, a tomato ethylene response transcription factor, into maize produces high lysine and protein content in the transgenic seeds compared to those of the non-transgenic plants (Wang et al., 2013). Other genes, such as GhLRP from cotton (Yue et al., 2014) or AtMAP18 from *A. thaliana* (Chang et al., 2015), can be used in a similar way to increase lysine content. As a final example, milk proteins constructs made of a α-lactoalbumin sequence coupled to a specific promoter such as Ubi-1 or 7-kDa γ-zein and an ER retention sequence, also increase lysine levels in transgenic plant seeds (Yang et al., 2002; Bicar et al., 2008).

In addition to zein gene promoters, signal-mRNA can also stimulate the accumulation of alternative heterologous proteins. The addition of prolamin mRNA 3′UTR to transgenic constructs will redirect them towards PB-ER (Hamada et al., 2003), as well as coupling the starch-binding domain of cyclodextrin glycosyltransferase from *Bacillus circulans* to guide this glycosyltransferase inside starch granules (Ji et al., 2003). Nevertheless, alterations affecting normal seed development physiology can be generated by the introduction of foreign, modified proteins, that is, 20-kDa γ-zein proteins whose sequences exhibited high level of modifications (substitutions of Proline-Xaa motifs by Proline-Lysine) are expressed and co-localized to native 20-kDa γ-zein in maize PBs (Torrent et al., 1997).

4 Improving the starch content of maize using mutant lines

Different parts of the maize grain have different carbohydrate profiles (Nuss and Tanumihardjo, 2010). Overall, starch is the predominant carbohydrate, constituting 70% of kernel dry weight. The endosperm consists of 90% starch. Fibre is present mainly in the pericarp, where it is integrated with some complex carbohydrates such as hemicellulose (75%), cellulose (25%) and lignin (0.1%) (Burge and Duensing, 1989). Simple sugars (1–5%) such as glucose, sucrose, fructose, maltose and raffinose are stored in the germ (Boyer and Shannon 1987).

Starch is composed of two glucans: amylose and amylopectin. The amylose molecule is linear, formed by chains of glucose units joined by α-(1–4) glycosidic linkages and containing limited α-(1–6) branching points. Amylopectin is produced when α-(1–4) amylose linear chains produce a heavily α-(1–6) branched structure (Tester and Karkalas, 2002). Ordinary maize starch has a 25–30/70–75 amylose to amylopectin ratio, which can be altered by genetic modifications to create a wide range of maize structural types.

In a plant's seed, the starch is present in the endosperm as water-insoluble granules. Inside these granules, amorphous amylose and crystalline amylopectin form alternate lamellas, constituting a matrix. Additionally, the crystallinity of different starches depends on the genetically controlled amylose/amylopectin ratio, amylose chain length and amylopectin branching patterns. Amylopectin-rich starches are waxy and show more crystallinity than ordinary starches, while ordinary starches show higher crystallinity than high amylose starches (Singh et al., 2003). Amylose is usually located at the granule periphery, while amylopectin is found in the core, as amylose is formed at the later stages of starch accumulation (Pan and Jane, 2000).

Maize is classified into three different groups according to its starch structure and composition (Nelson and Pan, 1995): (i) waxy maize (100% amylopectin); (ii) high amylose maize (40–70% amylopectin) and (iii) sugary maize (85% sucrose). Starch granule structures can then be classified into types A, B and C. In type A, polymorphous starch granules are loosely packed, with numerous channels and gaps inside them. A-amylopectin molecules exhibit predominantly short-branched chains. Their branching points are dispersed through both the amorphous and crystalline regions and the molecule's central core contains an additional helix. Both of these characteristics make them easily digestible by amylases and resistant to acid hydrolysis. Type B starch granules have smooth surfaces without pinholes. B-amylopectin molecules have just a few branched chains, which are large and linear. These molecules of amylopectin are responsible for type B starch condense and homogeneous appearance. Branching points form discrete clusters in the amorphous region, and the molecule's central core contains water molecules. B-amylopectin molecules are highly susceptible to acid hydrolysis digestion. Finally, type C amylopectin has intermediate features between A and B (Srichuwong and Jane, 2007).

Biosynthesis of starch depends on the enzymes, which are needed to metabolize it, namely AGPases, SSs, and starch-branching and debranching enzymes (SBEs and DBEs). As a starting point, sucrose is converted to uridine diphosphate glucose (UDP-glucose) and fructose by sucrose synthase in the endosperm cell cytosol. This is then transformed into glucose-1-P by ADP-glucose pyrophosphorylase, which is transported from the cytosol to the amyloplasts by an adenylate translocator (Tester et al., 2004). Once inside the amyloplasts, amylose is synthetized by granular-bound starch synthases (GBSSI), while amylopectin formation is driven by soluble starch synthases (GBSI and GBSII), branching enzymes (SBEI, SBEIIa and SBEIIb) and debranching enzymes (isoamylase and pullulanase) (Zhang et al., 2008). Most of these activities were discovered by isolation and characterization of several starch maize mutants at different stages of development (Nelson and Pan, 1995).

The starch in maize kernels is not only an important source of carbohydrates in human and animal diets, but also an important raw material for industrial and manufacturing applications. To produce starch with properties tailored to food, fuel, fibre or other applications, marker-assisted selection (MAS) is an alternative, efficient strategy for improving starch quantity and quality when QTLs or genes have been identified. Multiple traits have been improved by MAS in maize, such as head smut resistance (Zhao et al., 2012), pro-vitamin A content (Azmach et al., 2013), kernel oil content (Hao et al., 2014) and haploid induction rate (Dong et al., 2014). Interestingly, the top two QTLs, qSTA4-1 on chromosome 4 and qSTA10- 1 on chromosome 10, are stable across environments and are available for introgression of their favourable alleles to improve kernel starch content using MAS. Both QTLs explain ~10% of the starch variation and have additive effects of ~0.5% in the CI7/K22 RIL population. Several examples of maize mutants have therefore been successfully used to isolate the genes encoding key enzymes in starch metabolism, such as Shrunken1 (sh1), Shrunken2 (sh2), Brittle2 (bt2), agpsemzm, agpllzm, Waxy1 (wx1), SS1,

Sugary2 (su2), Dull1 (du1), SS2b-2, SS2c, SS3b-1, SS3b-2, SS4, SBEIa, SBEIIa, Amylose extender1 (ae1) and Sugary1 (su1) (Hennen-Bierwagen et al., 2013; Huang et al., 2014).

The sh1 gene encodes the major isoform of sucrose synthase and provides an important link in sucrose–starch conversion reactions, as sucrose synthase catalyses the reversible reaction between sucrose and UDP-glucose (Chourey et al., 1976). sh2, bt2, agpsemzm and agpllzm encode the large or small subunits of AGPase, which converts glucose-1-phosphate to ADP-glucose, the precursor for starch synthesis (Huang et al., 2014). wx1, encoding granule-bound SS I, is solely responsible for amylose production, whereas SS1, Su2, Du1, SS2b-2, SS2c, SS3b-1, SS3b-2 and SS4, encoding four types of soluble SS, are responsible for amylopectin production (Nelson et al., 1962; Harn et al., 1998). SBEIa, SBEIIa and ae1 encode the SBE isoforms Ia, IIa and IIb, respectively (Baba et al., 1991; Fisher et al., 1993; Gao et al., 1997), which are all responsible for amylopectin production. su1 encodes a DBE of the isoamylase type, and mutant su1 kernels contain the highly branched, water-soluble phytoglycogen and constitute the original sweet corns (James et al., 1995).

Several examples of recessive mutants with modified amylose/amylopectin ratios have been described. The amylose extender (Ae) mutant contains an anomalous Ae gene, which elevates expression to 80/20, producing a glassy, tarnished endosperm phenotype. This gene encodes a starch-branching enzyme IIb (SBEIIb), producing amylopectin structures with fewer α-(1–6) branched points than normal (Kim et al., 1998). The waxy (*wx*) mutant, also known as highly comestible starch, contains only amylopectin starch (0/100 amylose/amylopectin ratio). This occurs through homozygosity. An altered Wx locus encodes GBSSI, which determines the amylose content of endosperm tissue (Klösgen et al., 1986).

Other mutants induce highly sugary and low starch endosperms. For example, the sugary1 (*su1*) mature kernel has a glassy, dried, translucent and shrunken appearance, because the debranching enzyme α-(1–6) glucosidase is damaged and produces an accumulation of phytoglycogen and simple sugars at the expense of amylopectin (James et al., 1995) The sugary2 (*su2*) mutant contains amylopectin with more short glucan chains and fewer medium-length ones. This effect is produced by the aberrant SSIIa gene (Zhang et al., 2004). The dull (*du*) mutant phenotype has tarnished, glassy, dull kernels. This phenotype exhibits a reduction in the activity of SSII and SBEIIa, although only SBII is directly affected, while the reduced productivity of SBEIIa results primarily from the associated synthase deficiency (Gao et al., 1998). Finally, the brittle2 (*bt2*) and shrunken2 (*sh2*) mutants also have reduced levels of starch in their endosperms.

In addition, the transport of ADP-glucose into the amyloplast stroma is disrupted in mutants such as brittle1 (*bt1*). Such an irregular adenylate translocator substantially reduces the starch content of the resulting corn, producing matured kernels with a collapsed, angular shape (Shannon et al., 1998).

Most of the mutants discussed above have been included in breeding programmes to develop maize hybrids with altered nutritional attributes related to starch and sugar content (starch modification). For example, the name 'Sweet maize' is given to corn derived from varieties with abnormalities in endosperm carbohydrate composition. Such corn accumulates two to three times more simple sugars, and may include the sugary, brittle, dull, shrunken, waxy and amylose extender mutants. The Golden Bantam, Stowell´s Evergreen, Golden Rod, Cream-o-Gold, Black Mexican and Country Gentleman varieties are some of the first self-pollinated lines to emerge from these breeding programmes. This type of corn is also harvested 20–24 days after pollination when the sugar content is highest. As the selection period only lasts between five and seven days, the choice of hybrid traits must be quick (Pajic et al., 2000). In addition, in *su* types, sugar content decreases to nearly 50% at around day 28 because of the conversion of sugars to phytoglycogen (Tsai and Glover, 1974).

Three strategies have been used to obtain these starch modifications: recurrent selection, backcross breeding and pedigree breeding. First, recurrent selection is based on a four-year cycle followed by phenotypic selection. It has been used to produce important YUZPI*su* and YUZPII*su* elite inbred lines in programmes in Yugoslavia (Pajic, 1990). Backcross breeding focuses on improving a specific characteristic in a particular strain, by introducing a new gene into elite inbred lines, even in both parental lines of crossing. This method was used to develop Illini Xtra Sweet, in which the first *ae*, *du*, *sh2* and *wx* hybrids were created through introgression of AE, DU, SH2 and WX genes on Ia453 and Ia5152 inbred lines. Backcross breeding was also used to develop Florida Stay Sweet, Iobelle and the *su* hybrid Jubilee through introgression of the SH2 gene on Ia2312, Ia2256 and C13 (Pajic et al., 2000). The third strategy, pedigree selection, aims to develop new inbred lines through heterotic pattern crossing of parents selected for commercial utility. This method of inbreeding depends on evaluating successive generations and hybrids on the basis of early appearance, plant phenotype and growing season duration. It has the advantage of allowing the improvement of both additive and non-additive effects (Lee and Tracy, 2009).

'Amylose maize' varieties have been widely used in the paper, textile, adhesive and food processing industries. Breeding programmes have involved introducing AE genes into good agronomic inbred lines through alternate backcrossing and inbreeding. The complex heredity of the AE gene, involving epistatic interactions, makes this a very difficult task. Recurrent selection method cycles generate high amylose inbred lines and hybrids with amylase content of up to 94% (compared to an initial amylase content of 70%). Visual recognition of the desired phenotype was easy because the amylose endosperm appeared tarnished (Pajic et al., 2000).

Finally, 'Waxy maize' mutants have an important role in the food industry because they improve the uniformity, texture and stability of food. They are also used in animal feed mixtures. The WX gene is responsible for the special features of waxy maize, and in homozygous maize creates starch granules made of 100% amylopectin. Recurrent selection was applied to ensure that WX genes were expressed in elite inbred lines, allowing the development of waxy hybrids. One of the most important hybrids is the YUZP 704*wx* variety, as this yields 1.7% more energy than other hybrids (Pajic et al., 2000).

5 Improving the digestibility of maize protein and starch

Maize zein is quite resistant to digestion in comparison to other storage proteins. This resistance is probably due to the disulphide bonds between prolamin molecules, especially β and γ, generating a compact structure with low superficial area for hydrolytic enzyme accession (Holding, 2014).

It has been observed that maize and sorghum prolamin digestibility increases when reducing agents are added (Oria et al., 1995). In addition, although α-, β- and γ-zeins behave as resistant prolamins after seed digestion, α-prolamin isolates are easily hydrolysed. It therefore seems that only β- and γ-prolamins have bad digestibility properties, due to the formation of the disulphide interaction reticule on the PB's surface. In contrast, α-prolamins are not hydrolysed because of their interior location (Oria et al., 2000).

The digestibility of prolamins also decreases when seed products are cooked by boiling or pressure. β-prolamins form a highly compact grid, because more disulphide bonds are formed during the heating process of cooking (Emmambux and Taylor, 2009). However, cysteine residues of 27-kDa γ-zeins are the main components which form the disulphide bond matrix. Sequence deletion of these by directed mutagenesis increases digestibility rates (Lee and Hamaker, 2006).

Some mutants with lowered zein levels are both more nutritious and more easily digestible because of the improved access to PB inner zones for degradation enzymes. In the *high digestibility high lysine* (hdhl) *mutant* of sorghum, irregular PBs with invaginations allow easy access to α-zeins by hydrolysis enzymes (Oria et al., 2000). Abnormal protein is caused by a point mutation in a conserved signal peptide of the α-prolamin sequence, disturbing the cleavage and accumulation of the protein precursors (Wu et al., 2013). A similar phenotype is present in the maize mutants fl2 and De-B30 discussed above, where the ER signal sequence of 22-kDa and 19-kDa α-zeins are both affected (Coleman et al., 1997; Kim et al., 2006).

Starch digestibility is affected by the plant species, the extent of starch–protein interaction, the physical form of the granule, inhibitors such as tannins and the type of starch. The hard peripheral endosperm layer's resistance to digestive action is largely responsible for the overall digestibility of the starch.

Starch digestion is carried out by three main intestinal hydrolytic enzymes: α-amylase, maltase and isomaltase (Tester et al., 2004). Starch can then be classified into three groups, depending on how glucose is released and absorbed from it in the gastrointestinal tract (Englyst et al., 1992): (i) rapidly digestible starch (RDS) is hydrolysed quickly by digestive enzymes, producing an immediate increase in plasma glucose levels and an insulin response; (ii) slowly digestible starch (SDS) is partially digested with a minor glycaemic and insulin replay; and finally (iii) resistant starch (RS) cannot be digested by amylases in the small intestine, and thus plays a similar role to fibre.

RS can be further divided into four different subcategories, depending on its digestion resistance. Starches in the RS1 group include those physically inaccessible for digestion because they are inside granules encapsulated by carbohydrates or proteins. The RS2 group of starches contains those with a particular granule structure, namely those classified as types B and C in Section 4. Type A polymorphous starch granules, on the other hand, are easily digestible. RS3 starch is that which displays gelatinization and retrogradation phenomena. This occurs because amylopectin and amylose re-associate after being heated in the presence of water, forming enzymatic hydrolysis aggregates. In these cases, the original helical structure is disturbed by the loss of H bonds, producing type B polymorphous starch granules. In addition to these three natural categories, there is also the RS4 group, which consists of chemically modified starches (Tester et al., 2004).

The digestibility properties of starch also depend on several other parameters. Granular size is significant because smaller granules possess a high surface/volume ratio for enzyme action, and so larger granules are digested more slowly. Starches with a high amylose/amylopectin ratio are less digestible because amylopectin has a higher surface area than amylose (Carre et al., 2004). Waxy maize can therefore be digested more easily than high amylose starch (Jane et al., 2003). As indicated above, the type of polymorphism exhibited by starch is important. Type A starch is more accessible to digestible enzymes than are B and C polymorphs, because the channels and branching points in both the amorphous and crystalline regions of type A starch facilitate digestion agent attack (Jane et al., 2003).

The manufactured food industry subjects food to several production techniques such as milling, fermentation, steam cooking, baking, parboiling, extrusion, autoclaving, and heat and moisture treatment. Changes in starch proprieties will appear after gelatinization, depolymerization, retrogradation and recrystallization of starch. When crystalline zones are formed, these are not susceptible to enzymatic hydrolysis. This fact is being used to design experimentally controlled conditions to create more digestion-resistant starch (Birt et al., 2013).

Habitual consumption of high-glycaemic foods triggers cellular stress, because cells must confront variable conditions. High blood glucose and insulin levels, altered hormone secretion and elevated serum free fatty acid levels could produce beta cell dysfunction, dyslipidaemia and endothelial problems. A link has been suggested between excessive intake of RDS and increased susceptibility to obesity, type 2 diabetes and some heart disorders (Ludwig et al., 2002). Current efforts therefore focus on developing foods with high SDS and RS levels and lower glycaemic index (Birt et al., 2013). Through biotechnology, researchers are pursuing the creation of high amylose maize lines through the inhibition of the amylopectin branching enzymes SBEIIa and SBEIIb. Similar results have already been obtained for potato and durum wheat using antisense methodologies (Schwall et al., 2002; Sestili et al., 2010). In addition, transgenic RNAi lines which inhibit all SBEs have been investigated, producing resistant C-type starch granules in rice (Wei et al., 2010).

6 Assessing and reducing the anti-nutritional properties of maize

The properties of maize that are beneficial to health have been attributed to high antioxidant and antiradical activities, and too many other mechanisms such as anti-mutagenesis, anti-carcinogenesis, oestrogenic activities, inhibition of enzymes and induction of detoxification enzymes (Garodia et al., 2007).

Anthocyanins found in the pericarp and aleurones determine the colour of maize, and these water-soluble compounds are potent natural antioxidants due to their ability to trap free radicals (Stavric, 1994). Anthocyanin-rich foods and anthocyanin pigments have been suggested as potential agents to reduce the risk of colon cancer by inhibiting the proliferation of human colon cancer cells *in vitro* (Lila, 2004). Anthocyanins have been described as a 'functional food factor' (Tsuda et al., 2003), and could play a nutritional, biochemical role in helping to prevent diabetes and obesity. Maize has a higher total phenolic content and a higher antioxidant capacity than other grains (Adom et al., 2002).

However, maize grain components have also been reported to show clinical allergenicity, although the factors that may influence allergy elicitation are still unknown. This phenomenon means that the relevant macromolecules are anti-nutritional components of maize grains (Moneret-Vautrin et al., 1998). Maize allergy can occur after the ingestion of maize or maize derivatives, or by the inhalation of or contact with maize flour or pollen. Allergy to maize can cause different clinical manifestations according to the kind of environmental exposure, dietary habits and pre-existing sensitizations. Thus, maize allergy can be manifested in maize-pollen-induced respiratory symptoms (Gonzalo et al., 2004), occupational asthma (Weichel et al., 2006a) or dermatitis (Cristaudo et al., 2004) due to inhaled or handled maize flour, or by food allergy due to maize ingestion (Scibilia et al., 2008).

So far, 22 maize allergen proteins have been identified (Jimenez-Lopez et al., 2012). The major maize allergens in pollen, Zea m 13 (Cosgrove et al., 1997), which is a polygalacturonases, and Zea m 12, which is a profilin (Petersen et al., 2006) are also pollen allergens. Zea m 14 is the only major food allergen to be identified so far (Pastorello et al., 2000), but Zea m 25 is a minor maize food allergen (Weichel et al., 2006b; Fasoli et al., 2009). Zea m 14 was identified as a heat-resistant lipid transfer protein (LTP) with a molecular weight of 9 kDa and identified as a major food allergen of maize mediating an immunoglobulin E (IgE) response (Pastorello et al., 2009).

Precedent studies on individuals with food allergies have been reported to show clinical allergenicity (Pasini et al., 2002). However, it should be noted that *in vitro* cross-reactivity does not, in itself, guarantee that the individual from whom the serum was procured would react clinically upon ingestion of the cross-reactive protein.

Pastorello et al. (2000) performed one of the first studies on maize allergens, and recognized the LTP as a major maize allergen. LTP had first been identified as an important allergen in fruits, but it is also present in nuts, various vegetables and cereals. LTP is an extremely stable protein, and maintains IgE-binding activity even after cooking at 100°C (Pastorello et al., 2003). These properties make LTP a strong food allergen that can cause severe reactions.

More recently, a rapid identification and quantification method for the detection of LTP in 14 maize lines has been developed (Kuppannan et al., 2011). This operates through the use of liquid chromatography-ultraviolet/mass spectrometry, allowing the accurate quantification of LTP in maize samples for the assessment of the food allergen. Fasoli et al. (2009) identified broadly ten protein fractions in maize extract: vicilin-like embryo storage proteins, globulin-2 precursors, 50 kDa γ-zein, endo-chitinase, thioredoxin and trypsin inhibitor 4 low-molecular-weight proteins of 28–48 kD (which could possibly be classified as zeins), and three higher molecular weight seed storage proteins which should be classified as vicilins.

Interestingly, it has been also found that two major classes of allergens, belonging to the family of profilins and expansins (Petersen et al., 2006) belong to group 1 of the cross-reactive allergens of grass pollens. Furthermore, thioredoxins are a novel family of cross-reactive allergens that might contribute to the symptoms of baker's asthma and might in addition be related to grass pollen allergy, as indicated by the reactivity of grass pollen allergic patients to cereal thioredoxins (Mittag et al., 2004).

Maize has been a prominent example in discussions about whether genetic modification (GM) could introduce allergens, and in efforts to use GM to remove intrinsic allergens. New and diverse plant varieties have been developed to be more pest and/or disease resistant, more productive and to have improved nutritional quality, flavour and shelf life (Goodman et al., 2011).

In the EU, MON810 maize is the only cultivated, genetically modified maize used for food purposes. This maize was developed through a specific GM: the truncated cry1Ab gene containing a variation of two amino acid residues with the holotype sequence, Cry1Ab1, that encodes for the insecticidal crystal protein(s) delta-endotoxin of *Bacillus thuringiensis* (Bt), acting as a potent insecticide were inserted. This may cause the production of new proteins or the formation of either new metabolites or altered levels of existing metabolites, and these could compromise safety (Sicherer et al., 2014). This includes the potential production of new allergens or toxins.

Recently, maize flour IgE-binding proteins have been characterized using proteomic tools (Fonseca et al., 2012) in an attempt to understand whether natural proteomic differences between maize varieties result in different IgE-binding protein profiles among maize-allergic individuals. The characterization of maize IgE-binding proteins emphasizes the idea that the allergenicity of plants is not static, but potentially influenced by the plant genotype and growth conditions.

The expression of five known maize allergens in MON 810 has been evaluated using quantitative real-time PCR techniques, and this has been compared to the expression in the mutant's non-transgenic counterpart. Such investigations were undertaken throughout the seed's development because one of the main concerns about GM foods is their potential impact on human health. In this study, fourteen new IgE-binding proteins were identified as present in both transgenic and non-transgenic maize (Fonseca et al., 2012).

Barros et al. (2010) used non-targeted molecular profiling to provide insight into the extent of variation in maize transcriptome, proteome and metabolome by analysing three maize genotypes. The comparison between GM and non-GM crops usually comprises agronomic/phenotypic features, feed performance studies and crop composition.

The study of effects occurring in transgenic maize seeds as a result of GMs showed that environment has the greatest influence on the proteomic profiles of transgenic seeds, but also that a transformation technique such as particle bombardment can induce additional genome alterations which cause a different protein expression (Zolla et al., 2008).

Kauffman et al. (1983) analysed 39 cases of maize-specific IgE. Nine patients (23%) showed greater than 30% binding to the GM and non-GM maize antigens and were considered to have significantly raised maize-specific IgE. Later in the analysis, the sera of five people with clinical sensitization to maize were used to identify seven protein fractions in the range 88–28 kDa, which played a role in IgE binding to non-GM maize antigens.

Batista et al. (2005) studied the allergenicity of three GM maize varieties expressing different levels of the Cry 1Ab protein, in two groups of sensitive members of the population. The study proceeded on the assumption that the participating population was consuming maize-derived products from a GM maize source. This study concluded that Cry 1Ab was safe because of the absence of detectable differences in IgE reactivity to GM and non-GM maize. Buzoianu et al. (2012) and Walsh et al. (2012) reported a non-significant effect on the growth and immune responses of animals fed with GM maize (Bt-MON810) expressing Cry 1Ab.

Bt protein sequences have been shown to produce crystal proteins known as Cry proteins or inclusion bodies that are specifically effective in controlling certain orders and species of insect pests. Mathur et al. (2015) assessed GM maize seeds containing Bt-Cry 1Ab, Cry 1Ac and Cry 1C protein sequences for allergenicity. They used *in silico* bioinformatics searches and *in vitro* methods involving the determination of specific IgE-binding patterns to GM and non-GM maize antigens in the sera of maize-sensitized patients. Using the *in silico* tools, they confirmed that Cry 1Ab, Cry 1Ac and Cry 1C protein sequences are non-allergenic sequences with no cross-reactivity to known allergens. The incorporation of these transgene protein sequences produces no appreciable changes in endogenous protein expression.

Using an animal model (Andreassen et al., 2015), it was suggested that Cry1Ab proteins from three different sources did not act as an adjuvant. The contents of MON810 maize tissues in their study could represent long-term exposures of plant Cry1Ab as well as purified plant proteins to mimic the total exposures experienced in real-life situations.

γ-zein is successfully expressed in several important crops such as soya bean, barley and alfalfa (Krishnan et al., 2011). The maize γ-zein has also been introduced into soya bean to improve its nutritional quality. However, the use of the 27-kDa γ-zein as a means of improving the nutritional value of soya bean protein by improving its amino acid profile induces a strong immunological response in young pigs, suggesting that an allergen may have been introduced (Krishnan et al., 2011). In this study, it was demonstrated that the same protein may also be a potential food allergen.

In order for GM to be applied to crops more broadly, there is an urgent need for increased knowledge and better biochemical and molecular methods, including animal models, to test for food allergens experimentally. The supporting data can be provided to evaluate newly proposed and actual GM products. The use of bioinformatics tools is an important step for novel protein safety assessment.

7 Conclusions

Maize has great importance since it occupies the third place after wheat and rice as basic and accessible food for large population worldwide, particularly in developing countries.

Massive efforts have been made towards improvement of maize nutraceutical properties (mainly at genetic manipulation and processing levels), because of factors such as large variability in the maize chemical composition, and its poor protein value, because of its deficiency in two essential amino acids (tryptophan and lysine) in the protein composition. Thus, enhancing nutritional quality of maize for human health has been possible through appropriate manipulation of mutant lines used in breeding programmes, as well as in the more efficient molecular marker systems, giving as a result the QPM. Furthermore, efforts have also been directed on carbohydrate composition, which is genetically controlled, and as starch is a major nutritional component of maize kernel, variations of the amylose to amylopectin ratio provided nutritional positive significance.

As maize quality depends on kernel components of different nature, interdisciplinary and multidisciplinary approaches harnessing research expertise in proteins biochemistry, genetics, plant breeding and nutrition may be important and necessary for the discovery of new genes, their expression patterns and the identification of rate-limiting steps of metabolic and regulatory pathways to decide new and future strategies regarding overall maize improvement.

8 Where to look for further information

Allergome, http://www.allergome.org/

Echarte L, Nagore L, Di Matteo J, et al. 2013. Grain yield determination and resource use efficiency in maize hybrids released in different decades. In *Agricultural Chemistry*, InTech, ISBN 978-953-51-1026-2

Hartings H, Fracassetti M and Motto, M. 2012. Genetic enhancement of grain quality-related traits in maize. In *Transgenic Plants – Advances And Limitations*, InTech, ISBN 978-953-51-0181-9

Hossein A, Schulin R, Chaney RL, et al. 2010. Micronutrient-efficient genotypes for crop yield and nutritional quality in sustainable agriculture – A review. *Agron. Sustain. Dev.*, 30, 83–107.

Huang S, Adams WR, Zhou Q, et al. 2004. Improving nutritional quality of maize proteins by expressing sense and antisense zein genes. *J Agric. Food Chem.*, 52(7), 1958–64

Huffman RD, Edwards JW, Pollak LM, et al. 2016. Interaction of Genetic Mechanisms Regulating Methionine Concentration in Maize Grain. *Crop Sci.*, 56, 1–11

Jimenez-Lopez, J. C., Kotchoni, S. O., Gachomo, E. W., Castro, A. J., Rodríguez-García, M. I. and Alché, J. D. 2012. Molecular features of maize allergens and their implications in human health. In Jimenez-Lopez, J. C (Ed), *Maize: Cultivation, Uses And Health Benefits*. Nova Science Publishers Inc., pp.137–78. ISBN: 978-1-62081-514-4.

Khan R and Dubey RB. 2015. Combining ability analysis for nutritional quality and yield in maize (*Zea mays* L.). *The Bioscan*, 10(2), 785–8

Liu H, Shib J, Sun C, et al. 2016. Gene duplication confers enhanced expression of 27-kDa γ-zein for endosperm modification in quality protein maize. *Proc. Natl. Acad. Sci. USA*, 113(18), 4964–9

Mavengahama, S., McLachlan M and De Clercq, W. 2013. The role of wild vegetable species in household food security in maize based subsistence cropping systems. *Food Secur.*, 5(2), 227–33

Mladenovic S, Andjelkovic V and Ignjatovic, D. 2012. Genetic diversity of maize landraces as sources of favorable traits. In *The Molecular Basis of Plant Genetic Diversity*, InTech, ISBN: 978-953-51-0157-4

Nuss ET and Tanumihardjo SA. 2010. Maize: a paramount staple crop in the context of global nutrition. *Compr. Rev. Food Sci. Food Saf.*, 9(4), 417–36

Nkongolo KK, Daniel G, Mbuya, K., et al. 2015. Characterization of new quality protein maize (qpm) varieties from a breeding program: analysis of amino acid profiles and development of a variety-diagnostic molecular marker. *British Biotech. J.*, 6(3), 101

Pérez-Massot E, Banakar R, Gómez-Galera, S., et al. 2013. The contribution of transgenic plants to better health through improved nutrition: opportunities and constraints. *Genes Nutr.*, 8, 29

SDAP, https://fermi.utmb.edu/

Tang M, He X, Luo Y, Ma L, Tang X and Huang, K. 2013. Nutritional assessment of transgenic lysine-rich maize compared with conventional quality protein maize. *J. Sci. Food Agric.*, 93, 1049–54

TWumasi-Afriyie S, Friesen D and Pixley, K. 2011. Quality protein maize: Progress and prospects. *Plant Breed Rev.*, 34, 83.

Wilkinson, M. 2015. Maize cultivars for anaerobic digestion and animal nutrition in Europe. *Int. J. Agricultural Manag.*, 4(4), 144–51

Wu Y and Messing, J. 2014. Proteome balancing of the maize seed for higher nutritional value. *Front. Plant Sci.*, 5, 240.

9 Acknowledgements

This work has been funded by the European Research Program MARIE CURIE (FP7-PEOPLE-2011-IOF), grant ref. number PIOF-GA-2011-301550 to JCJ-L; and by the Spanish Ministry of Economy and Competitiveness, grant ref. number RYC-2014-16536 (Ramon y Cajal Research Program) to JCJ-L.

10 References

Adom, K. K. and Liu, R. H. (2002), 'Antioxidant activity of grains', *J. Agric. Food Chem.*, 50(21), 6182–7.

Andreassen, M., Bøhn, T., Wikmark, O.-G., Van den Berg, J., Løvik, M., et al. (2015), 'Cry1Ab protein from Bacillus thuringiensis and MON810 cry1Ab-transgenic maize exerts no adjuvant effect after airway exposure', *Scand. J. Immunol.*, 81(3), 192–200.

Azmach, G., Gedil, M., Menkir, A. and Spillane, C. (2013), Marker-trait association analysis of functional gene markers for provitamin A levels across diverse tropical yellow maize inbred lines. *BMC Plant Biol.*, 13(1), 227.

Baba, T., Kimura, K., Mizuno, K., Etoh, H., Ishida, Y., Shida, O., et al. (1991), Sequence conservation of the catalytic regions of anylolytic enzymes in maize branching enzyme-I. *Biochem Bioph Res. Co.*, 181(1), 87–94.

Barrangou, R., Fremaux, C., Deveau, H., Richards, M., Boyaval, P., Moineau, S., Romero, D. A. and Horvath, P. (2007), CRISPR provides acquired resistance against viruses in prokaryotes. *Science* 315, 1709–12.

Bhave, M. R., Lawrence, S., Barton, C. and Hannah, L. C. (1990), Identification and molecular characterization of shrunken-2 cDNA clones of maize. *Plant Cell*, 2(6), 581–8.

Bicar, E. H., Woodman-Clikeman, W., Sangtong, V., Peterson, J. M., Yang, S. S., et al. (2008), 'Transgenic maize endosperm containing a milk protein has improved amino acid balance', *Transgenic Res.*, 17, 59–71.

Barros, E., Lezar, S., Anttonen, M. J., van Dijk, J. P., Röhlig, R. M., et al. (2010), 'Comparison of two GM maize varieties with a near-isogenic non-GM variety using transcriptomics, proteomics and metabolomics', *Plant Biotechnol. J.*, 8(4), 436–51.

Batista, R., Nunes, B., Cardoso, C., Jose, H. S., de Almeida, A. B., et al. (2005), 'Lack of detectable allergenicity of transgenic maize and soya samples', *J. Allergy Clin. Immunol.*, 116, 403–10.

Bressani, R. (1991), 'Protein quality of high lysine maize for humans', *Cereal Food World*, 36, 806–11.

Bjarnason, M. and Pollmer, W. G. (1972), 'The maize germ: Its role as contributing factor to protein quantity and quality', *Z. Pflanzenzucht.*, 68, 83–9.

Boyer, C. D. and Shannon, J. C., (1983), 'The use of endosperm genes for sweet corn improvement', *Plant Breed. Rev.*, 1, 139–61.

Buleon, A., Colonna, P., Planchot, V. and Ball, S. (1998), 'Starch granules: structure and biosynthesis', *Int. J. Biol. Macromol.*, 23, 85–112.

Burge, R. M. and Duensing, W. J. (1989), 'Processing and dietary fiber ingredient applications of com bran', *Cereal Foods World*, 34, 535–8.

Buzoianu, S. G., Walsh, M. C., Rea, M. C., Cassidy, J. P., Ross, R. P., et al. (2012), 'Effect of feeding genetically modified Bt MON810 maize to 40 day old pigs for 110 days on growth and health indicators', *Animal*, 6(10), 1609–19.

Carre, B. (2004), 'Causes for variation in digestibility of starch among feedstuffs', *World's Poult. Sci. J.*, 60, 76–89.

Chang, Y., Shen, E., Wen, L., Yu, J., Zhu, D. and Zhao, Q. (2015), 'Seed-Specific Expression of the Arabidopsis AtMAP18 Gene Increases both Lysine and Total Protein Content in Maize', *PLoS ONE*, 10(11), e0142952. doi:10.1371/journal.pone.0142952

Chen, W., Chen, H., Zheng, T., Yu, R., Terzaghi, W. B., Li, Z., et al. (2014). Highly efficient genotyping of rice biparental populations by GoldenGate assays based on parental resequencing. *Theor. Appl. Genet.*, 127(2), 297–307.

Choulika, A., Perrin, A., Dujon, B. and Nicolas, J. F. (1995), 'Induction of homologous recombination in mammalian chromosomes by using the I-SceI system of Saccharomyces cerevisiae', *Mol. Cell Biol.*, 15, 1968–73.

Coleman, C. E., Lopes, M. A., Gillikin, J. W., Boston, R. S. and Larkins, B. A. (1995), 'A defective signal peptide in the maize high-lysine mutant floury 2', *Proc. Natl. Acad. Sci. USA*, 92, 6828–31.

Coleman, C. E., Clore, A. M., Ranch, J. P., Higgins, R., Lopes, M. A. and Larkins, B. A. (1997), 'Expression of a mutant alpha-zein creates the floury2 phenotype in transgenic maize', *Proc. Natl. Acad. Sci. USA*, 94(13), 7094–7.

Coleman, C. E. and Larkins, B. A. (1999), 'The prolamins of maize'. In Shewry, P. R. and Casey, R. (Eds), *Seed Proteins*. Academic Publishers, pp.109–39.

Cosgrove, D. J., Bedinger, P. A. and Durachko, D. M. (1997), 'Group I allergens of grass pollen as cell wall-loosening agents', *Proc. Natl. Acad. Sci. USA*, 94, 6559–64.

Cook, J. P., McMullen, M. D., Holland, J. B., Tian, F., Bradbury, P., Ross-Ibarra, J., et al. (2012), 'Genetic architecture of maize kernel composition in the nested association mapping and inbred association panels', *Plant Physiol.*, 158(2), 824–34.

Cristaudo, A., Simonato, B., Pasini, G., De Rocco, M., Curioni, A. and Giannattasio, M. (2004), 'Contact urticaria and protein contact dermatitis from corn in a patient with serum IgE specific for a salt-soluble corn protein of low molecular weight', *Contact. Dermatitis*, 51(2), 84–7.

Dong, X., Xu, X. W., Li, L., Liu, C. X., Tian, X. L., Li, W., et al. (2014), 'Marker-assisted selection and evaluation of high oil in vivo haploid inducers in maize', *Mol. Breed.*, 34(3), 1147–58.

Doudna, J. A. and Charpentier, E. (2014), 'Genome editing. The new frontier of genome engineering with CRISPR-Cas9', *Science*, 346, 1258096.

Dudley, J. W., Dijkhuizen, A., Paul, C., Coates, S. T. and Rocheford, T. R. (2004), 'Effects of random mating on marker–QTL associations in the cross of the Illinois high protein × Illinois low protein maize strains', *Crop Sci.*, 44(4), 1419–28.

Emmambux, M. N and Taylor, R. N. (2009), 'Properties of heat-treated sorghum and maize meal and their prolamin proteins', *J. Agric. Food Chem.*, 57(3), 1045–50.

Englyst, H. N., Kingman, S. M. and Cummings, J. H. (1992), 'Classification and measurement of nutritionally important starch fractions', *Eur. J. Clin. Nutr.*, 46, S33–S50.

FAO. (2013), FAOSTAT, Food Supply, http://www.fao.org/publications/2013/sofi/en (Accessed 3 June 2015)

Fasoli, E., Pastorello, E. A., Farioli, L., Scibilia, J., Aldini, G., et al. (2009), 'Searching for allergens in maize kernels via proteomic tools', *J. Proteomics*, 72, 501–10.

Fisher, D. K., Boyer, C. D. and Hannah, L. C. (1993), 'Starch branching enzyme II from maize endosperm', *Plant Physiol.*, 102(3), 1045.

Fonseca, C., Planchon, S., Renaut, J., Oliveira, M. M. and Batista, R. (2012), 'Characterization of maize allergens—MON810 vs. its non-transgenic counterpart', *J. Proteomics.*, 75(7), 2027–37.

Gao, M., Fisher, D. K., Kim, K. N., Shannon, J. C. and Guiltinan, M. J. (2014), 'Independent genetic control of maize starch-branching enzymes IIa and IIb (isolation and characterization of a Sbe2a cDNA)', *Plant Physiol.*, 114(1), 69–78.

Gao, M., Wanat, J., Stinard, P. S., James, M. G. and Myers, A. M. (1998), 'Characterization of dull1, a maize gene coding for a novel starch synthase', *Plant Cell*, 10, 399–412.

Galili, G. and Amir, R. (2013), 'Fortifying plants with the essential amino acids lysine and methionine to improve nutritional quality', *Plant Biotech. J.*, 11(2), 211–22.

Garodia, P., Ichikawa, H., Malani, N., Sethi, G. and Aggarwal, B. B. (2007), 'From ancient medicine to modern medicine: ayurvedic concepts of health and their role in inflammation and cancer', *J. Soc. Integr. Oncol.*, 5, 25–37.

Geevers, H. O. and Lake, J. K. (1992), 'Development of modified opaque-2 maize in South Africa'. In Mertz (Ed.), *Quality Protein Maize*. AACC, pp. 49–78.

Gibbon, B. C. and Larkins, B. A. (2005), 'Molecular genetic approaches to developing quality protein maize', *Trends Genet.*, 21(4), 227–33.

Gonzalo-Garijo, M. A., Pérez-Calderón, R., Muñoz-Rodriguez, A., Tormo-Molina, R. and Silva-Palacios, I. (2004), 'Hypersensitivity to maize pollen', *Allergy*, 59(3), 365.

Goodman, R. E. and Affua, O. T. (2011), 'Suggested improvements for the allergenicity assessment of genetically modified plants used in foods', *Curr. Allergy Asthma Rep.*, 11, 317–24.

Hamada, S., Ishiyama, K., Sakulsingharoj, C., Choi, S. B., Wu, Y., et al. (2003), 'Dual regulated RNA transport pathways to the cortical region in developing rice endosperm', *Plant Cell*, 15, 2265–72.

Hao, X. M., Li, X. W., Yang, X. H. and Li, J. S. (2014), 'Transferring a major QTL for oil content using marker-assisted backcrossing into an elite hybrid to increase the oil content in maize', *Mol. Breed.*, 34(2), 739–48.

Harn, C., Knight, M., Ramakrishnan, A., Guan, H. P., Keeling, P. L. and Wasserman, B. P. (1998), 'Isolation and characterization of the zSSIIa and zSSIIb starch synthase cDNA clones from maize endosperm', *Plant Mol. Biol.*, 37(4), 639–49.

Hennen-Bierwagen, T. A. and Myers, A. M. (2003), 'Genomic specification of starch biosynthesis in maize endosperm', *Seed Genomics.*, 233, 123–37.

Henley, E. C., Taylor, J. R. N. and Obukosia, S. D. (2010), 'The importance of dietary protein in human health: combating protein deficiency in sub-saharan africa through transgenic biofortified sorghum', *Adv. Food Nutr. Res.*, 60, 21–52.

Holding, D. R., et al. (2007), 'The maize floury1 gene encodes a novel endoplasmic reticulum protein involved in zein protein body formation', *Plant Cell*, 19, 2569–82.

Huang, B. Q., Hennen-Bierwagen, T. A. and Myers, A. M. (2014), 'Functions of multiple genes encoding ADP-glucose pyrophosphorylase subunits in maize endosperm, embryo, and leaf', *Plant Physiol.*, 164(2), 596–611.

Hsu, P. D., Lander, E. S. and Zhang, F. (2014), 'Development and applications of CRISPR-Cas9 for genome engineering', *Cell*, 157, 1262–78.

Holding, D. R., Otegui, M. S., Li, B. L., Meeley, R. B., Dam, T., et al. (2007), 'The maize floury1 gene encodes a novel endoplasmic reticulum protein involved in zein protein body formation', *Plant Cell*, 19, 2569–82.

Holding, D. R., Hunter, B. G., Chung, T., Gibbon, B. C., Ford, C. F., et al. (2008), 'Genetic analysis of opaque2 modifier loci in quality protein maize', *Theor. Appl. Genet.*, 117, 157–70.

Holding, D. R., Meeley, R. B., Hazebroek, J., Selinger, D., Gruis, F., et al. (2010), 'Identification and characterization of the maize arogenate dehydrogenase gene family', *J. Exp. Bot.*, 61, 3663–73.

Holding, D. R. (2014), 'Recent advances in the study of prolamin storage protein organization and function', *Front. Plant Sci.*, 5, 276.

Huang, S., Adams, W. R., Zhou, Q., Malloy, K. P., Voyles, D. A., et al. (2004), 'Improving nutritional quality of maize proteins by expressing sense and antisense zein genes', *J. Agric. Food Chem.*, 52(7), 1958–64.

Jane, J.-l., Ao, Z., Duvick, S. A., Wiklund, M., Yoo, S.-H., et al. (2003), 'Structures of amylopectin and starch granules: How are they synthesized?', *J. Appl. Glycosci.*, 50, 167–72.

James, M. G., Robertson, D. S. and Myers, A. M. (1995), 'Characterization of the maize gene sugary1, a determinant of starch composition in kernels', *Plant Cell*, 7(4), 417–29.

James, M. and Myers, A. (2009), 'Seed starch synthesis'. In Jeff L. Bennetzen and Sarah C. Hake (Eds), *Handbook of Maize: Its Biology.*, 233, 439–56.

Ji, Q., Vincken, J. P., Suurs, L. C. and Visser, R. G. (2003), 'Microbial starch-binding domains as a tool for targeting proteins to granules during starch biosynthesis', *Plant Mol. Biol.*, 51, 789–801.

Jimenez-Lopez, J. C., Kotchoni, S. O., Gachomo, E. W., Castro, A. J., Rodríguez-García, M. I. and Alché, J. D. (2012), 'Molecular features of maize allergens and their implications in human health'. In Jimenez-Lopez, J. C. (Ed.), *Maize: Cultivation, Uses And Health Benefits.* Nova Science Publishers Inc., pp. 137–78. ISBN: 978-1-62081-514-4.

Kim, C. S., et al. (2004), 'A defective signal peptide in a 19-kD alpha-zein protein causes the unfolded protein response and an opaque endosperm phenotype in the maize De*-B30 mutant', *Plant Physiol.*, 134, 380–7.

Kim, C. S., et al. (2006), 'The maize Mucronate mutation is a deletion in the 16-kDa gamma-zein gene that induces the unfolded protein response', *Plant J.*, 48, 440–51.

Kauffman, H. F., Beaumont, F., Meurs, H., Heidi, S. V. and de Vries, K. (1983), 'Comparison of antibody measure- ment against Aspergillus fumigatus by means of double—diffusion and enzyme linked immunosorbent assay (ELISA)', *J. Allergy Clin. Immunol.*, 72, 255–61.

Kim, K., Fisher, D. K., Gao, M. and Guiltinan, M. J. (1998), 'Molecular cloning and characterization of the Amylose-Extender gene encoding starch branching enzyme IIB in maize', *Plant Mol. Biol.*, 38, 945–56.

Kim, C. S., Hunter, B. G., Kraft, J., Boston, R. S., Yans, S., et al. (2994), 'A defective signal peptide in a 19-kD alpha-zein protein causes the unfolded protein response and an opaque endosperm phenotype in the maize De*-B30 mutant', *Plant Physiol.*, 134, 380–7.

Kim, C. S., Gibbon, B. C., Gillikin, J. W., Larkins, B. A., Boston, R. S. and Jung, R. (2006), 'The maize Mucronate mutation is a deletion in the 16-kDa gamma- zein gene that induces the unfolded protein response', *Plant J.*, 48, 440–51.
Kirma, M., Araujo, W. L., Fernie, A. R. and Galili, G. (2012), 'The multifaceted role of aspartate-family amino acids in plant metabolism', *J. Exp. Bot.*, 63(14), 4995–5001.
Klösgen, R. B., Gierl, A., Schwarz-Sommer, Z. and Saedler, H. (1986), 'Molecular analysis of the waxy locus of *Zea mays*', *Mol. Gen. Genet.*, 203, 237–44.
Krishnan, H. B., Jang, S., Kim, W. S., Kerley, M. S., Oliver, M. J. and Trick, H. N. (2011), 'Biofortification of soybean meal: immunological properties of the 27 kDa γ-zein', *J. Agric. Food Chem.*, 59(4), 1223–8. doi: 10.1021/jf103613s
Kuppannan, K., Albers, D. R., Schafer, B. W., Dielman, D. and Young, S. A. (2010), 'Quantification and characterization of maize lipid transfer protein, a food allergen, by liquid chromatography with ultraviolet and mass spectrometric detection', *Anal. Chem.*, 83, 516–24.
Lai, J. S. and Messing, J. (2002), 'Increasing maize seed methionine by mRNA stability', *Plant J.*, 30, 395–402.
Lang, Z., Zhao, Q., Yu, J., Zhu, D. and Ao, G. (2004), 'Cloning of potato SBgLR gene and its intron splicing in transgenic maize', *Plant Sci.*, 166(5), 1227–33.
Lee, E. A. and Tracy, W. F. (2009), 'Modern maize breeding'. In Bennetzen, J. L. and Hake, S. C. (Eds), *The Maize Handbook–Volume 2: History And Practice Of Genetics, Genomics And Improvement*. Springer-Verlag, pp. 141–60.
Lee, S. H. and Hamaker, B. R. (2006), 'Cys155 of 27 kDa maize γ-zein is a key amino acid to improve its in vitro digestibility', *FEBS Lett.*, 580, 5803–6.
Li, K., Yan, J. B., Li, J. S. and Yang, X. H. (2014), 'Genetic architecture of rind penetrometer resistance in two maize recombinant inbred line populations', *BMC Plant Biol.*, 14(1), 152.
Liang, Z., Zhang, K., Chen, K. and Gao, C. (2014), 'Targeted mutagenesis in *Zea mays* using TALENs and the CRISPR/Cas system', *J. Genet. Genomics*, 41, 63–8.
Lila, M. A. (2004), 'Anthocyanins and human health: an in vitro investigative approach', *J. Biomed. Biotechnol.*, 5, 306–13.
Lohmer, S., Maddaloni, M., Motto, M., Fonzo, N. D., Hartings, H., et al. (1991), 'The maize regulatory locus Opaque-2 encodes a DNA-binding protein which activates the transcription of the b-32 gene', *EMBO J.*, 10, 617–24.
Lopes, M. A. and Larkins, B. A. (1995), 'Genetic analysis of opaque2 modifier gene activity in maize endosperm', *Theor. Appl. Genet.*, 91, 274–81.
Ludwig, D. S. (2002), 'The glycemic index physiological mechanism relating to obesity, diabetes, and cardiovascular disease', *J. Am. Med. Assoc.*, 287, 2414–23.
Maccaferri, M., Ricci, A., Salvi, S., Milner, S. G., Noli, E., Martelli, P. L., et al. (2015), 'A highdensity, SNPbased consensus map of tetraploid wheat as a bridge to integrate durum and bread wheat genomics and breeding', *Plant Biotechnol J.*, 13(5), 648–63.
Maddaloni, M., Donini, G., Balconi, C., Rizzi, E., Gallusci, P., et al. (1996), 'The transcriptional activator Opaque-2 controls the expression of a cytosolic form of pyruvate orthophosphate dikinase-1 in maize endosperms', *Mol. Gen. Genet.*, 250, 647–54.
Mathur, C., Kathuria, P. C., Dahiya, P. and Singh, A. B. (2015), 'Lack of Detectable Allergenicity in Genetically Modified Maize Containing "Cry" Proteins as Compared to Native Maize Based on In Silico & In Vitro Analysis', *PLoS ONE*, 10(2), doi: 10.1371/journal.pone.0117340
Mertz, E. T., Bates, L. S. and Nelson, O. E. (1964), 'Mutant gene that changes protein composition and increases lysine content of maize endosperm', *Science*, 145(3629), 279–80.
Miclaus, M., Wu, Y. R., Xu, J. H., Dooner, H. K. and Messing, J. (2011), 'The maize high-lysine mutant opaque7 is defective in an Acyl-CoA synthetase-like protein', *Genetics*, 189, 1271–80.
Mittag, D., Niggemann, B., Sander, I., Reese, I., Fiedler, E. M., et al. (2004), 'Immunoglobulin E-reactivity of wheat-allergic subjects (baker's asthma, food allergy, wheat-dependent, exercise-induced anaphylaxis) to wheat protein fractions with different solubility and digestibility', *Mol. Nutr. Food Res.*, 48, 380–9.

Moneret-Vautrin, D. A., Kanny, G. and Beaudouin, E. (1998), 'L'allergie alimentaire au mais existe-t-elle?', *Allerg. Immunol.*, 30, 230.

Myers, A. M., James, M. G., Lin, Q. H., Yi, G., Stinard, P. S., et al. (2011), 'Maize opaque5 encodes monogalactosyldiacylglycerol synthase and specifically affects galactolipids necessary for amyloplast and chloroplast function', *Plant Cell*, 23, 2331–47.

Nelson, O. E. and Rines, H. W. (1962), 'The enzymatic deficiency in the waxy mutant of maize', *Biochem. Bioph. Res. Co.*, 9(4), 297–300.

Nelson, O. E. and Pan, D. (1995), 'Starch synthesis in maize endosperms', *Annu. Rev. Plant Physiol. Plant Mol. Biol.*, 46, 475–96.

Nuss, E. T. and Tanumihardjo, S. A. (2010), 'Maize: a Paramount staple crop in the context of global nutrition', *Compr. Rev. Food Sci. Food Saf.*, 9, 417–36.

Oria, M. P., Hamaker, B. R. and Shull, J. M. (1995), 'Resistance of sorghum αandγ kafirins to pepsin digestion', *J. Agric. Food Chem.*, 43, 2148–53.

Oria, M. P., Hamaker, B. R., Axtell, J. D. and Huang, C. P. (2000), 'A highly digestible sorghum mutant cultivar exhibits a unique folded structure of endosperm protein bodies', *Proc. Natl. Acad. Sci. USA*, 97, 5065–70.

Pajic, Z., Dumanovic, J., Misovic, M. and Rosulj, M. (2000), 'Breeding maize for nutritional value and industrial use', *Genetika*, 32(3), 255–82.

Pajic, Z. (1990), 'Popcorn and sweet corn breeding'. *Maize '90, Maize Breeding, Production, Processing and Marketing in Mediterranean Countries*. Belgrade, pp. 79–97.

Pan, D. D. and Jane, J.-l. (2000), 'Internal structure of normal maize starch granules revealed by chemical surface gelatinization', *Biomacromolecules*, 1, 126–32.

Pasini, G., Simonato, B., Curioni, A., Vincenzi, S., Cristaudo, A., et al. (2002), 'IgE- mediated allergy to corn: a 50 kDa protein, belonging to the reduced soluble proteins, is a major protein', *Allergy*, 57, 98–106.

Pastorello, E. A., Farioli, L., Pravettoni, V., Ispano, M., Scibola, E., et al. (2000), 'The maize major allergen, which is responsible for food-induced allergic reactions, is a lipid transfer protein', *J. Allergy Clin. Immunol.*, 106(4), 744–51.

Pastorello, E. A., Pompei, C., Pravettoni, V., Farioli, L., Calamari, A. M., et al., (2003), 'Lipid-transfer protein is the major maize allergen maintaining IgE-binding activity after cooking at 100°C, as demonstrated in anaphylactic patients and patients with positive double-blind, placebo-controlled food challenge results', *J. Allergy Clin. Immunol.*, 112, 775–83.

Pastorello, E. A, Farioli, L., Pravettoni, V., Scibilia, J., Conti, A., et al., (2009), 'Maize food allergy: lipid-transfer proteins, endochitinases, and alpha-zein precursor are relevant maize allergens in double-blind placebo-controlled maize-challenge- positive patients', *Anal. Bioanal. Chem.*, 395, 93–102.

Petersen, A., Dresselhaus, T., Grobe, K. and Becker, W. M. (2006), 'Proteome analysis of maize pollen for allergy-relevant components', *Proteomics*, 6, 6317–25

Prasanna, B. M., Vasal, S. K., Kassahun, B. and Singh, N. N. (2001), 'Quality protein maize', *Curr. Sci.*, 81, 1308–19.

Puchta, H., Dujon, B. and Hohn, B. (1993), 'Homologous recombination in plant cells is enhanced by in vivo induction of double strand breaks into DNA by a site-specific endonuclease', *Nucleic Acids Res.*, 21, 5034–40.

Puchta, H., Dujon, B. and Hohn, B. (1996), 'Two different but related mechanisms are used in plants for the repair of genomic double-strand breaks by homologous recombination', *Proc. Natl. Acad. Sci. USA*, 93, 5055–60.

Segal, G., Song, R. and Messing, J. (2003), 'A new opaque variant of maize by a single dominant RNA-interference-inducing transgene', *Genetics*, 165, 387–97.

Schmidt, R. J., Burr, F. A., Aukerman, M. J. and Burr, B. (1990), 'Maize regulatory gene opaque-2 encodes a protein with a "leucine-zipper" motif that binds to zein DNA', *Proc. Natl. Acad. Sci. USA*, 87, 46–50.

Schwall, G. P., Safford, R., Westcott, R. J., Jeffcoat, R., Tayal, A., et al. (2000), 'Production of very-high-amylose potato starch by inhibition of SBE A and B', *Nat. Biotechnol.*, 18, 551–4.

Scibilia, J., Pastorello, E. A., Zisa, G., Ottolenghi, A., Ballmer-Weber, B., et al. (2008), 'Maize food allergy: a double-blind placebo-controlled study', *Clin. Exp. Allergy*, 38(12), 1943–9.
Segal, G., Song, R. and Messing, J. (2003), 'A new opaque variant of maize by a single dominant RNA-interference-inducing transgene', *Genetics*, 165, 387–97.
Sestili, F., Janni, M., Doherty, A., Botticella, E., D'Ovidio, R., et al. (2010), 'Increasing the amylose content of durum wheat through silencing of the SBEIIa genes', *BMC Plant. Biol.*, 10, 144.
Shannon, J. C., Pien, F., Cao, H. and Liu, K. (1998), 'Brittle-1, an adenylate translocator, facilitates transfer of extraplastidial synthesized ADP-glucose into amyloplast of maize endosperms', *Plant Physiol.*, 117, 1235–52.
Sicherer, S. H. and Sampson, H. A. (2014), 'Food allergy: Epidemiology, pathogenesis, diagnosis and treatment', *J. Allergy Clin. Immunol.*, 133, 291–307.
Singh, N., Singh, J., Kaur, L., Sodhi, N. S. and Gill, B. S. (2003), 'Morphological, thermal and rheological properties of starches from different botanical sources', *Food Chem.*, 81, 219–31.
Singleton, W. R. (1939), 'Recent linkage studies in maize: V. opaque endosperm-2 (o2)', *Genetics*, 24, 61–3.
Smih, F., Rouet, P., Romanienko, P. J. and Jasin, M. (1995), 'Double-strand breaks at the target locus stimulate gene targeting in embryonic stem cells', *Nucleic Acids Res.*, 23, 5012–19.
Srichuwong, S. and Jane, J. (2007), 'Physicochemical properties of starch affected by molecular composition and structures', *Food Sci. Biotechnol.*, 16, 663–74.
Stavric, B. (1994), 'Biological significance of trace levels of mutagenic heterocyclic aromatic amines in human diet: a critical review', *Food Chem. Toxicol.*, 32(10), 977–94.
Tester, R. F. and Karkalas, J. (2002), 'Starch'. In Steinbüchel, A. (Ed.), Biopolymers, vol. 6. Polysaccharides. Wiley–VCH, pp. 381–438.
Tester, R. F. and Karkalas, J., Qi, X. (2004), 'Starch structure and digestibility Enzyme- Substrate relationship', *World's Poult. Sci. J.*, 60, 186–95.
Torrent, M., Alvarez, I., Geli, M. I., Dalcol, I. and Ludevid, D. (1997), 'Lysine-rich modified gamma-zeins accumulate in protein bodies of transiently transformed maize endosperms', *Plant Mol. Biol.*, 34, 139–49.
Tsai, C. Y. and Glover, D. V. (1974), 'Effect of brittle-l sugary-l double mutant combination on carbohydrate and postharvest quality or sweet corn', *Crop Sci.*, 14, 808–10,
Tsuda, T., Horio, F., Uchida, K., Aoki, H. and Osawa, T. (2003), 'Dietary cyanidin 3-O-β-D-glucoside-rich purple corn color prevents obesity and ameliorates hyperglycemia in mice', *J. Nutr.*, 133, 2125−30.
Vasal, S. K. (2000), 'Quality protein maize story', *Food Nutr. Bull.*, 21, 445–50.
Walsh, M. C., Buzoianu, S., Gardiner, G., Rea, M., Ross, R., et al. (2012), 'Effects of short-term feeding of Bt MON810 maize on growth performance, organ morphology and function in pigs', *Br J Nutr.*, 107(3), 364–71.
Wang, G., Wang, F., Wang, G., Wang, F., Zhang, X. W., et al. (2012), 'Opaque1 encodes a myosin XI motor protein that is required for endoplasmic reticulum motility and protein body formation in maize endosperm', *Plant Cell*, 24, 3447–62.
Wang, M., Liu, C., Li, S., Zhu, D., Zhao, Q. and Yu, J. (2013), 'Improved nutritive quality and salt resistance in transgenic maize by simultaneously overexpression of a natural lysine-rich protein gene, SBgLR, and an ERF transcription factor gene, TSRF1', *Int. J. Mol. Sci.*, 14(5), 9459–74.
Wang, T., Wang, M., Hu, S., Xiao, Y., Tong, H., et al. (2015), 'Genetic basis of maize kernel starch content revealed by high-density single nucleotide polymorphism markers in a recombinant inbred line population', *BMC Plant Biology*, 15, 288.
Wei, C., Xu, B., Qin, F., Yu, H., Chen, C., Meng, X., et al. (2010), 'C-type starch from high-amylose rice resistant starch granules modified by antisense RNA inhibition of starch branching enzyme', *J. Agri.c Food Chem.*, 58, 7383–8.
Weichel, M., Glaser, A. G., Ballmer-Weber, B. K., Schmid- Grendelmeier, P. and Crameri, R. (2006a), 'Wheat and maize thioredoxins: a novel cross-reactive cereal allergen family related to baker's asthma', *J. Allergy Clin. Immunol.*, 117(3), 676–81.

Weichel, M., Vergoossen, N. J., Bonomi, S., Scibilia, J., Ortolani, C., et al. (2006b), 'Screening the allergenic repertoires of wheat and maize with sera from double-blind, placebo-controlled food challenge positive patients', *Allergy*, 61(1), 128–35.

WHO/FAO/UNU Expert Consultation (2007), 'Protein and Amino Acid Requirements in Human Nutrition', World Health Organization, Geneva Report of a Joint WHO/FAO/UNU Expert Consultation. World Health Organization Technical Report No. 935.

Woo, Y. M., Hu, D. W. N., Larkins, B. A. and Jung, R. (2001), 'Genomics analysis of genes expressed in maize endosperm identifies novel seed proteins and clarifies patterns of zein gene expression', *Plant Cell*, 13, 2297–317.

Wu, X. R., Kenzior, A., Willmot, D., Scanlon, S., Chen, Z., et al. (2007), 'Altered expression of plant lysyl tRNA synthetase promotes tRNA misacylation and translational recoding of lysine', *Plant J.*, 50(4), 627–36.

Wu, Y., Holding, D. and Messing, J. (2010), 'γ-zein is essential for endosperm modification in quality protein maize', *Proc. Natl. Acad. Sci. USA.*, 107, 12810–15.

Wu, Y. and Messing, J. (2012), 'RNA interference can rebalance the nitrogen sink of maize seeds without losing hard endosperm', *PLoS ONE*, 7(2), e32850. doi: 10.1371/journal.pone.0032850.

Wu, Y., Yuan, L., Guo, X., Holding, D. R. and Messing, J. (2013), 'Mutation in the seed storage protein kafirin creates a high-value food trait in sorghum', *Nat. Commun.*, 4, 2217.

Young, V. R., Scrimshaw, N. S., Pellet, P. L. (1998), 'Significance of dietary protein source in world nutrition: animal and/or plant protein?'. In Waterlow, J. C., Armstrong, D. G., Fowden, L. and Riley, R. (Eds), *Feeding a World Population of More than 8 Billion People: A Challenge to Science*. Oxford University Press, pp. 205–12.

Yu, J., Peng, P., Zhang, X., Zhao, Q., Zhy, D., et al. (2004), 'Seed specific expression of lysine rich protein Sb401 gene significantly increases both lysine and total protein in maize seeds', *Mol. Breed.*, 14, 1–7.

Yuan, L., Dou, Y., Kianian, S. F., Zhang, C. and Holding, D. R. (2014), 'Deletion mutagenesis identifies a haploinsufficient role for g-zein in opaque2 endosperm modification', *Plant Physiol.*, 164, 119–30.

Yue, J., Li, C., Zhao, Q., Zhu, D. and Yu, J. (2014), 'Seed-specific expression of a lysine-rich protein gene, GhLRP, from cotton significantly increases the lysine content in maize seeds', *Int. J. Mol. Sci.*, 15(4), 5350–65.

Zhang, X., Colleoni, C., Ratushna, V., Sirghie-Colleoni, M., James, M. G. and Myers, A. M. (2004), 'Molecular characterization demonstrates that the *Zea mays* gene sugary2 codes for the starch synthase isoform SSIIa', *Plant Mol. Biol.*, 54, 865–79.

Zhang, G., Ao, Z. and Hamaker, B. (2008), 'Nutritional property of endosperm starches from maize mutants: a parabolic relationship between slowly digestible starch and amylopectin fine structure', *J. Agric. Food Chem.*, 56(12), 4686–94.

Zhao, X., Tan, G., Xing, Y., Wei, L., Chao, Q., Zuo, W., et al. (2012), 'Marker-assisted introgression of qHSR1 to improve maize resistance to head smut', *Mol. Breed.*, 30(2), 1077–88.

Zolla, L., Rinalducci, S., Antonioli, P. and Righetti, P. G. (2008), 'Proteomics as a complementary tool for identifying unintended side effects occurring in transgenic maize seeds as a result of genetic modifications', *J. Proteome Res.*, 7(5), 1850–61.

Zou, G. H., Zhai, G. W., Feng, Q., Yan, S., Wang, A. H., Zhao, Q., et al. (2012), 'Identification of QTLs for eight agronomically important traits using an ultra-high-density map based on SNPs generated from high-throughput sequencing in sorghum under contrasting photoperiods', *J. Exp. Bot.*, 63(15), 5451–62.

Chapter 10

Analysing maize grain quality

Glen P. Fox and Tim J. O'Hare, The University of Queensland, Australia

1 Introduction

Maize is the world's most produced cereal and supplies a large portion of the daily calories for billions of people. It also contributes to farmers' income and food supply in third-world countries. Table 1 shows some of the uses for maize.

Various physical properties of maize can be assessed visually. Maize has a very large germ (embryo) compared to other cereals. Another obvious physical attribute is its large

Table 1 Types of products from milling processes

Processing	Products
Dry milling	• Grits • Polenta • Semolina • Flour
Wet milling	• Starches – Regular – Waxy – High amylose – Modified • By-products

http://dx.doi.org/10.19103/AS.2016.0001.14

seed, composed mostly of the starchy endosperm. The large endosperm is full of starch and protein (the storage protein called zein). In most maize hybrids, the starch is typical of most cereals, with a composition of around 70–80% amylopectin and 20–30% amylose.

Since starch is the most abundant component in maize, we begin by focusing on the latest methods to understand the variation in starch structure as well as the amylose and amylopectin structures that comprise the starch endosperm. We review methodologies to test aspects of maize quality that are of interest for human health, such as protein content, phenol content (in particular, levels of flavonoids and carotenoids) and levels of myotoxins. Understanding of the hardness and texture of maize is important for commercial milling processes, and so we also review how these qualities can be determined.

2 The starch content of maize

Maize, like all cereals, has high starch content. In the case of maize, the kernel consists of approximately 75% starch (dry basis). Starch provides energy, and during germination it would provide a source of sugar for the developing germ. When used as a source of energy for animals (including humans), maize can be prepared in a number of different ways, through cracking or milling and then cooking into a range of food products. These foods vary between regions and countries, and range from ugali (a starchy mash) in eastern Africa to porridges and breads in South American countries. Maize flour is also used as a thickening agent in cooking and in industrial products, because gelatinised starch (heated and solubilised) develops a high viscosity when heated.

The starch in maize is predominantly composed of amylopectin, which makes up around 75% of the starch content. The level of amylopectin varies between 70% and 80% in normal maize, but a genetic variation in the starch synthesis pathway can induce a kernel with starch composed of 100% amylopectin, sometimes called waxy corn. Amylose is the other starch polymer, making up around 25% of the starch content of maize. There can also be a genetic variation in the synthesis pathway to produce high amylose maize, which has an amylose content of up to 40%. Even in this case, however, amylopectin will be by far the largest simple polymer in maize, making up around 40% of the total kernel mass (dry basis).

There are major structural differences between amylopectin and amylose. Amylopectin is a highly branched polymer, similar to a tree (Fig. 1). There can be differences in the number of branches as well as the size of branches. In contrast, amylose is a relatively straight chain polymer with very few branches. Although there are differences in the size and shape of these two polymers, they form starch granules together (Fig. 2).

Measuring total starch, amylose and/or amylopectin content is relatively straightforward with high-throughput commercial kits, or with simple assays with iodine for amylose/amylopectin. This is usually the first step taken in considering the starch content of maize. Currently, the most common method is the Megazyme method. Previous methods included iodine binding and other colourimetric assays. However, starch structure can vary between different maize samples and possibly even within a heterogeneous maize kernel. Therefore, measuring total starch often provides little information on starch functionality in processing. It is only through recent advances in separation and detection technology that we have been able to elucidate the structure of starch, and advance our understanding of some of the structural differences that affect processing.

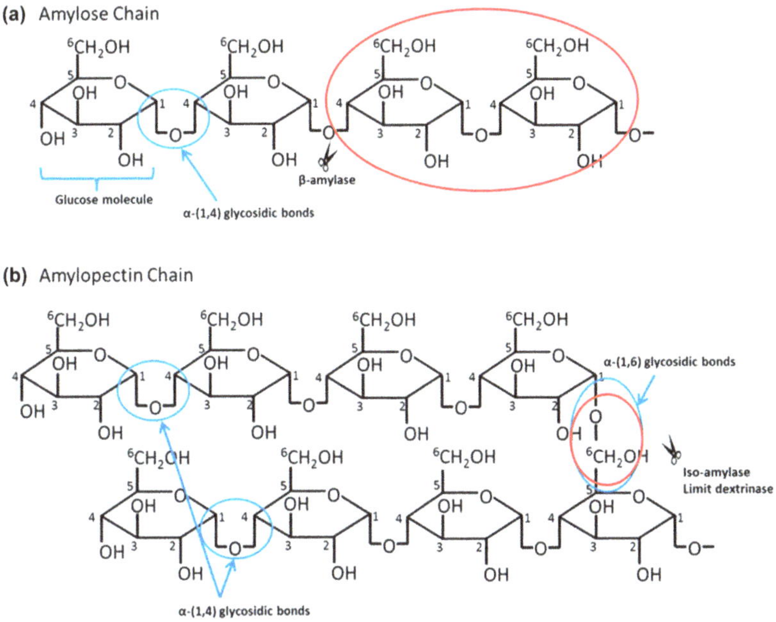

Figure 1 (a) Glucose molecules linked together with α-(1→4) linkages to form amylose, which is cut by β-amylase to form maltose (glucose + glucose) from the non-reducing end. (b) An amylopectin chain made up of chains formed by α-(1→4) linkages and branched by α-(1→6) linkages. These branches are cut by debranching enzymes isoamylase and limit dextrinase, resulting in more straight chains for hydrolysis by α-amylase and β-amylase.

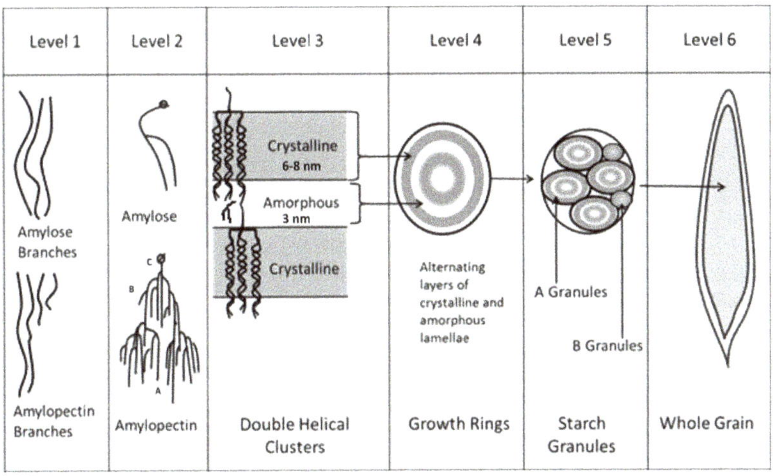

Figure 2 The six levels of starch structure in cereal grain (modified from Gilbert 2014). Modified with A, B and C chains labelled on the amylopectin polymer and the width of the crystalline and amorphous layers in nanometers (nm).

The next step in investigating starch is to consider the amount of amylopectin and amylose present, as there are a number of food uses for high amylose maize. Commercial kits are available as well as simple colourimetric assays, all of which provide information on the total amount of amylose and, by default, amylopectin, since the amount of each polymer is expressed as a percentage of total starch. This information could allow users such as cereal chemists or food technologists to select the maize samples most suited to their intended application and/or process. However, there is once again a limitation here, because these methods provide no information on the complexity of the structure, especially of the amylopectin molecules.

The next step in investigating the starch content of maize is to quantify the actual size, mass and/or level of branching of amylopectin and amylose in the sample. Recent developments using separation techniques such a high performance liquid chromatography (HPLC) can provide information about the size, mass or charge of the individual chains of amylose or amylopectin, and the distribution of the chains can be plotted. However, there is no strong relationship between the amount of amylose or amylopectin and the size of the polymer.

One key factor in accurately determining the starch structure of a sample is to avoid shearing affects (starch damage) when milling samples. Cryomilling is therefore preferred over other milling processes, which cause heating of the sample.[1] Another important factor is the appropriate solubilisation of starch, again without causing damage.

One example of an improvement in structural analysis is the use of size exclusion chromatography (SEC), also known as gel permeation chromatography. SEC is a chromatographic method to separate molecules by their size, or more accurately, on their hydrodynamic volume. This involves the assumption of universal calibration, which means the separation is solely by V_h and does not depend on the nature of polymers.[2] In SEC, the sample is dissolved and the solution is then passed through a column packed with a material containing pores. As molecules are forced through the column under pressure, smaller molecules diffuse into the column pores, thus taking a longer pathway. They are therefore eluted more slowly than larger molecules.

After the separation, a number of detectors can be used to quantify the starch chain lengths. However, SEC is only suitable to analyse molecules with small or medium size such as amylose and debranched starch molecules. The analysis of larger molecules such as amylopectin by SEC is not very accurate due to the shearing action on large molecules when being separated in the column.[2] A particular problem here is that SEC characterisation requires the molecular-level dissolution of starch without intermolecular association and without damage being caused by the dissolution process. Schmitz et al. highlighted improvements in the methodology involving dimethyl sulphoxide and lithium bromide as a combined solvent, and improvements to the equipment, including detectors which can handle this harsh solvent system while still giving acceptable signal-to-noise ratios.[3]

Once a starch sample has been prepared, there are a range of detection options including

- multiple angle laser light scattering,
- capillary electrophoresis (CE) and fluorophore-assisted CE,
- nuclear magnetic resonance and
- X-ray.

It is beyond the scope of this chapter to compare the various advantages and disadvantages of these different detection methods in detecting the size, length and mass of starch molecules. Gilbert and Sullivan have provided a useful review.[4]

With regard to specific applications in analysing the starch in maize, several reports provide detailed information on variation in maize starch structure, either as a raw product or once processed for a range of products such as films or extruded products.[5–11] The results of these studies have provided information about the activity of starch branching and debranching enzymes, and on the basis of these results a model has been proposed for both the activity of individual enzymes and the interaction between these enzymes.[12]

3 The protein content of maize

Proteins supply amino acids, which are an important part of an animal's diet. However, like most cereals, maize is quite low in protein content. The proteins in maize are similar to those found in other cereals, with storage proteins (zein in maize) being the most abundant. In addition, there are structural proteins as well as free and bound proteins within the endosperm, and all other tissue groups. The dominant protein is the zein fraction and, as in other cereals, this storage protein is found in a matrix around the starch granules, but comprises small protein body somehow acting as a bridge between the starch and protein matrix.

The amino acid profile in maize is slightly different from that in other cereals. If used for food, then it is important to consider both the balance of amino acids and whether a foodstuff is high in the essential amino acids, such as lysine and methionine.

Measuring total protein content has been a routine practice for over 100 years. It has usually followed a method originally developed by Kjeldahl, which requires digesting a sample in a strong acid, and then following a series of steps to distil the ammonia and then quantify the NH bonds present. However, in the last two decades, a more environmental-friendly, faster method has been adopted by many industries. This method is based on combustion, where total nitrogen, sulphur and carbon can be quantified. This method is based on a principle outlined prior to Kjeldahl in 1836, but the required technology was not available at that time.

During the nineteenth century, a nitrogen conversion factor was proposed. The idea was that by multiplying the nitrogen content of a foodstuff by the factor, one could arrive at the protein content of the foodstuff. Following this, foodstuff-specific nitrogen conversion factors were developed for a multitude of commodities, including maize. In 1969, Tkachuk detailed different factors for individual grain species.[13]

The analysis of proteins in maize has followed a similar path to protein analysis in other cereals, with sodium dodecyl sulphate polyacrylamide gel electrophoresis (SDS-PAGE) and isoelectric focusing (IEF) used to separate proteins based on molecular weight and charge, respectively. Figure 3 shows an SDS gel separating maize storage proteins using non-reducing and reducing reagents.

Two-dimensional electrophoresis is an innovation, where both SDS-PAGE and IEF are combined to provide a two-dimensional separation of proteins on both charge and molecular weight. However, as individual methods, these are not very accurate, even when molecular weight or pH markers are used. HPLC provides a robust quantitative method to more accurately measure the size of proteins. These methods can be used to show differences between hybrids, identify protein mutants, track changes during grain filling and monitor changes during the processing of maize. Figure 4 shows an HPLC chromatogram of maize storage proteins.

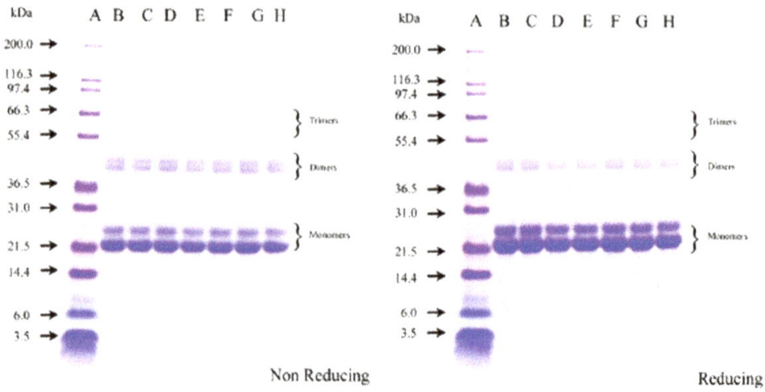

Figure 3 SDS-PAGE zein from dough (Sly et al.[14]).

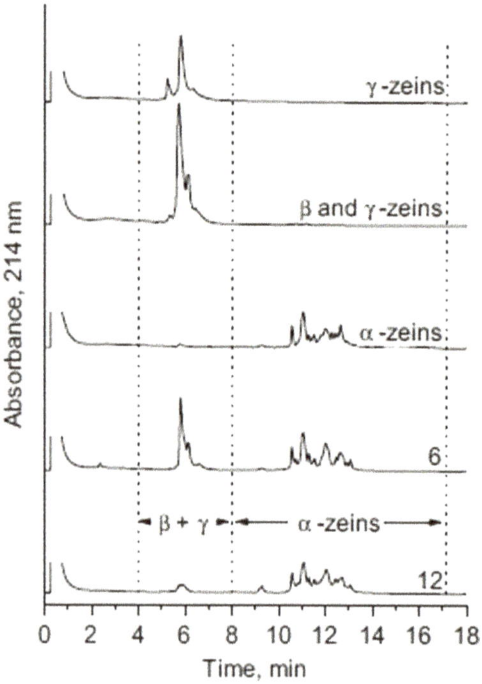

Figure 4 RPHPLC of zein proteins (Schober et al.[15]).

Table 2 An example of proteomics results from four different maize endosperm types. The proteins identified are the most abundant for that particular hybrid.

Function	UniProtID	Description
Up_*o2*	C5XX52	Glyceraldehyde-3-phosphate dehydrogenase
	B6SHW9	Ubiquitin fusion protein
	B4FAL9	Fructose-biphosphate aldolase
	B4FT23	14-3-3-like protein
	C0PHR4	Adenosylhomocysteinase
Down_*o2*	B6SIX6	Prolamin PPROL 17
	P04698	Zein-α PZ22.3
	B6SI09	Aquaporin TIP3.1
	B6TIK6	Sarcosine oxidase
Up_*fl2*	B6TNF1	Calnexin
	P24067	Luminal-binding protein 2
	A545E7	Protein disulphide isomerase
	B4FT23	14-3-3-like protein
	Q5EUD5	Protein disulphide isomerase
Down_*fl2*	Q09HU3	Trypsin inhibitor (fragment)
	Q946V2	Legumin
	Q43706	Sucrose synthase
	B6TIK6	Sarcosine oxidase
	C0LNQ9	UDP-glucosyltransferase

As already mentioned, zein is the major storage protein in maize. It can be divided into four main classes: α-, β-, γ-[9] and δ-zein.[16,17] During kernel development, zein proteins are deposited in a specific order inside the spherical body.[18] Zein comprises a family of proteins that differ in solubility characteristics, isoelectric points and molecular weights and can make up to 70% of total protein in maize.[19] Zein proteins have been connected to endosperm hardness,[20,21] and contribute anti-nutritional properties (in terms of amino acid balance, in particular low lysine and tryptophan).[22] They can be extracted for use as a biodegradable polymer.[23] Hence, the identification and classification of zein proteins would be beneficial to various industries.

SDS-PAGE has conventionally been used for characterisation of cereal storage proteins, including zein.[24–26] With SDS-PAGE, it has been shown that α-zein comprises two protein groups with molecular weights of 19 and 22 kDa; γ-zein comprises three proteins with molecular weights of 16, 27 and 50 kDa and δ-zein comprises two proteins with molecular weights of 10 and 18 kDa.[27] However, SDS-PAGE has limitations in resolution, which inhibits accurate calculation of protein mass and the differentiation of proteins with similar masses.

Mass spectrometry (MS) is a more powerful technique for providing quantitative identification of individual proteins. Zein proteins can be analysed with four MS techniques: matrix-assisted laser desorption ionisation time-of-flight MS (MALDI-TOF MS),[28,29] capillary electrophoresis MS (CE-MS),[30] reversed-phase HPLC-electrospray ionisation MS (RP-HPLC-ESI-MS)[31] and HPLC time-of-flight MS (HPLC-TOF-MS).

MALDI-TOF MS is a relatively fast technique, providing increased molecular weight accuracy and differentiation of proteins compared to SDS-PAGE. MALDI-TOF-MS has been used to determine the heritability of zein classes[29,32] and has also been used to evaluate differences between the zein profiles of high lysine transgenic parental lines and crosses.[32]

In recent years, a new technique called proteomics has been developed, which uses two-dimensional PAGE or LC-MS/MS techniques to separate hundreds of individual proteins. In combination with a MS database, proteomic methods can be used to identify these individual proteins. Proteomics has been used to identify metabolomic proteins such as 14-3-3 proteins,[33] heat shock proteins,[34] proteins produced in response to abiotic stress during kernel development[35] and proteins produced in response to *Aspergillus* infection.[36] This technique is now the best option when exploratory experiments are conducted to identify the hundreds of individual proteins present in a sample. This also allows researchers to understand protein expression in multiple metabolic pathways simultaneously. While the initial investment for the equipment is very high, high-throughput methods provide extensive data about protein levels and composition (possibly more than the researcher needs from a single experiment).

4 Phenolic compounds in maize

4.1 Overview of phenols in maize

Phenols are compounds that occur naturally in cereals, vegetables and fruits. Approximately 8000 phenolic compounds have been identified in different plant species.[37] Phenolic compounds are those that contain a benzene ring with one or more hydroxyl groups attached. These compounds can exist as soluble or insoluble phenols; soluble phenols

Figure 5 Health benefits associated with dietary polyphenols.

Source: Adapted from Pandey and Rizvi[37]

are found in cell vacuoles, while insoluble phenols are present in cell walls.[38] Phenolic compounds can be categorised as phenolic acids, condensed tannins, flavonoids, coumarins and resorcinols.[39] These different phenols affect appearance, odour, taste and oxidative stability.

When considering maize as a food, the range and level of phenol compounds matters little unless these compounds can be absorbed. Polyphenol bioavailability varies according to each polyphenol compound, and there is also no correlation between the level of polyphenols in a food and the bioavailability of polyphenols in that food once ingested. In addition, different polyphenols have different absorption sites in the human body. Some are absorbed in the gastrointestinal digestive tract, while a few are absorbed in other parts of the body.[37] Some phenol compounds are widespread between plant species, while others are specific to certain species, plant organs or developmental stages.[40]

Phenolic compounds are considered to be strong antioxidants, and they are associated with promoting human health and reducing the risk of chronic illnesses such as diabetes, cancer and cardiovascular diseases (Fig. 5). A number of epidemiological and meta-analysis studies have shown a strong protective effect of diets rich in plant polyphenols.[37]

4.2 Flavonoids

Flavonoids are compounds made up of three carbon atoms bonded by two aromatic rings. They can be classed into flavanols, flavonols, anthocyanins and flavones (Fig. 6). Maize contains numerous flavonoids, found mostly in the pericarp.

Anthocyanins

One of the most important flavonoids is anthocyanins. Anthocyanins are pigments that are soluble in water and occur naturally in plants. They are also responsible for the blue, purple and red colours in plant foods. Currently, more than 635 anthocyanins have been identified.[39] Anthocyanins are 'harmless' and easily incorporated in aqueous media, and so they are often used as water-soluble natural colourants.[41] Anthocyanins are also among the most used colourants in the food industry, especially in the beverage industry.[39] Isolated anthocyanins are very unstable and susceptible to degradation. Their stability is influenced by numerous factors, including storage temperature, pH,

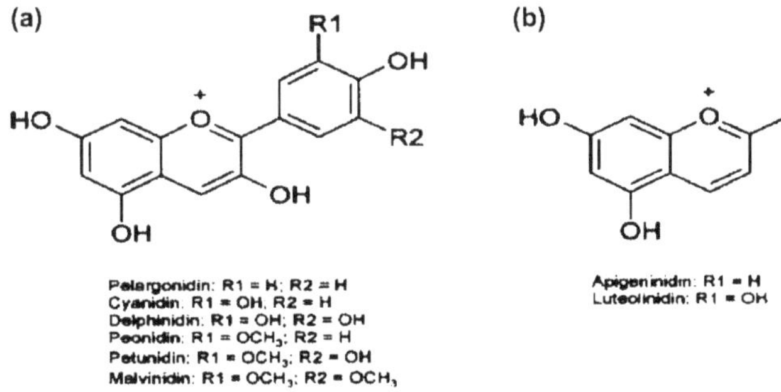

Figure 6 Structures of (a) common anthocyanins and (b) 3-deoxyanthocyanins.

Source: Adapted from Dykes and Rooney[42]

chemical structure, light, oxygen, the presence of proteins, enzymes, other flavonoids and metallic ions.[41]

Anthocyanins also have antioxidant, anti-mutagenic and anti-tumour properties. These properties play an important role in the prevention of cardiovascular and neuronal diseases, diabetes and cancer, among others.[41,43] Anthocyanins have intense colour under acidic conditions. They exist in a red flavylium cation form at pH <2, while they are colourless at pH values between 3 and 6.[39] Anthocyanins degrade at pH >7.[41] Their absorption is between 465 and 550 nm in the visible range and between 270 and 280 nm in the ultraviolet (UV) range.[44] Total anthocyanin content differs significantly across cultivars and plant species. In addition, environmental factors influence a plant's total anthocyanin content significantly.[39]

Effective anthocyanin quantification is hindered by poor separation techniques, the use of different extracting solvents and a lack of suitable, generally accepted standards for measurement procedures. Recent developments in HPLC-MS/MS techniques can quantify multiple anthocyanins present in a single sample.[45]

Carotenoids

Carotenoids are lipid-soluble pigments that naturally occur in flowers, plants, grains, vegetables, fruits, prawns and bird feathers. They are responsible for red, yellow and orange colours.[46,47] Humans and animals do not synthesise carotenoids, and must therefore acquire them from their diet.[48]

Carotenoids can be classified into two groups, namely carotenes (non-oxygenated carotenoid) and xanthophylls (oxygenated derivatives). Carotenes consist of α- and β-carotenes, whereas xanthophylls consist of β-cryptoxanthin, zeaxanthin and lutein[49,50] (Fig. 7). More than 600 carotenoids have been identified, and they have been linked to disease prevention because of their ability to scavenge free radicals and quench singlet oxygen.[51]

Carotenoid antioxidant activity is dependent on several factors,[51] including absorption, concentration, cellular distribution and dietary bioavailability. In order for them to have a positive effect, it is essential that carotenoids be released from the food matrix and combined with dietary lipids and bile salts. Lipids are therefore essential for the absorption of carotenoids and for subsequent antioxidant activity. The absorption of carotenoids is also influenced by dietary fibre, an individual's health status and the physical form of carotenoids.[52] Food processing such as particle reduction (cutting) and cooking is suggested to improve the release of carotenoids and consequently availability.

Since the turn of the twenty-first century, effort has been made to increase carotenoid concentrations in maize and sweetcorn. In the case of maize, this effort has been directed towards increasing the concentration of the pro-vitamin A carotenoid pigment β-carotene[53], while in sweetcorn the focus has been on increasing the concentration of zeaxanthin[54], one of the principal macular pigments associated with slowing macular degeneration[55,56].

Vitamin A deficiency is prevalent in many countries of sub-Saharan Africa[55]. In many of these countries, white maize is widely eaten as a principal component of the diet[55]. Although vitamin A precursor pigments (pro-vitamin A) such as β-carotene and β-cryptoxanthin exist in small amounts in yellow maize (<1 μg/g DW), they are virtually non-existent in white maize[56]. Dry weight is indicated as DW. Consequently, increasing β-carotene in maize has been targeted as a means of overcoming vitamin A deficiency. β-Carotene can be enzymatically cleaved to form two retinol molecules following consumption, while

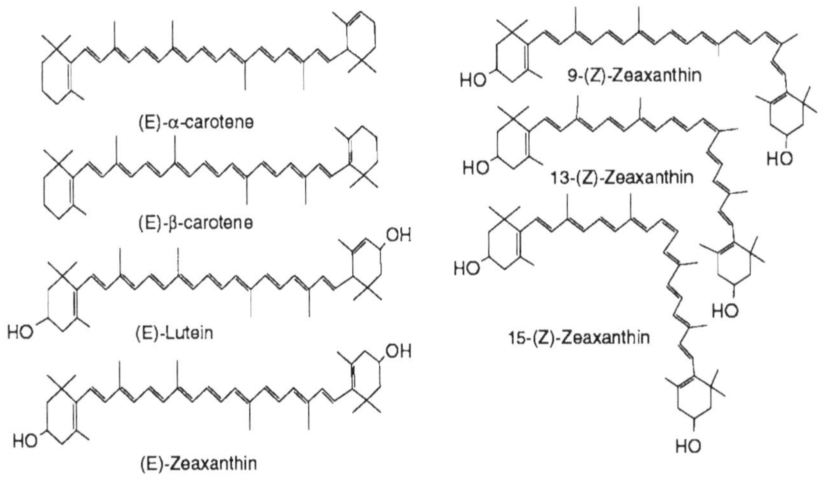

Figure 7 Major carotenoids found in maize.

Source: Adapted from Pixley et al.[57]

cleavage of β-cryptoxanthin forms only one molecule[58,59]. Consequently, effort has been focused on maximising β-carotene concentration, with a target kernel concentration of 15 μg/g DW[60]. This target has already been achieved in many breeding lines[53].

At the same time, efforts have been directed towards increasing the concentration of the carotenoid pigment, zeaxanthin, in sweetcorn, in order to provide an ample dietary source of this pigment[53] for sufferers of age-related macular degeneration, the leading cause of blindness in Australia, Europe and the United States[61]. Sweetcorn was selected, rather than maize, because it is seen as a readily adoptable part of a normal Western meal. Sweetcorn has been previously identified as one of the best sources of zeaxanthin in our diets (2–5 μg/g FW)[62,63]. Fresh weight is indicated as FW. However, in order to achieve a dosage equivalent to that used in clinical studies such as AREDS2 (2 mg/person/day), a person would need to consume 4–11cobs per day[64]. The target concentration for sweetcorn was therefore set at 20 μg/g FW,[65] so that consumption of 100 g kernels would supply sufficient zeaxanthin as part of a normal meal. This target has been achieved with experimental hybrids released for commercialisation.

The principal carotenoids present in yellow maize and sweetcorn are lutein and zeaxanthin (in that order), followed by β-cryptoxanthin, zeinoxanthin, antheraxanthin and β-carotene[66–68]. All of the carotenoids of interest for biofortification are located in the β-arm of the carotenoid biosynthesis pathway (Fig. 7)[69]. β-Carotene is the precursor of β-cryptoxanthin, while β-cryptoxanthin is the precursor of zeaxanthin. Lutein and zeinoxanthin are on the α-arm of the pathway and compete for resources for biosynthesis with the β-arm[70]. It should be stated that although both zeaxanthin and lutein are considered to be important for slowing macular degeneration, lutein concentration in sweetcorn is considerably lower than that found in dark-green leafy vegetables, by a factor of 40 to 60[63]. Lutein levels in sweetcorn were therefore not chosen as a target for biofortification.

In order to identify maize or sweetcorn germplasm with higher carotenoid levels, the principal tool used for screening is HPLC[66,71]. This involves milling of samples (at the immature eating-stage for sweetcorn, and at full kernel maturity for maize), followed by lipophilic solvent extraction (repeated several times, depending on carotenoid concentration), solvent removal and resuspension of the carotenoids in a carrier suitable for HPLC analysis. Initial studies used C18 columns for analysis, but C30 columns were found to better separate the lutein peak from the zeaxanthin peak[72] The same columns are used for analysis of β-carotene (*trans* and two *cis* isomers) and β-cryptoxanthin. The order of peak elution can vary between columns, but a typical example is shown in Figure 8. Individual carotenoid concentration is determined by peak area as compared to pure standards of each target compound[66].

Because solvent extraction and HPLC analysis are time-consuming and expensive, prescreening of germplasm with quicker and cheaper techniques is very useful[66]. Sorting kernels by colour is one method that has proved very useful. Despite many publications detailing that the difference in spectrophotometry absorbance wavelength between lutein and zeaxanthin was minor when measured in hexane or other lipophilic solvents[73,74], kernels containing higher levels of the β-arm carotenoids, zeaxanthin, β-carotene and β-cryptoxanthin are more orange than kernels with lower levels[75]. The latter are more

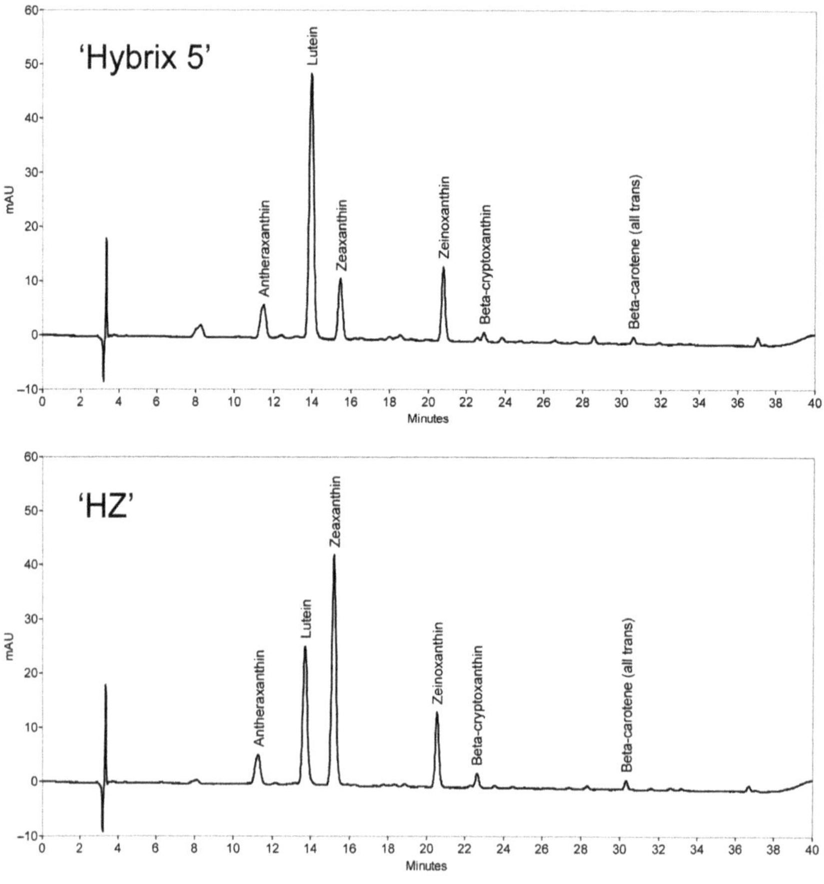

Figure 8 Typical HPLC chromatogram (YMC C30 column) showing the principal carotenoid peaks identified in 'Hybrix5' and 'HZ' sweet-corn (from Gallon et al.[69]).

yellow, and predominantly tend to contain lutein and zeinoxanthin. In general, β-arm carotenoids are orange and α-arm carotenoids are yellow, although the minor β-arm carotenoid antheraxanthin is yellow[75].

The colour of kernel has been reported to be very useful for prescreening sweetcorn germplasm (coefficient of determination of r^2 = 0.84), but less successful (r^2 = 0.24) with mature maize kernels[54,70]. This is possibly due to the use of different techniques for each biofortification programme, with the former using objective colour measurement (hue angle) with a chromameter, and the latter using subjective colour estimates by eye. It is probable that measurements by eye may be influenced by relative colour of kernel, as well as the incident light under which the kernels are sorted, which can influence the apparent colour of a kernel. By contrast, chromameter measurements are objective and produce their own standardised light source. The relationship between hue angle and zeaxanthin (or a combination of zeaxanthin, β-carotene and β-cryptoxanthin) was curvilinear such that as hue angle decreased, concentration of zeaxanthin increased. Further increase in zeaxanthin concentration was associated with less colour change, presumably due to pigment saturation near the kernel surface. A minimum hue angle cut-off of 82 degrees, however, was successful at rejecting kernels with low zeaxanthin concentrations. In both programmes for zeaxanthin and β-carotene, however, the biofortified germplasm is a deeper orange than standard yellow kernels[66,76].

Increasing kernel maturity can also be associated with an increase in the colour of kernel and an associated increase in carotenoid concentrations[69]. It is therefore important to measure kernels at the appropriate maturity for their end use. In the case of sweetcorn, kernels should be measured at optimum eating quality, which is an immature kernel stage, approximately 18–25 days after pollination, depending on moisture content[77]. Kernels should be analysed as soon as possible to avoid changes in the carotenoid profile, although these have been found to be minor under postharvest refrigeration of sweetcorn[78]. Postharvest drying of mature maize kernels can also impact on carotenoid levels, particularly (-carotene when dried at high temperatures[79].

Owing to the perishability of sweetcorn, the procedure of carotenoid analysis is often aided by freezing samples for later analysis. Freezing, however, can alter the colour of kernel, with kernels dropping in hue angle up to 5 degrees[54]. Interestingly, zeaxanthin-biofortified sweetcorn is more affected than standard yellow sweetcorn, but the relationship between fresh and frozen hue angle is very linear. This effect is not related to a change in carotenoid profile, but appears to be a physical effect brought about by ice-crystal formation. Rapid freezing using liquid nitrogen has less impact on colour, and may be associated with the formation of smaller ice crystals. It is therefore recommended that sweetcorn be measured for hue angle prior to freezing, if prescreening by the colour of kernel is to be used.

5 Maize kernel hardness/texture

Maize hardness (texture) is a physical trait where a kernel remains intact after applications of applied force.[80–82] As mentioned earlier, the endosperm makes up the largest part of the kernel, that is, approximately 85% (dry basis). The endosperm comprises starch, which makes up approximately 90% of the endosperm (dry basis), and the protein bodies embedded in a protein matrix, which make up approximately 7% of the endosperm (dry basis).[83–86] Variation in hardness is related to the ratio of the two endosperm types (floury and vitreous) present within a maize kernel. The ratio of these two types contributes to kernel hardness.[87–94]

Table 3 Examples of ranges of carotenoid content in maize kernels (μg/g DW)

Germplasm adaption	Lutein	Zeaxanthin	β-crypto	β-carotene	α-carotene	Total carotenoids	Reference
Temperate	0.0–27.5	0.01–7.7	0.07–2.4	0.07–7.6	0.0–0.7	0.15–33.1	[66]
	0.0–31.0	0.76–43.9	0.16–10.8	0.07–13.6	0.01–2.0	8.5–47.2	[64]
Tropical/subtropical	1.3–32.3	0.3–34.8	0.0–6.13	0.7–5.8	0–2.31	2.42–81.3	[63]
	0.4–19	0.3–21.5	0.3–4.6	0.3–4.3	0.0–0.6	1.3–49.5	[65]

Since maize is the most produced grain globally and it is primarily used for food, it is essential that maize kernel integrity be maintained to avoid losses for farmers, millers and consumers. For a number of food products, the maize is reduced to flour through either simple or complex milling methods and the hardness of the maize kernels will influence the milling procedure and efficiency of the process. Measuring maize hardness is important, but it is not a routine or simple process.

A number of methods and processes to measure maize hardness were reviewed by Fox and Manley (2008). Destructive methods such as particle size index (PSI) or the tangential abrasive dehulling device (TADD) can be used to assess the hardness of bulk samples.[88,95,96] PSI is a common method used to fractionate ground maize kernels through sieves of different mesh sizes. Hard maize kernels give rise to coarser particles, while soft maize kernels give rise to finer particles.[81,97]

Near-infrared (NIR) hyperspectral imaging, which combines NIR spectroscopy with digital imaging, has been used to investigate food and agricultural products non-destructively (see[98,99]). The NIR hyperspectral images provide chemical composition and spatial information about a sample, using a number of NIR wavelengths.[92,100–102] Advantages of this method include the ability to depict chemical composition and location of the chemical compounds, minimal sample preparation, no use of chemical substances and fast acquisition times.[103–107] Interpretation of hyperspectral images is possible using chemometric methods. Chemometric methods such as principal component analysis (PCA) are used for qualitative or exploratory data analysis with the help of preprocessing techniques such as multiplicative scatter correction to remove scatter effects.[100]

The NIR hyperspectral imaging has been used to classify maize[89,90,108] and wheat[109 kernels based on hardness. Other applications included detection of pregerminated wheat, barley and sorghum grains[110,111] investigation of food quality[112] and testing for contamination[113] to ensure food safety.[100,102,114–119]

A more recent technique, X-ray micro-computed tomography (μCT), is fast becoming the technique of choice in food research, because it enables three-dimensional visualisation of a sample's internal structure non-destructively and non-invasively.[120] X-ray μCT has been used to investigate various food products such as meat,[121,122] fruits and vegetables,[123,124] emulsions and foam products[125] and coffee.[126] The technique has also been used to investigate the crispiness of cornflakes produced using different mechanical and processing (i.e. batch and extruded) conditions.[127] Texture is important in terms of cornflake quality, and X-ray μCT results correlated with the destructive piston method. In another study, sensory characteristics such as the crispiness of products made from maize starch extrudates (e.g. tortilla chips) were correlated with microstructural data such as air cell wall thickness and cell diameter.[128] X-ray μCT has been used to provide information on texture variability between biscuits made from different solvent-coupled flavours.[129]

6 Mycotoxins in maize

6.1 Identification methods

All the above information describes methodologies to test the major components within maize. However, it would be remiss not to review another very important aspect of maize quality, although this is not a physical or biochemical feature of maize itself. This final aspect of maize quality is the existence and degree of fungal contamination and the possible mycotoxins produced. This is an important aspect of food safety and much attention has been paid to two major fungal species: *Fusarium* and *Aspergillus*, both of which produce mycotoxins harmful to humans and animals. Many books and reviews have been written about these fungal species, their respective toxins and their harmful impacts. The aflatoxins produced by *Aspergillus* species are especially harmful, and in some instances deadly to animals and humans.

Fungi can be identified through the usual microbiological process of plating on a specific agar, followed by incubation. Genetic analysis can identify species and real-time PCR can quantify the amount of fungus present. Quantification of the toxins requires LC-MS/MS for accurate measurement, but a number of simpler tests such as ELISA and other kits can provide an indicative level. LC-MS/MS has been shown to identify multiple toxins and/or precursors produced by multiple fungi present in a single sample. For toxins from *Fusarium* and *Penicillium* such as DON and ochratoxin, respectively, ELISA or other detection methods are sufficiently accurate as the threshold level is in parts per million. NIR has also been calibrated for these toxins, but the level of accuracy is much lower.

Aflatoxin is a major global food safety risk, which could threaten millions of people, and maize is particularly susceptible to the *Aspergillus* fungus, which produces it. For the most accurate measure, LC-MS/MS is required as the threshold is 10 parts per billion (ppb), around 1000th level of *Fusarium* toxins. There are simpler assays which can give an indication of aflatoxin levels, that is, $<$300 ppb. However, use of these methods would be questionable for grading maize for human consumption.

6.2 Sampling

For all other biochemical traits, a normal distribution of values between single kernels is expected, and this has been shown for protein.[118] However, in the case of mycotoxins, the sample is skewed: a small portion of kernels have very high levels of toxins, and these influence the overall result by significantly raising the average toxin level, usually to an acceptable level. A number of studies have shown the difference between the levels of toxins in individual kernels of maize, usually for *Fusarium* toxins. These results highlight the importance of appropriate sampling for such skewed samples.

6.3 Single kernel sorting

As mentioned above, for some attributes, sorting individual kernels with either high desirable characteristics, or in the case of mycotoxins, highly undesirable levels, may be the best option to ensure samples are safe for human consumption. In recent years, a number of physical, NIR and hyperspectral infrared applications have explored the kernel-to-kernel variation for protein,[104] hardness,[130] *Fusarium* infection[131–133] and DON levels.[134,135] A number of these applications were recently reviewed.[136] However, in order for single kernel sorting to move into full scale production, its speed and accuracy will need to be increased. When using this technology, smallholder farmers will need to address the question of what to do with rejected kernels. There are options to treat these kernels to make them acceptable, or to use them to make briquettes-type products, so that no kernels are wasted.

7 Conclusions: current uses and future trends in maize grain analysis

7.1 How research can contribute to enhanced and sustainable crop production

Current research into understanding starch structure, combined with proteomics (and other 'omics' platforms), can provide important data on functionality. While these technologies could enhance production for large commercial processors, they are likely to remain out of reach and possibly irrelevant for smallholders, who grow most of their maize for family consumption and sell a small proportion at local markets. Total daily calories is the most pressing issue for smallholders in the short term.

If it can be demonstrated that high amylose hybrids provide improved digestion, then understanding amylose structure can be seen as relevant to ensuring that 'healthy' starch is available for consumers.

The development of handheld devices for quantifying starch structure will be of value to large-scale processors, in combination with data about the improved efficiencies of certain starch structures. There will then be opportunities for farmers who produce a high yield of the relevant hybrids to sell their products to these processors.

7.2 Future trends in analysing maize grain quality

Recent developments in technology such as proteomics and metabolomics, hyperspectral imaging and field-based measurements can provide a wealth of data to researchers. These data now require specialised bioinformatic and chemometric analysis. Technologies with genome-wide selection can potentially help us develop hybrids with more food functionality, rather than just high-calorie loads. Maize hybrids with high levels of polyphenols such as anthocyanins and carotenoids will contribute to improve health. Where possible, whole maize hybrids should be replaced with coloured maize hybrids that have similar yields.

We should continue to seek a better understanding of single kernel variation, and whether and how this is important for specific processes. This research should be supported by technology companies. There are obvious benefits in mycotoxin analysis and understanding the kernel-to-kernel variation in either fungal infection or toxin accumulation. High-speed sorting devices for field and large-scale applications will ensure safe maize is always available. In addition, this type of technology could assist in optimising

Table 4 Summary of technologies available to measure maize quality

Compound group	Methods
Starch	RVA
• Amylose	Iodine
• Amylopectin	SEC
	FACE
	Colourimetric kits
Proteins	SDS-PAGE
	HLPC
	LS-MS/MS
	MALDI-TOF_MS
	Proteomics
	NIR to predict total protein
Phytochemicals	HPLC
Anthocyanins	LC-MS/MS
Carotenoids	
Hardness	Particle size index
	TADD
	RVA
	X-ray imaging

quality for high-value markets or ensuring food safety through the removal of defective kernels for humans and animals.

8 Where to look for further information

There are a number of recommended reference text which provides good detail on all aspects of maize production and quality.

P. J. White and L. A. Johnson (Eds) (2007), *Corn – Chemistry and Technology*, 2nd Edition. American Association of Cereal Chemists, St Paul, MN, USA. ISBN 1-891127-33-0

C. Wrigley, H. Corke and C. E. Walker (Eds) (2004), *Encyclopaedia of Grain Science*. Elsevier, Amsterdam. ISBN 0-12-765490-9

R. C. Hoseney (Ed.) (1998), *Principles of Cereal Science and Technology*. American Association of Cereal Chemists, 2nd Edition, St Paul, MN, USA 0-913250-79-1

For collaborations, most maize breeding companies welcome collaborations where there are mutual benefits in IP sharing.

9 References

1. Hasjim, J., Cesbron-Lavau, G., Gidley, M. J. and Gilbert, R. G. (2010), In vivo and in vitro starch digestion: are current in vitro techniques adequate? *Biomacromolecules 11*, 3600–8.
2. Cave, R. A., Seabrook, S. A., Gidley, M. J. and Gilbert, R. G. (2009), Characterization of starch by size-exclusion chromatography: the limitations imposed by shear scission, *Biomacromolecules 10*, 2245–53.
3. Schmitz, S., Dona, A. C., Castignolles, P., Gilbert, R. G. and Gaborieau, M. (2009), Assessment of the Extent of Starch Dissolution in Dimethyl Sulfoxide by H-1 NMR Spectroscopy, *Macromolecular Bioscience 9*, 506–14.
4. Gilbert, R. G. and Sullivan, M. A. (2014), The molecular size distribution of glycogen and its relevance to diabetes, *Australian Journal of Chemistry 67*, 538–43.
5. Shrestha, A. K., Blazek, J., Flanagan, B. M., Dhital, S., Larroque, O., Morell, M. K., Gilbert, E. P. and Gidley, M. J. (2015), Molecular, mesoscopic and microscopic structure evolution during amylase digestion of extruded maize and high amylose maize starches, *Carbohydrate Polymers 118*, 224–34.
6. Powell, P. O., Sullivan, M. A., Sweedman, M. C., Stapleton, D. I., Hasjim, J. and Gilbert, R. G. (2014), Extraction, isolation and characterisation of phytoglycogen from su-1 maize leaves and grain, *Carbohydrate Polymers 101*, 423–31.
7. Vilaplana, F., Hasjim, J. and Gilbert, R. G. (2012), Amylose content in starches: Toward optimal definition and validating experimental methods, *Carbohydrate Polymers 88*, 103–11.
8. Witt, T., Gidley, M. J. and Gilbert, R. G. (2010), Starch digestion mechanistic information from the time evolution of molecular size distributions, *Journal of Agricultural and Food Chemistry 58*, 8444–52.
9. Shrestha, A. K., Ng, C. S., Lopez-Rubio, A., Blazek, J., Gilbert, E. P. and Gidley, M. J. (2010), Enzyme resistance and structural organization in extruded high amylose maize starch, *Carbohydrate Polymers 80*, 699–710.
10. Hasjim, J., Lavau, G. C., Gidley, M. J. and Gilbert, R. G. (2010), In vivo and in vitro starch digestion: are current in vitro techniques adequate?, *Biomacromolecules 11*, 3600–8.
11. Hernandez, J. M., Gaborieau, M., Castignolles, P., Gidley, M. J., Myers, A. M. and Gilbert, R. G. (2008), Mechanistic investigation of a starch-branching enzyme using hydrodynamic volume SEC analysis, *Biomacromolecules 9*, 954–65.
12. Wu, A. C. and Gilbert, R. G. (2010), Molecular weight distributions of starch branches reveal genetic constraints on biosynthesis, *Biomacromolecules 11*, 3539–47.
13. Tkachuk, R. (1970), L-Cysteineglycine: its occurrence and and identification, *Canadian Journal of Biochemistry*, *48*,1029–1036.
14. Sly, A., Taylor, J. and Taylor, J. (2014), Improvement of zein dough characteristics using dilute organic acids, *Journal of Cereal Science 60*,157–63.
15. Schober, T. J., Bean, S. R., Tilley, M., Smith, B. M. and Loeger, B. P. (2011), Impact of different isolation procedures on the functionality of zein and kafirin, *Journal of Cereal Science 54*, 241–9.
16. Prat, S., Perezgrau, L. and Puigdomenech, P. (1987), Multiple variability in the sequence of a family of maize endosperm proteins, *Gene 52*, 41–9.
17. Kirihara, J. A., Petri, J. B. and Messing, J. (1988), Isolation and sequence of a gene encoding a methionine-rich 10-kDa zein protein from maize, *Gene 71*, 359–70.
18. Lending, C. R., Kriz, A. L., Larkins, B. A. and Bracker, C. E. (1988), Structure of maize protein bodies and immunocytochemical localization of zeins, *Protoplasma 143*, 51–62.
19. Prasanna, B. M., Vasal, S. K., Kassahun, B. and Singh, N. N. (2001), Quality protein maize, *Current Science 81*, 1308–19.
20. Dombrink-Kurtzman, M. A. and Bietz, J. A. (1993), Zein composition in hard and soft endosperm of maize, *Cereal Chemistry 70*, 105–8.
21. Fox, G. and Manley, M. (2009), Hardness methods for testing maize kernels, *Journal of Agricultural and Food Chemistry 57*, 5647–57.

22. Shewry, P. R. and Tatham, A. S. (1990), The prolamin storage proteins of cereal seeds – structure and evolution, *Biochemical Journal 267*, 1–12.
23. Shukla, R. and Cheryan, M. (2001), Zein: the industrial protein from corn, *Industrial Crops and Products 13*, 171–92.
24. Esen, A. (1987), A proposed nomeclature for the alcohol-soluble proteins (zeins) of maize (Zea Mays L.), *Journal of Cereal Science 5*, 117–28.
25. Esen, A. (1986), Separation of alcohol-soluble proteins (zeins) from maize into 3 fractions by differential solubility, *Plant Physiology 80*, 623–7.
26. Shewry, P. R. and Halford, N. G. (2002), Cereal seed storage proteins: structures, properties and role in grain utilization, *Journal of Experimental Botany 53*, 947–58.
27. Woo, Y. M., Hu, D. W. N., Larkins, B. A. and Jung, R. (2001), Genomics analysis of genes expressed in maize endosperm identifies novel seed proteins and clarifies patterns of zein gene expression, *Plant Cell 13*, 2297–317.
28. Wang, J. F., Geil, P. H., Kolling, D. R. J. and Padua, G. W. (2003), Analysis of zein by matrix-assisted laser desorption/ionization mass spectrometry, *Journal of Agricultural and Food Chemistry 51*, 5849–54.
29. Adams, W. R., Huang, S. S., Kriz, A. L. and Luethy, M. H. (2004), Matrix-assisted laser desorption ionization time-of-flight mass spectrometry analysis of zeins in mature maize kernels, *Journal of Agricultural and Food Chemistry 52*, 1842.
30. Erny, G. L., Marina, M. L. and Cifuentes, A. (2007), CE-MS of zein proteins from conventional and transgenic maize, *Electrophoresis 28*, 4192–201.
31. Lopez, M. C. G., Garcia-Canas, V. and Alegre, M. L. M. (2009), Reversed-phase high-performance liquid chromatography-electrospray mass spectrometry profiling of transgenic and non-transgenic maize for cultivar characterization, *Journal of Chromatography A 1216*, 7222–8.
32. Huang, S. S., Kruger, D. E., Frizzi, A., D'Ordine, R. L., Florida, C. A., Adams, W. R., Brown, W. E. and Luethy, M. H. (2005), High-lysine corn produced by the combination of enhanced lysine biosynthesis and reduced zein accumulation, *Plant Biotechnology Journal 3*, 555–69.
33. Dou, Y., Liu, X. G., Yin, Y. J., Han, S. P., Lu, Y., Liu, Y. and Hao, D. Y. (2015), Affinity chromatography revealed insights into unique functionality of two 14-3-3 protein species in developing maize kernels, *Journal of Proteomics 114*, 274–86.
34. Lazaro-Mixteco, P. E., Nieto-Sotelo, J., Swatek, K. N., Houston, N. L., Mendoza-Hernandez, G., Thelen, J. J. and Dinkova, T. D. (2012), The absence of heat shock protein HSP101 affects the proteome of mature and germinating maize embryos, *Journal of Proteome Research 11*, 3246–58.
35. Kandianis, C. B., Michenfelder, A. S., Simmons, S. J., Grusak, M. A. and Stapleton, A. E. (2013), Abiotic stress growth conditions induce different responses in kernel iron concentration across genotypically distinct maize inbred varieties, *Frontiers in Plant Science 4*.
36. Chen, Z. Y., Brown, R. L., Damann, K. E. and Cleveland, T. E. (2007), Identification of maize kernel endosperm proteins associated with resistance to aflatoxin contamination by Aspergillus flavus, *Phytopathology 97*, 1094–103.
37. Pandey, K. B. and Rizvi, S. I. (2009), Plant polyphenols as dietary antioxidants in human health and disease, *Oxidative Medicine and Cellular Longevity 2*, 270–8.
38. Naczk, M. and Shahidi, F. (2006), Phenolics in cereals, fruits and vegetables: Occurrence, extraction and analysis, *Journal of Pharmaceutical and Biomedical Analysis 41*, 1523–42.
39. He, J. A. and Giusti, M. M. (2010), Anthocyanins: Natural colorants with health-promoting properties, in Doyle, M. P. and Klaenhammer, T. R. (Eds), *Annual Review of Food Science and Technology, Vol 1*, pp. 163–87.
40. Cheynier, V. (2012), Phenolic compounds: from plants to foods, *Phytochemistry Reviews 11*, 153–77.
41. Castaneda-Ovando, A., Pacheco-Hernandez, M. D., Paez-Hernandez, M. E., Rodriguez, J. A. and Galan-Vidal, C. A. (2009), Chemical studies of anthocyanins: A review, *Food Chemistry 113*, 859–71.

42. Dykes, L. and Rooney, L. (2007), Phenolic compounds in cereal grains and their health benefits, *Cereal Foods World 52*, 105–11.
43. Bustos, D. V., Riegel, R. and Calderini, D. F. (2012), Anthocyanin content of grains in purple wheat is affected by grain position, assimilate availability and agronomic management, *Journal of Cereal Science 55*, 257–64.
44. Glover, B. J. and Martin, C. (2012), Anthocyanins, *Current Biology 22*, R147–R150.
45. Netzel, M., Netzel, G., Tian, Q. G., Schwartz, S. and Konczak, I. (2006), Sources of antioxidant activity in Australian native fruits. Identification and quantification of anthocyanins, *Journal of Agricultural and Food Chemistry 54*, 9820–6.
46. Menkir, A., Liu, W. P., White, W. S., Mazlya-Dixon, B. and Rocheford, T. (2008), Carotenoid diversity in tropical-adapted yellow maize inbred lines, *Food Chemistry 109*, 521–9.
47. Jamieson, H. (2013), Carotenoids and health, *Annals of Nutrition and Metabolism 62*, 23.
48. Fraser, P. D. and Bramley, P. M. (2004), The biosynthesis and nutritional uses of carotenoids, *Progress in Lipid Research 43*, 228–65.
49. Kean, E. G., Hamaker, B. R. and Ferruzzi, M. G. (2008), Carotenoid bioaccessibility from whole grain and degermed maize meal products, *Journal of Agricultural and Food Chemistry 56*, 9918–26.
50. Hammond, B. R. and Renzi, L. M. (2013), Carotenoids, *Advances in Nutrition 4*, 474–6.
51. Rao, A. V. and Rao, L. G. (2007), Carotenoids and human health, *Pharmacological Research 55*, 207–16.
52. Voutilainen, S., Nurmi, T., Mursu, J. and Rissanen, T. H. (2006), Carotenoids and cardiovascular health, *American Journal of Clinical Nutrition 83*, 1265–71.
53. Pixley, K., Rojas, N. P., Mutale, R., Surles, R. and Simpungwe, E. (2013), Chapter 17: Biofortification of maize with provitamin A carotenoids, in Sherry A. Tanumihardjo (Ed.), *Carotenoids and Human Health*, pp. 271–92. Humana Press, New York.
54. O'Hare, T. J., Fanning, K. J. and Martin, I. F. (2015a), Zeaxanthin biofortification of sweet-corn and factors affecting zeaxanthin accumulation and colour change, *Archives of Biochemistry and Biophysics 572*, 184–7.
55. Ma, L., Dou, H. L., Wu, Y. Q., Huang, Y. M., Huang, Y. B., Xu, X. R., Zou Z. Y. and Lin, X. M. (2012). Lutein and zeaxanthin intake and the risk of age-related macular degeneration: a systematic review and meta-analysis, *British Journal of Nutrition 107*, 350–9.
56. Chew, E. Y., Clemons, T. E., SanGiovanni, J. P., Danis, R., Ferris 3rd, F. L., Elman, M., Antoszyk, A., Ruby, A., Orth, D., Bressler, S., Fish, G., Hubbard, B., Klein, M., Chandra, S., Blodi, B., Domalpally, A., Friberg, T., Wong, W., Rosenfeld, P., Agron, E., Toth, C., Bernstein, P. and Sperduto,R. (2013), Lutein + zeaxanthin and omega-3 fatty acids for age-related macular degeneration - the age-related eye disease study 2 (AREDS2) randomized clinical trial, *J.A.M.A. 309*, 2005–15.
57. WHO (2009), WHO global database on Vitamin A deficiency. Geneva, World Health Organisation, 2009.
58. Li, S., Tayie, F. A., Young, M. F. and Rocheford,T. (2007), Retention of provitamin A carotenoids in high beta-carotene maize (Zea mays) during traditional African household processing *Journal of Agricultural and Food Chemistry 55*, 10744–750.
59. Menkir, A., Liu, W., White, W. S., Maziya-Dixon, B. and Rocheford, T. (2008), Carotenoid diversity in tropical-adapted yellow maize inbred lines, *Food Chemistry 109*, 521–9.
60. Bouis, H. E. and Welch,R. M. (2010), Biofortification – a sustainable agricultural strategy for reducing micronutrient malnutrition in the global south, *Crop Science 50*, S20–32.
61. Pascolini, D., Mariotti, S. P., Pokharel, G. P., Pararajasegaram, R., Etya'ale, D., Négrel, A. D. and Resnikoff,S. (2004), 2002 global update of available data on visual impairment: a compilation of population-based prevalence studies, *Ophthalmic Epidemiology 11*, 67–115.
62. Holden, J. M., Eldridge, A. L., Beecher, G. R. et al. (1999), Carotenoid content of U.S. foods: an update of the database, *Journal of Food Composition and Analysis 12*, 169–96.
63. Perry, A., Rasmussen, H. and Johnson,E. J. (2009), Xanthophyll (lutein, zeaxanthin) content in fruits, vegetables and corn and egg products, *Journal of Food Composition and Analysis 22*, 9–15.

64. O'Hare, T. J., Calvo-Brenes, P., Fanning, K. and Pun, S. (2015b), Freezing induces reversible colour-change in both zeaxanthin-biofortified and yellow sweet-corn, *Acta Horticulturae* (in press).
65. O'Hare, T. J., Martin, I., Fanning, K. J., Holton, T. A., Innes, D., Wong, L. S., Keating, V., Pun, S., Fekybelu, S. and Zeppa, A. (2010), Enhancing zeaxanthin in sweetcorn for age-related macular degeneration, *Sabrao Journal of Breeding and Genetics* 41, Special suppl., August 2009.
66. Fanning, K. J., Martin, I., Wong, L., Keating, V., Pun, S. and O'Hare, T., (2010), Screening sweetcorn for enhancing zeaxanthin concentration, *Journal of the Science of Food and Agriculture 90*, 91–6.
67. Kopsell, D. A., Armel, G. R., Mueller, T. C., Sams, C. E., Deyton, D. E., McElroy J. S. and Kopsell, D. E. (2009), Increase in nutritionally important sweet corn kernel carotenoids following Mesotrione and Altrazine applications, *Journal of Agricultural and Food Chemistry 57*, 6362–8.
68. Gallon, C. Z, Fuller, S. C., Fanning, K. J., Smyth, H. E., Pun, S., Martin, I. F. and O'Hare, T. J. (2013), Increase in ß-ionone, a carotenoid-derived volatile in zeaxanthin-biofortified sweetcorn, *Journal of Agricultural and Food Chemistry 61*, 7181–7.
69. Weber, E. J. (1987), Carotenoids and tocols of corn grain determined by HPLC, *Journal of the American Oil Chemists' Society 64*, 1129–34.
70. Harjes, C. E., Rocheford, T. R., Bai, L., Brutnell, T. P., Kandianis, C. B., Sowinski, S. G., Stapleton, A. E., Vallabhaneni, R., Williams, M., Wurtzel, E. T., Yan, J. and Buckler,E. S. (2008), Natural genetic variation in lycopene epsilon cyclase tapped for maize biofortification, *Science 319*, 330–3.
71. Kurilich, A. C. and Juvik, J. A. (1999), Quantification of carotenoid and tocopherol antioxidants in Zea mays, *Journal of Agricultural and Food Chemistry 47*, 1948–55.
72. Howe, J. A. and Tanumihardjo, S. A. (2006), Evaluation of analytical methods for carotenoid extraction from biofortified maize (Zea mays sp.), *Journal of Agricultural and Food Chemistry 54*, 7992–7.
73. Wrolstad, R. E., Decker, E. A. and Schwartz,S. J. (2004), *Handbook of Food Analytical Chemistry: Pigments, Colorants, Flavors, Texture, and Bioactive Food Components*. Eds. R. E. Wrolstad, T. E. Acree, E. Decker, M. H. Penner, D. S. Reid, S. J. Schwartz, C. F. Shoemaker and P. Sporns. John Wiley & Sons, Hoboken, New Jersey, 624p.
74. Zang, L. Y., Sommerburg, O. and van Kuijk, F. J. (1997), Absorbance changes of carotenoids in different solvents, *Free Radical Biology and Medicine 23*, 1086–9.
75. Melendez-Martinez, A. J., Britton, G., Vicario, I. M. and Heredia F. J. (2007), Relationship between the colour and the chemical structure of carotenoid pigments, *Food Chemistry 101*, 1145–50.
76. Heying, E. K., Grahn, M., Pixley, K. V., Rocheford, T. and Tanumihardjo, S. A. (2013), High-provitamin A carotenoid (orange) maize increases hepatic vitamin A reserves of offspring in a vitamin A-depleted sow-piglet model during lactation, *Journal Nutrition 143*, 1141–46.
77. Beckingham, C. (2007), Commodity growing guides – sweet corn. NSW Department of Primary Industries. http://www.dpi.nsw.gov.au/agriculture/horticulture/vegetables/commodity/sweet-corn.
78. O'Hare, T. J., Fanning, K., Kirchoff,S. and Tyler,P. (2015c), Developing 'superyellow' enhanced pigment sweet-corn for eye health. HIA Final Project Report, Project VT07081, 110pp.
79. Ortiz, D., Rocheford, T., Hamaker, B. R., Dufour, D. and Ferruzzi,M. G. (2014), Changes in the carotenoid profiles of biofortified corn cultivars at pre-harvest and storage stages. Abstracts of 17th Intern. Sym. on Carotenoids, 29 June to 4 July 2014, Park City, Utah, p. 141.
80. Chandrashekar, A. and Mazhar, H. (1999), The biochemical basis and implications of grain strength in sorghum and maize, *Journal of Cereal Science 30*, 193–207.
81. Pomeranz, Y., Czuchajowska, Z. and Lai, F. (1986), Comparison of methods for determination of hardness and breakage susceptibility of commercially dried corn, *Cereal Chemistry 63*, 39–43.
82. Pomeranz, Y., Hall, G., Czuchajowska, Z. and Lai, F. (1986), Test weight, hardness and breakage susceptibility of yellow dent corn hybrids, *Cereal Chemistry 63*, 349–51.

83. Evers, T. and Millar, S. (2002), Cereal grain structure and development: some implications for quality, *Journal of Cereal Science 36*, 261–84.
84. Mestres, C. and Matencio, F. (1996), Biochemical basis of kernel milling characteristics and endosperm vitreousness of maize, *Journal of Cereal Science 24*, 283–90.
85. Robutti, J., Borrás, F., González, R., Torres, R. and De Greef, D. (2002), Endosperm properties and extrusion cooking behavior of maize cultivars, *LWT – Food Science and Technology 35*, 663–9.
86. Aldrich, S. R. and Leng, E. R. (1965), Modern corn production, *Modern corn production.*
87. Dombrink-Kurtzman, M. and Bietz, J. A. (1993), Zein composition in hard and soft endosperm of maize, *Cereal Chemisty 70*, 105–8.
88. Dombrink-Kurtzman, M. and Knutson, C. (1997), A study of maize endosperm hardness in relation to amylose content and susceptibility to damage, *Cereal Chemistry 74*, 776–80.
89. Manley, M., Williams, P., Nilsson, D. and Geladi, P. (2009), Near infrared hyperspectral imaging for the evaluation of endosperm texture in whole yellow maize (Zea maize L.) kernels, *Journal of Agricultural and Food Chemistry 57*, 8761–9.
90. Williams, P., Geladi, P., Fox, G. and Manley, M. (2009), Maize kernel hardness classification by near infrared (NIR) hyperspectral imaging and multivariate data analysis, *Analytica Chimica Acta 653*, 121–30.
91. Almeida-Dominguez, H. D., Suhendro, E. L. and Rooney, L. W. (1997), Factors affecting rapid visco analyser curves for the determination of maize kernel hardness, *Journal of Cereal Science 25*, 93–102.
92. Lee, K. M., Herrman, T. J., Lingenfelser, J. and Jackson, D. S. (2005), Classification and prediction of maize hardness-associated properties using multivariate statistical analyses, *Journal of Cereal Science 41*, 85–93.
93. Lee, K.-M., Bean, S. R., Alavi, S., Herrman, T. J. and Waniska, R. D. (2006), Physical and biochemical properties of maize hardness and extrudates of selected hybrids, *Journal of Agricultural and Food Chemistry 54*, 4260–9.
94. Blandino, M., Mancini, M. C., Peila, A., Rolle, L., Vanara, F. and Reyneri, A. (2010), Determination of maize kernel hardness: comparison of different laboratory tests to predict dry-milling performance, *Journal of the Science of Food and Agriculture 90*, 1870–8.
95. Huber, K. and BeMiller, J. (2000), Channels of maize and sorghum starch granules, *Carbohydrate Polymers 41*, 269–76.
96. O'Kennedy, K. (2011), Characterisation of zein from South African maize of varying endosperm texture, In *MSc in Food Science*, University of Stellenbosch: Stellenbosch.
97. Abdelrahman, A. and Hoseney, R. (1984), Basis for hardness in pearl millet, grain sorghum, and corn, *Cereal Chemistry 61*, 232–5.
98. Elmasry, G., Kamruzzaman, M., Sun, D.-W. and Allen, P. (2012), Principles and applications of hyperspectral imaging in quality evaluation of agro-food products: a review, *Critical Reviews in Food Science and Nutrition 52*, 999–1023.
99. Manley, M. (2014), Near-infrared spectroscopy and hyperspectral imaging: non-destructive analysis of biological materials, *Chemical Society Reviews 43*, 8200–14.
100. Gowen, A. A., O'Donnell, C. P., Cullen, P. J., Downey, G. and Frias, J. M. (2007), Hyperspectral imaging – an emerging process analytical tool for food quality and safety control, *Trends in Food Science and Technology 18*, 590–8.
101. Reich, G. (2005), Near-infrared spectroscopy and imaging: Basic principles and pharmaceutical applications, *Advanced Drug Delivery Reviews 57*, 1109–43.
102. Mehl, P. M., Chen, Y.-R., Kim, M. S. and Chan, D. E. (2004), Development of hyperspectral imaging technique for the detection of apple surface defects and contaminations, *Journal of Food Engineering 61*, 67–81.
103. Orman, B. A. and Schumann Jr, R. A. (1991), Comparison of near-infrared spectroscopy calibration methods for the prediction of protein, oil, and starch in maize grain, *Journal of agricultural and food chemistry 39*, 883–6.

104. Baye, T. M., Pearson, T. C. and Settles, A. M. (2006), Development of a calibration to predict maize seed composition using single kernel near infrared spectroscopy, *Journal of Cereal Science 43*, 236–43.
105. Huang, H., Yu, H., Xu, H. and Ying, Y. (2008), Near infrared spectroscopy for on/in-line monitoring of quality in foods and beverages: A review, *Journal of Food Engineering 87*, 303–13.
106. Büning-Pfaue, H. (2003), Analysis of water in food by near infrared spectroscopy, *Food Chemistry 82*, 107–15.
107. Geladi, P. (2002), Some recent trends in the calibration literature, *Chemometrics and Intelligent Laboratory Systems 60*, 211–24.
108. McGoverin, C. M. and Manley, M. (2012), Classification of maize kernel hardness using near infrared hyperspectral imaging, *Journal of Near Infrared Spectroscopy 20*, 529–35.
109. Du Toit, G. (2009), Near infrared hyperspectral imaging and chemometrics for exploration and classification of whole wheat kernels, Stellenbosch: University of Stellenbosch, Stellenbosch.
110. McGoverin, C. M., Engelbrecht, P., Geladi, P. and Manley, M. (2011), Characterisation of non-viable whole barley, wheat and sorghum grains using near-infrared hyperspectral data and chemometrics, *Analytical and Bioanalytical Chemistry*, 1–7.
111. Arngren, M., Hansen, P. W., Eriksen, B., Larsen, J. and Larsen, R. (2011), Analysis of pre-germinated barley using hyperspectral image analysis, *Journal of Agricultural and Food Chemistry 21*, 11385–94.
112. Qiao, J., Ngadi, M. O., Wang, N., Gariépy, C. and Prasher, S. O. (2007), Pork quality and marbling level assessment using a hyperspectral imaging system, *Journal of Food Engineering 83*, 10–16.
113. Siripatrawan, U., Makino, Y., Kawagoe, Y. and Oshita, S. (2011), Rapid detection of *Escherichia coli* contamination in packaged fresh spinach using hyperspectral imaging, *Talanta 85*, 276–81.
114. Dale, L. M., Thewis, A., Boudry, C., Rotar, I., Dardenne, P., Beaten, V. and Pierra, J. A. F. (2013), Hyperspectral imaging applications in agriculture and agro-food product quality and safety control: a review, *Applied Spectroscopy Reviews 48*, 142–59.
115. Di, W. and Sun, D.-W. (2013), Advanced applications of hyperspectral imaging technology for food quality and safety analysis and assessment: a review-part II: applications, *Innovative Food Science and Emerging Technologies 19*, 15–28.
116. Moghaddam, T. M., Razavi, S. M. and Taghizadeh, M. (2013), Applications of hyperspectral imaging in grains and nuts quality and safety assesment: a review, *Journal of Food Measurement and Characterization 7*, 129–40.
117. Fernández Pierna, J. A., Vermeulen, P., Amand, O., Tossens, A., Dardenne, P. and Baeten, V. (2012), NIR hyperspectral imaging spectroscopy and chemometrics for the detection of undesirable substances in food and feed, *Chemometrics and Intelligent Laboratory Systems 117*, 233–9.
118. Del Fiore, A., Reverberi, M., Ricelli, A., Pinzari, F., Serranti, S., Fabbri, A. A., Bonifazi, G. and Fanelli, C. (2010), Early detection of toxigenic fungi on maize by hyperspectral imaging analysis, *International Journal of Food Microbiology 144*, 64–71.
119. Williams, P. J. (2013), Near infrared (NIR) hyperspectral imaging and X-ray computed tomography combined with statistical and multivariate data analysis to study Fusarium infection in maize, Stellenbosch: University of Stellenbosch.
120. Bull, C. (1993), A review of sensing techniques which could be used to generate images of agricultural and food materials, *Computers and electronics in agriculture 8*, 1–29.
121. Frisullo, P., Laverse, J., Marino, R. and Nobile, M. D. (2009), X-ray computed tomography to study processed meat microstructure, *Journal of Food Engineering 94*, 283–9.
122. Young, M., Lewis, R., Jones, H. and Wolf, B. T. (2002), The use of X-ray computer tomography for measuring the muscularity of live sheep, *Animal Science*.
123. Mendoza, F., Verboven, P., Mebatsion, H. K., Kerckhofs, G., Wevers, M. and Nicolaï, B. (2007), Three-dimensional pore space quantification of apple tissue using X-ray computed microtomography, *Planta 226*, 559–70.

124. Léonard, A., Blacher, S., Nimmol, C. and Devahastin, S. (2008), Effect of far-infrared radiation assisted drying on microstructure of banana slices: an illustrative use of X-ray microtomography in microstructural evaluation of a food product, *Journal of Food Engineering 85*, 154–62.
125. Lim, K. and Barigou, M. (2004), X-ray micro-computed tomography of cellular food products, *Food Research International 37*, 1001–12.
126. Laverse, J., Mastromatteo, M., Frisullo, P. and Del Nobile, M. A. (2012), X-ray microtomography to study the microstructure of mayonnaise, *Journal of Food Engineering 108*, 225–31.
127. Chaunier, L., Della Valle, G. and Lourdin, D. (2007), Relationships between texture, mechanical properties and structure of cornflakes, *Food Research International 40*, 493–503.
128. Agbisit, R., Alavi, S., Cheng, E., Herald, T. and Trater, A. (2007), Relationships between microstructure and mechanical properties of cellular cornstarch extrudates, *Journal of texture studies 38*, 199–219.
129. Yang, N., Fisk, I. D., Linforth, R., Brown, K., Walsh, S., Mooney, S., Sturrock, C. and Hort, J. (2012), Impact of flavour solvent on biscuit micro-structure as measured by X-ray micro-Computed Tomography and the distribution of vanillin and HMF (HPLC), *European Food Research and Technology 235*, 1083–91.
130. Williams, P., Geladi, P., Fox, G. and Manley, M. (2009), Maize kernel hardness classification by near infrared (NIR) hyperspectral imaging and multivariate data analysis, *Analytica Chim Acta 653*, 121–30.
131. Williams, P., Manley, M., Fox, G. and Geladi, P. (2010), Indirect detection of Fusarium verticillioides in maize (Zea mays L.) kernels by near infrared hyperspectral imaging, *Journal of Near Infrared Spectroscopy 18*, 49–58.
132. Williams, P. J., Geladi, P., Britz, T. J. and Manley, M. (2012), Near-infrared (NIR) hyperspectral imaging and multivariate image analysis to study growth characteristics and differences between species and strains of members of the genus Fusarium, *Analytical and Bioanalytical Chemistry 404*, 1759–69.
133. Williams, P. J., Geladi, P., Britz, T. J. and Manley, M. (2012), Investigation of fungal development in maize kernels using NIR hyperspectral imaging and multivariate data analysis, *Journal of Cereal Science 55*, 272–8.
134. Pearson, T. C. and Wicklow, D. T. (2006), Detection of corn kernels infected by fungi, *Transactions of the Asabe 49*, 1235–45.
135. Pearson, T. C., Wicklow, D. T. and Pasikatan, M. C. (2004), Reduction of aflatoxin and fumonisin contamination in yellow corn by high-speed dual-wavelength sorting, *Cereal Chemistry 81*, 490–8.
136. Fox, G. and Manley, M. (2014), Applications of single kernel conventional and hyperspectral imaging near infrared spectroscopy in cereals, *Journal of the Science of Food and Agriculture 94*, 174–9.

Part 3

Translating research into practice: improving maize cultivation in the developing world

Chapter 11

Constraints in adopting improved technologies for maize cultivation: the case of Africa

T. Abdoulaye, The International Institute of Tropical Agriculture (IITA), Nigeria; A. S. Bamire and A. A. Akinola, Obafemi Awolowo University, Nigeria; and A. Alene, A. Menkir and V. Manyong, The International Institute of Tropical Agriculture (IITA), Nigeria

1 Introduction

Improved agricultural technology is essential for increasing agricultural productivity, achieving food self-sufficiency and alleviating poverty and food insecurity in Africa. Although Hayami and Ruttan (1985) stated that the process of agricultural technology and growth has remained outside the concern of most development economists, the government has played a critical role in raising productivity through the promotion of new technologies (Schultz, 1964). The adoption of improved agricultural technologies such as improved seeds, fertilizers, herbicides, pesticides, zero tillage, other soil fertility management practices (e.g. animal manure, legume intercrop, crop residue, ridging, mounding, earth bunding, contour ploughing and live fence) and irrigation systems has been associated with higher earnings and lower poverty (Bamire et al., 2002; Kassie et al., 2011; Idrisa et al., 2014; CIMMYT, 2015), improved nutritional status (Kumar and Quisumbing, 2010), lower staple food prices (de Janvry and Sadoulet, 2002; Karanja et al., 2003) and increased

http://dx.doi.org/10.19103/AS.2016.0001.18

employment opportunities as well as earnings for landless labourers (Binswanger and von Braun, 1991). Indeed, the adoption of improved technologies is heralded as a major driver of the success of the Green Revolution in Asia (Ravallion and Chen, 2004).

Improved technology is, nonetheless, particularly relevant for smallholder farmers who are disadvantaged in many ways, thereby making them a priority for development efforts. However, technologies are of little value unless farmers judge them to be appropriate and subsequently adopt them. An improved technology may increase yields or agricultural output, but that does not necessarily mean that it should be adopted or selected for use by an individual or an organization. For example, some crops may have higher yields, but they may also be more sensitive to drought, thereby requiring large investments in irrigation infrastructure for them to be profitable (La Rovere et al., 2010). These investments may be very costly in some cases since the labour and capital costs of infrastructure development are real costs. Developing an understanding of what farmers are doing, how they are doing it and why they are doing it the way it is done is highly important in order to identify factors that influence farmers' decisions on farm cropping systems and factors that must be considered to improve cropping systems.

Understanding the process of technology adoption in maize production is, therefore, imperative in identifying the technologies in use, predicting adoption patterns of the technologies, supporting adopters to sustain the process and providing information on the most favourable way of developing and marketing new technologies for food security and economic well-being of people. For example, a study of the maize production trends in East Africa (Ethiopia Kenya and Tanzania) shows that maize yields increased from 1 tonne/ha in the early 1960s to 1.5 tonnes/ha in the mid-1980s due partly to the increase in use of improved maize seed and fertilizer, which many farmers have adopted over the years (De Groote et al., 2003). The adoption process has, however, been truncated in Kenya and Tanzania because despite the liberalization of the seed and fertilizer sectors in these countries, sales have not increased. On the other hand, the maize sector in Ethiopia is still largely under government control and a strong extension service with credit programme increased the use of improved maize seed and fertilizer in the late 1990s.

Thus, apart from developing new agricultural technologies, understanding the underlying forces of adoption and identifying the drivers of adoption, as well as the tools and methods for promoting their adoption by smallholder farmers, become important. The adoption of improved technologies has been documented at farm and national levels in Africa (Bamire et al., 2010; Abdoulaye et al., 2012, 2013). However, very few studies have been carried out to synthesize the dynamics of adoption of improved maize production technologies in Africa. Information on technology in use and on the know-how of more successful farmers might aid in developing improved technology for other farmers. An understanding of existing farming methods and the agronomic, economic and social constraints facing farmers also helps in determining how new technology could be modified to be practical and acceptable to farmers. The information can be used as a benchmark to measure progress made in increasing agricultural production and in improving farmers' welfare.

This chapter qualitatively reviews studies on adoption of improved technologies for maize production in Africa by using a comprehensive search in the databases, such as Web of Science and Google Scholar. Apart from the introduction, the review examines the theory used to understand farmers' decisions on adoption of improved technologies and identifies the factors influencing adoption. It also highlights improved technologies used in maize production in Africa; examines the different variables and constraints to their adoption and develops a framework that integrates, rearranges and classifies the variables and constraints into groups and examines strategies for promoting the adoption of improved maize technologies in Africa.

2 Understanding farmers' decisions on improved technology adoption: the adoption theory

The economic literature on farmers' decisions is based on normative theory and on the assumption that decisions can be modelled only in terms of the individual's maximization of profit (Austin et al., 1998; Willock et al., 1999). Profit maximization is a standard assumption in neoclassical production economics and the farmer who is guided by the profit maximization rationale must choose between alternatives in combining inputs to generate outputs. Farmers' decisions and behaviour is not, however, driven only by the maximization of profit (Austin et al., 1998; Willock et al., 1999) but could be studied from two different approaches: purely economic models based on the expected utility theory and the social-psychology theory.

The expected utility theory presents a farmer that compares any improved technology with the traditional technology and adopts it if its expected utility (based on his/her perceptions of benefits and costs) is greater than that of the traditional technology (Batz et al., 1999). Although the utility function is unobserved, the relationship between the expected utility for each of the two technologies is hypothesized to be a function of the vector of some observed variables and an error term (Adesina and Zinnah, 1993; Batz et al., 1999). Empirical studies have analysed the impact of different variables on each farmer's adoption decisions using econometric models such as logit, probit and tobit (Batz et al., 1999).

The social-psychology theory scientifically involves the use of psychological constructs such as farmer's thoughts, feelings and behaviours to explain their decision to adopt an improved technology. Within this context, Fishbein and Ajzen (1991) developed the theory of reasoned action to understand farmers' behaviour and this was extended to the theory of planned behaviour. Reimer et al. (2012) extended the theory of planned behaviour by using reasoned action approach (Fishbein and Ajzen, 2010). The theories tend to overlap in terms of individual attitudes, using psychological constructs to frame human behaviour with the assumption that individual behaviour originates from his/her intentions to perform a specific behaviour (Hansson et al., 2012). Martínez-García et al. (2013) defined farmers' intention as being determined by his attitude and the traditional way of doing things based on his objectives and goals. For example, a farmer who believes that an improved technology increases profitability will evaluate this outcome better if his objective is to increase profit. Thus, a positive attitude is developed towards adoption when a person evaluates that his decision has a higher probability to lead to positive outcomes than to negative ones. Attitude and utility have, however, been interpreted to be the same by Lapple and Kelley (2013). This implies that whenever only one of these theories is used, some gaps are identified: farmers' objectives and goals are expressed differently in the two theories; background factors and perceptions are used mainly in expected utility theory, while beliefs and psychological constructs that consider the social pressure on farmers to adopt an improved technology are used in socio-psychology theory. Also, studies in socio-psychology do not consider the importance of information acquisition and

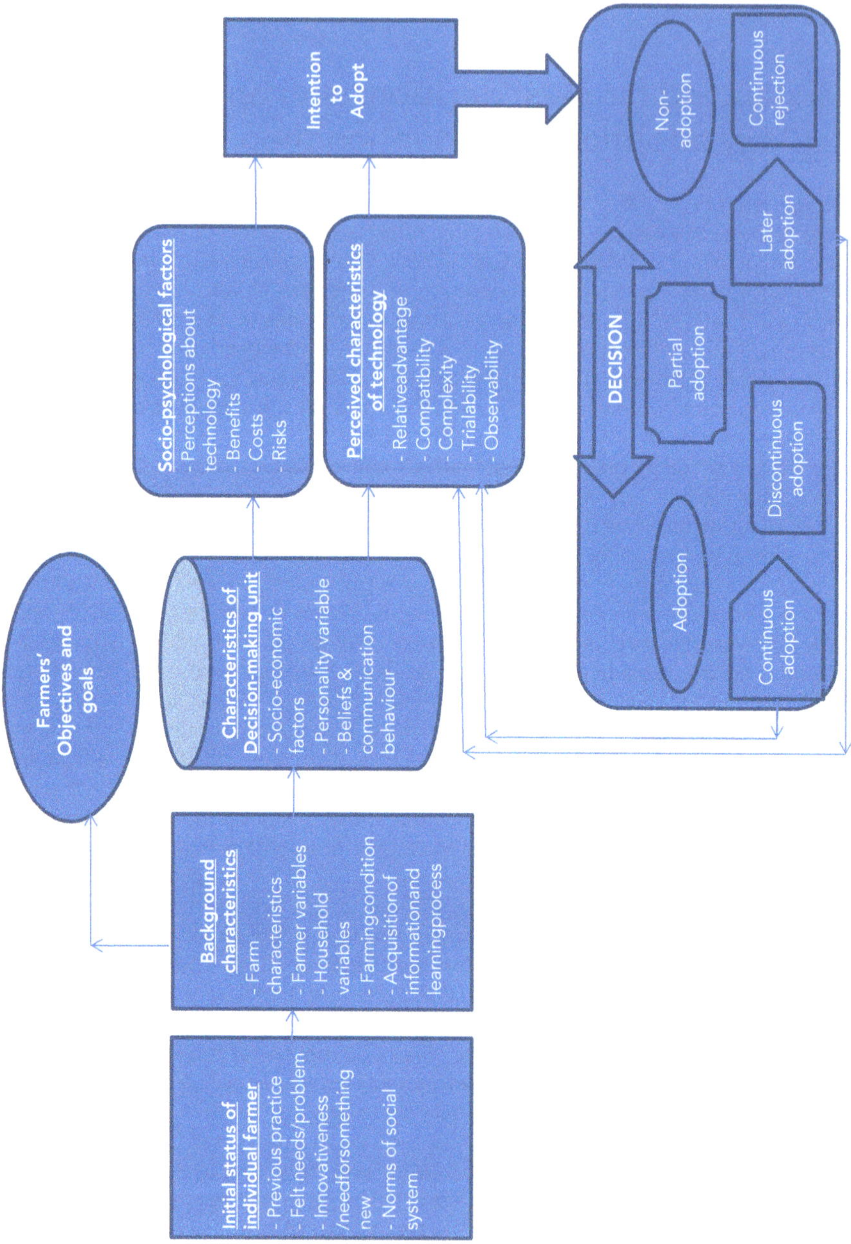

Figure 1 Framework for technology adoption decisions.
Source: Adapted from Joao et al. (2015).

the learning process. Since the adoption decision is influenced by an array of complex interrelated variables, it becomes imperative to consider the use of different theories rather than only one theory to have robust estimates from research studies. The combination of the expected utility theory and socio-psychology theory, therefore, overcomes the gaps identified in adoption decisions.

The general framework for adoption decisions of an improved technology could be based on how farmers make decisions in practice or on how they should make decisions (Öhlmér et al., 1998; Negatu and Parikh, 1999). The adoption decision framework is a dynamic process (see Fig. 1) which starts with the assumption that a farmer is aware of the new technology. Awareness here means that the farmer knows and has acquired relevant information on new technology and its practical relevance to his condition (Pannell et al., 2006; Adegbola and Gardebroek, 2007). For instance, farmer awareness of groundwater contamination increases the probability that a farmer will adopt innovations that help him solve this problem. On awareness, the farmer has the following options: immediately adopts, partially adopts and tries the technology, or rejects adoption.

3 Adoption of improved maize technologies

The backbone of any agricultural revolution is farmers' access to improved technologies. Maize (*Zea mays* L.) is one of the common and important food crops across Africa. It is a widely grown staple food crop that occupies over 33 million ha land every year, representing about 17% of the estimated 200 million ha cultivated land in sub-Saharan Africa (FAOSTAT, 2015). More than 208 million people in Africa depend on maize as a source of food and livelihood, with the top 19 countries, including South Africa, Nigeria, Ethiopia, Tanzania, Malawi, Kenya, Zambia, Uganda, Ghana, Mozambique, Cameroon, Mali, Burkina Faso, Benin, DR Congo, Angola, Zimbabwe, Togo and Cote d'Ivoire, accounting for 96% of the total maize production in SSA (FAOSTAT, 2015). The high demand of maize as a food crop is because of its high energy and nutritional value, rich in vitamins A, C and E; several essential minerals; and protein (Guang-Hai et al., 2007). It is also rich in dietary fibre and carbohydrates, which are a good source of energy. According to Nuss and Tanumihardjo (2011), maize accounts for about one-fifth of the calories and protein consumed in West Africa and almost half of the calories and protein consumed in East and Southern Africa, with nearly 73% of the 22 countries where maize constitutes the highest percentage of calorie intake in the national diet being in Africa. As its population continues to grow, the demand and consumption of maize in Africa will continue to increase rapidly over the coming years. Maize demand in Africa has been projected to increase by 79% from 29 million Metric Tons (MT) in 1997 to 52 million MT in 2020, hence the current increase in yields averaging only 1% presents a challenge for countries to meet this demand (James, 2003). The demand for maize is also expected to double by 2050 as a result of the increasing demand for feed and bio-energy (Rosegrant et al., 2007). To meet the growing maize demand, therefore, smallholder farmers in Africa have to increase maize productivity through increased adoption of improved technologies. The technologies range from improved seeds, fertilizers, conservation tillage, improved crop and soil fertility management practices, crop protection chemicals such as pesticides and herbicides, agricultural machinery, irrigation, to knowledge (Abdoulaye et al., 2012).

Over the decades, improved technologies have been developed for different agro-ecological zones to meet the needs at every stage of the production process. The majority of farmers in Africa partially adopted technologies out of a package of technologies. Adoption surveys as well as seed production and sales of improved varieties can be used to estimate technology adoption. For instance, adoption studies conducted on different crops in Kenya in 1998 showed high rates of adoption in the high potential areas; less than 20% adoption in the low potential (coast and dry transitional zones) and about 50% in the mid-altitudes, with a country-wide average of 74% (Wekesa et al., 2002; Ouma et al., 2002). Findings from Ogunsumi et al. (2010) showed that out of a package comprising nine maize technologies, only improved variety, seed rate/spacing, weed control and fertilizer were adopted. However, 36% adopters of these technologies eventually abandoned the technologies because of inadequate capital and high production cost. The Savings and Fertilizer Initiative (SAFI), which offered farmers subsidized fertilizer at harvest time as opposed to planting time, reported between 11 and 14% increase in fertilizer adoption (Duflo et al., 2008). The adoption rate for hybrid maize was 30% among smallholder farmers in the drought-prone central rift valley of Ethiopia (Bedru and Wegary, 2014) and only 3% for all districts in Ghana (Tripp and Mensah-Bonsu, 2013). Adoption studies in Ghana also had different estimates for fertilizer use across districts in the middle parts of the country, with 66% (Mensah-Bonsu et al., 2011) and 77% (IFPRI, 2011), while the 2012 nationwide study recorded 47% (Ragasa et al., 2013). Akinola et al. (2009, 2010), in their study of Balanced Nutrient Management Systems (BNMS), estimated an adoption rate of 40% for BNMS rotation and 48% for BNMS manure from an integrated soil fertility management (ISFM) technology package among maize farmers in Northern Guinea Savanna of Nigeria. In Malawi, only 20% of the maize land was covered with improved open-pollinated and hybrid maize varieties and more than 75% of the total seed purchase was of local varieties (Government of Malawi, 2004; World Bank, 2006), although the input subsidy programme of the Government of Malawi in the 2005 and 2006 seasons would have changed the adoption figures since then. Different high-yielding maize Open Pollinated Varieties (OPVs) and Hybrid of varying maturity have been released in Benin Republic, Ghana, Mali and Nigeria for boosting maize productivity (Abdoualye et al., 2009, 2012).

In Ghana, results from a study of 420 maize farmers on three Ghana Grains Development Project's generated maize technologies showed an adoption rate of 54% for modern varieties, 21% for fertilizer and 53% for the plant configuration recommendations (Morris et al., 1999). Only 12% of the farmers adopted all three technologies as a package. The coastal savannah zone of Ghana recorded the highest adoption rate (74%), whereas the forest zone had the lowest (33%). Male farmers recorded the highest adoption rate for modern varieties (59%), fertilizer (23%) and row planting (59%), while female farmers had 39% adoption for modern varieties, 16% for fertilizer and 38% for plant configuration recommendations. In Benin Republic, Baco et al. (2010) estimated 75% adoption for improved maize varieties, with females having the highest adoption (60%), while variations were found across agro-ecological zones. In a study of 150 maize farming households categorized into wealth groups in Mali, Fofana et al. (2010) recorded adoption rates of 58% for modern varieties, 50% for fertilizer, 59% for herbicides, 52% for insecticides and 42% for organic manure. In Ghana, variations exist in the adoption rates for herbicide use; Quinones and Diao (2011) reported 19%, whereas Mensah-Bonsu et al. (2011) reported 38%. Estimates for Africa south of the Sahara showed 3% adoption of herbicides among maize smallholders (Overfield et al., 2001), less than 5% in Africa (Gianessi and Williams, 2011) and 0.1% in Uganda (Magyembe, 1997). Adoption of zero tillage or no-till consisting of slashing, no ploughing, no burning and planting without mulch, which the Crops Research Institute, the Ministry of Food and Agriculture and other development partners actively promoted in the 1980s and 1990s was practised on only 4% of the maize area (Ekboir et al., 2002), while Mensah-Bonsu (2011) reported 38%. Similarly, 68% of maize area was not ploughed by either tractor or animal traction and 30% was not under slash-and-burn. Regasa et al. (2013) also reported that only 3% of maize area was applied with animal manure, 40% was intercropped mainly with cassava, 3% was intercropped with legumes, 16% was ploughed in with crop residue and 11% was planted in mulch. Bamire et al. (2012) studied farmers in the savanna zone of Nigeria and recorded adoption rates of 8% for local varieties, 10% for hybrids and 83% for OPVs.

The adoption rates estimated for improved maize varieties in Tanzania varied across zones in the country when considered as a percentage of the total: 8% in the central region, 19% in the eastern region, 13% in the Lake region, 19% in the northern region, 7% in the southern region, 24% in the southern highlands and 10% in the western region (Moshi, 1997). Also, based on seed sales, Hassan et al. (2001) estimated the total national maize area planted to improved maize varieties (OPVs and Hybrids) in Tanzania at only 4%. In Zambia, 83% of farmers were adopters of improved maize varieties in southern Zambia (Kalinda and Kuntashula, 2014) and 12% in Mozambique (Lopes, 2010), while Malawi recorded 55% (Katengeza et al., 2012). In DR Congo, the adoption rates were 10% for improved maize varieties, 6% for mineral fertilizers and 15% for row planting (Lambrecht et al., 2015).In general, the adoption of improved maize technologies can result in larger crop yields, increase household income and consumption and reduce poverty and inequality among smallholder farming households (Mendola, 2007; Kassie et al., 2011; Asfaw et al., 2012; Mathenge et al., 2014).

It is important to note that most adoption studies tend to be localized and this affects their quality and representativeness. For instance, Doss (2006) suggested that the use of microstudies to document adoption does not allow policy makers to get useful information on critical indicators that could be actionable. Therefore, since most adoption studies are localized in particular areas where the technologies are developed or first introduced, documenting lessons on the conditions leading to adoption of maize technologies in these areas become imperative. This will allow the designing and implementation of future programmes that will increase the uptake of improved varieties by many farmers.

4 Constraining factors in the adoption of improved maize technologies: economic and institutional factors

Despite research efforts in developing improved maize technologies, the adoption of improved crop varieties and input use remains generally low in many parts of Africa. In Uganda for instance, in spite of the resources spent on the public extension system, only 6% of farmers used improved seeds in 2006, whereas 3% used inorganic fertilizers (Uganda Bureau of Statistics, 2010). Even for farmers who initially adopted improved technologies, about 50% abandoned the variety within two years (Kijima et al., 2011; Akinbode and Bamire, 2015). Consequently, it is important to understand why adoption of agricultural technologies has remained very low in Africa despite the documented benefits of agricultural technical change. It is important to note that the factors that constrain technology adoption do not exist in isolation and the presence of one constraint may affect others.

4.1 Land availability

One of the most highlighted constraints to improved technology adoption is the availability of cultivable land (de Janvry et al., 2011; Carletto et al., 2007; Pingali et al., 2001). It is argued that availability of cultivable land helps reduce the liquidity constraints faced by households and also reduces risk aversion. Also, land availability through ownership of large plots of land can facilitate experimentation with improved technologies and also determine the pace of adoption as large landowners are more likely to be the early adopters (de Janvry et al., 2011). On the other hand, the limited availability of land may lead to the use of organic fertilizers in a poor resource setting (Bamire, 1995, 1999; Bamire et al., 2007; Akinola, 2009). Furthermore, the quality of land available may be a major factor in deciding the use of key inputs such as chemical fertilizers or adopting improved crop varieties expecting higher returns (Carletto et al., 2007; Bamire et al., 2012). In the case of Uganda, with overlapping and relatively unsecure property rights (Deininger and Ayalew Ali, 2008), availability of land alone may not lead to the adoption of improved technology. Furthermore, even in countries with secure property rights but poorly developed financial markets, land availability may not reduce the credit constraint.

4.2 Lack of capital

The cost of adopting an improved technology also constrains its adoption. Some authors argue that it is not the lack of liquidity per se but the timing of its requirement that constitutes a problem. Farmers often need cash to purchase fertilizers and seeds when rains start. However, they have sold most of their production since harvest time due to pressing needs (Abdoulaye and Sanders, 2006) leaving them without cash at planting time. In the absence of financing options, they are unable to raise the cash needed to purchase agricultural inputs. Also, the elimination of subsidies on seed and fertilizers since the 1990s due to the World Bank-sponsored structural adjustment programmes in most African countries has widened this constraint (Bamire et al., 2012; Muzari et al., 2012). In order to address the liquidity and supply constraints faced by resource-poor farmers related to technology adoption, a number of African countries, including Malawi, have implemented various forms of 'smart subsidies', an innovative delivery system that targets specific farmers to reduce common problems facing subsidy programmes (Minde et al., 2008). Also, 14 million farmers across Nigeria have received subsidized seeds and fertilizers through the e-wallet system (NATSA, 2014). The farmers have also redeemed 1.37 million MT of fertilizer worth N151 billion ($915 million) and 67 991 MT of improved maize seeds, worth N43 billion ($260 million). The fertilizer subsidy programme in Ghana encouraged more farmers to use fertilizer in their maize plots, based on the increased adoption rate from 21% (Morris et al., 1999), 25% based on the Ghana Living Standards Survey 5 conducted in 2005–6 (Quinones and Diao, 2011) and 47% in a most recent nationwide study (Ragasa et al., 2013). Based on extensive studies in Ethiopia, it has been shown that younger and much older household heads are risk averse and are less likely to adopt improved technologies. On the other hand, the availability of adult family members within households may facilitate the process of improved technology adoption because most farming households cannot easily acquire hired labour due to liquidity constraints

(Carletto et al., 2007). Furthermore, the continued availability of adult household members is an important determinant of whether households continue with the technology after adoption.

4.3 Changing prices

The relative price of agricultural output to inputs, for example, fertilizer has been declining in most countries and does not constitute an incentive to invest in input purchase (Abdoulaye and Sanders, 2005). Most cereal farmers have to sell more and more grain to be able to afford a kilogram of inorganic fertilizer. The changing prices for agricultural products also constrain improved maize technology adoption (Kijima et al., 2011). With an initial attraction by high product prices, farmers can abandon the improved technologies if the expected benefits from its adoption are lower than the prevailing costs. There are a number of ways through which profitability of products may be lowered. For crops like maize, changes in the international trade regime may negatively affect world prices and consequently reduce local prices. Most seed production systems are seasonal in nature, and this affects the annual cash flow pattern of the seed producers. For commercial seed companies, the costs of carrying large stocks for several months at high interest rates can be devastating. These problems are made worse by the need to dispose of some seed stocks as grain due to deterioration in store and/or excess production arising from poor demand estimates. The seasonality of agricultural enterprises and the changing profitability introduces a time dimension such that households may be able to adopt improved technologies for only some periods but not for all periods.

4.4 Labour

In a study conducted in West Africa (Niger), which evaluated the adoption of cereal technologies under weather/rainfall uncertainty, it was found that family labour was used to adopt labour-intensive technologies (Adesina and Sanders, 1991). This is due to rural–urban migration by both young and adult individuals within the farming community in search of a better living; while most youths now engage in fast income-generating business activities through the use of motor cycles popularly tagged 'Okada' in carrying passengers from one location to another. This has led to labour scarcity and consequent high cost of hired labour where available (Bamire et al., 2010). The study by Makokha et al. (2001) on determinants of fertilizer and manure use in maize production in Kiambu County, Kenya, reported the high cost of labour and other inputs, unavailability of demanded packages and untimely delivery as the main constraints to fertilizer adoption. Cost of hired labour was also reported by Ouma et al. (2002) as one among other factors constraining adoption of fertilizer and hybrid seed in Embu County, Kenya.

4.5 Non-functional seed sector

The farmers' systems of seed supply and crop development are the most important sources of seed in most farming systems of the world and in Africa. Despite the efforts of large seed programmes to replace the farmers' seed systems for a system in which farmers use seed as an external input, the major part of agricultural land in Africa is still sown with seed that is informally produced by farmers (Conny et al., 2002). Studies from the late 1990s and early

2000s suggest that the formal 'seed industry' in Africa provided less than 10% of the seeds needed by the farmers (Louwaars and Marrewijk, 1999; Kormawa et al., 2000). The maize seed industry in eastern and southern Africa, for example, has witnessed a proliferation of private seed companies following the liberalization and restructuring of the seed sector. However, despite the increased number of registered maize seed companies in major maize-producing countries between 1997 and 2007, the quantity of seed marketed barely doubled (Langyintuo et al., 2010). This suggests that the seed production and deployment environment is less than perfect. This is associated with the non-functional seed production and delivery systems in different countries. For example, the national seed system (NSS) in Nigeria plays a pivotal role in the development of the nation's seed industry, including the production of foundation seeds, supervision, monitoring and quality control. However, inadequate and delayed funding has made the performance of these functions ineffective. The NSS has representations at both state and regional levels. Its operations in quality control and seed certification have been drastically reduced due to inadequate numbers of trained staff and financial constraints (Badu-Apraku et al., 2014). These have led to low output, which makes it impossible to supply good quality-certified seeds to the farmers. Rather it has encouraged the sale of adulterated and unlabelled seeds in the market. For example, only 18% of maize area was planted with modern varieties and certified seed in Ghana (IFPRI, 2011). The use of the Growth Enhancement Scheme (GES) programme for reaching millions of farmers with seeds through their telephones or e-wallet system and input subsidies has, however, helped improve the seed delivery system in Nigeria and Malawi (NATSA, 2014). Targeting a formal seed sector that supplies 100% of the seed for planting is only realistic by linking formal (private and public) and farmers' seed systems as an effective strategy for improving national and local seed.

4.6 Inadequate extension delivery system

The extension services in most African countries that are entrusted with the training of manpower to handle the technical aspects of the seed industry are either constrained by inadequate finance, equipment and logistics or have collapsed. Staffs do not have training opportunities and therefore lack the knowledge and skill required to assist in seed testing, quality control and in providing technical assistance to contract growers. The inadequacy of extension officers makes it impossible to disseminate information on the seed industry, especially about the availability of improved seed varieties to the farmers.

4.7 Poor research funding

Long-term research, inputs, skill and expertise, materials and equipment are required in developing improved maize technologies. Governments in many African countries directly support scientific and technical research through budget allocations to the ministries, researches and other institutions. This is complemented with funding support from Consultative Group for International Agricultural Research (CGIAR) centres and other donor agencies that are active in the region. For example, the Bill and Melinda Gates Foundation, a founding member of Alliance for Green Revolution in Africa (AGRA), invests in basic research in a number of crops, including maize. However, these institutions can only provide about one-tenth of the overall funding required for research in improved technology in Africa that has been stagnating at that low level (Binswanger-Mkhize, 2009). Given the enormous challenges of climate change, low yields, pest and diseases, research activities in the development of improved maize technologies become imperative. However, the inadequate funding of the Ministries of Agriculture, research institutes and other scientists makes it difficult for them to engage effectively in research activities. In addition, when funds become available, they are usually disbursed late, causing distortions in the technology development and dissemination programmes while acquisition of the necessary technical materials and equipment are problematic, thereby demoralizing the researchers and scientists involved and low adoption by users of the technologies. There is, therefore, a need for African governments to have the political will to sharply increase funding for research in improved technologies. This could be through policies directed at protecting intellectual property rights regime by granting breeders/developers of the new ideas/technologies strong and long-lasting claims to the economic benefits of their ideas. The patent rights should not, however, prevent further research and the commercialization of the improved technologies. The promotion of research and development (R&D) is also essential to usher in innovation and technological change more directly.

4.8 Poor infrastructure

Improved seeds produced by the public sector are often sold to farmers through distribution channels like farmers' supply companies, agro-service centres, ADPs, cooperative societies and so on. The seed distribution system comprises all those activities involved in the flow of information about farmers' seed needs between the farmers themselves and the producing organizations. The seed distribution systems are important in the seed sector as they help in the timely distribution of improved seed to all the locations where it is required by the farmers. However, most rural areas are inaccessible due largely to poor road networks which hinder movement and performance of staff whose activities are required in the rural communities. This makes it difficult for transporters and impossible for farmers in such locations to enjoy the benefits of improved maize technology, and the few agro-dealers who find their ways into such rural areas often charge high prices for their stock. The farmers also run the risks of buying fake and adulterated seeds due to poor storage and handling and the unscrupulous behaviour of some seed traders (Abdoulaye et al., 2009). This is due to the poor seed policy environment and lack of organized institutions to handle seed distribution among farmers.

5 Constraining factors in the adoption of improved maize technologies: the importance of farmers' characteristics and perspectives

Different and diverse factors influence farmers' decisions associated with the adoption of improved technologies. Paramount is farmers' intention to adopt based on their perceptions about characteristics of the new technology. In this regard, the diffusion theory (Rogers, 2003) becomes relevant. This theory is based on five characteristics of a technology that affects its adoption: relative advantage, compatibility, complexity, observability and trialability. Relative advantage is the attitude or perceptions about benefits, costs and risks of the technology; the degree to which the technology is perceived as consistent with the existing values, past experiences and needs of potential adopters (compatibility); the degree to which the technology is perceived as relatively difficult to understand and use (complexity); the degree to which the results of the technology are visible to others (observability) and the degree to which a technology may be experimented with on a limited basis (trialability).

Apart from the intention to adopt based on perceptions about characteristics of the technology, other factors that influence farmers' adoption decisions are household characteristics, farmer characteristics or background, socio-demographic factors, farm characteristics and related institutional factors and acquisition of information (Ajayi and Oloruntoba, 2007). Furthermore, farmers' objectives and goals at the time of the decision, environmental factors and the costs and risks associated with the technology also affect technology adoption decisions. For example, Koundouri et al. (2006) showed that farmers' decision to adopt a new technology is constrained by risk factors, which are related to production risk and how the new technology can change the amount of production and profitability of the farmers.

In many studies, adoption behaviour has been explained in relation to personal characteristics and endowments, imperfect information, risk, uncertainty, institutional constraints, input availability and infrastructure (Feder et al., 1985; Koppel 1994; Rogers, 2003; Uaiene, 2009). However, a more recent strand of literature has included social networks and learning in the categories of factors determining adoption of technology (Uaiene, 2009). Some studies classify these factors into different categories. For example, Akudugu et al. (2012) grouped the determinant of agricultural technology adoption into three categories, namely, economic, social and institutional factors. Kebede et al. (1990) as cited by Lavison (2013) broadly categorized the factors that influence adoption of technologies into social, economic and physical categories; McNamara et al. (1991) categorized the factors into farmer characteristics, farm structure, institutional characteristics and managerial structure; Nowak (1987) grouped them into information, economic and ecological, while Wu and Babcock (1998) classified them under human capital, production, policy and natural resource characteristics.

A general and detailed review about some of the variables that may influence the adoption decision is given in the following subsections.

5.1 Farmers' perceptions about technology characteristics

Some studies have justified the use of perceptions about characteristics of the technology as influencing farmers' adoption decisions. Adesina and Zinnah (1993), Adesina and Baidu-Forson (1995), Negatu and Parikh (1999), Wubeneh and Sanders (2006) and Khan et al. (2008) tested the hypothesis that farmers' perceptions about characteristics or attributes of a new technology impact on the decision to adopt it. Their results support that farmers' perceptions of technology-specific attributes have a highly significant effect on adoption. Specific to maize, Bamire et al. (2003) and Justice and Abdoulaye (2012) showed that farm size and net farm income were highly significant factors that explained farmers' fertilizer use decisions in maize production in Osun state of Nigeria. According to Bamire (1999) and Abdoulaye et al. (2013), fertilizer enables the adoption of high-yielding varieties (HYVs) which can increase cereal yields several folds, and that without nutrients supply through fertilizer, HYVs cannot produce higher yields. In studying the determinants of adopting Imazapyr-resistant maize technology in Western Kenya, Mignouna et al. (2011) stated that the feature of the technology plays a critical role in the adoption decision process. They argued that farmers who perceive the technology being consistent with their needs and compatible with their environment are likely to adopt since they consider it a positive investment. The finding that farmer perceptions of technology-specific characteristics significantly condition technology adoption decisions is consistent with recent evidence in the literature, which suggests the need to go beyond the commonly considered socio-economic, demographic and institutional factors in adoption modelling (Feder et al., 1985; Feder and Umali, 1993).

5.2 Household characteristics

Here, the most important factor is to recognize that households are different in their characteristics, thus their adoption behaviour and risk-taking aptitude will be different. Abdoulaye and Lowenberg-DeBoer (2000) have classified farmers in three wealth groups according to their resource endowment and that led to different adoption patterns. Development programmes need to identify those groups depending on the context in which they operate and then tailor technology packages to each group in order to maximize uptake of their technologies. Household size is expected to have a positive effect on the adoption decision (Jara-Rojas et al., 2012). Households with greater number of relatives are more likely to adopt new technologies since they are able to share risks with relatives (Kassie et al., 2013). However, having more relatives may reduce incentives for hard work and induce inefficiency, such that farmers may exert less effort to invest in technologies. It is important to note that larger households connote increased home consumption and is expected to be inversely related with adoption (Jara-Rojas et al., 2012). Higher education levels of household members also increase the probability of adoption (Asfaw and Admassie, 2004). The off-farm employment by other household members has an indeterminate effect on adoption decisions. This is because labour outside the farm may allow better access to information about new technologies or the capacity to finance investments, but it may also divert time and effort away from agricultural activities, reducing investments in technologies and the availability of labour (Kassie et al., 2013). Off-farm income acts as an important strategy for overcoming credit constraints faced by the rural households in many developing countries (Reardon et al., 2007). It also acts as a substitute for borrowed capital in rural economies where credit markets are either missing or dysfunctional (Ellis and Freeman, 2004; Diiro, 2013). According to Diiro (2013), off-farm income is expected to provide farmers with liquid capital for purchasing productivity-enhancing inputs such as improved seed and fertilizers. However, not all technologies have shown positive relationship between off-farm income and their adoption. Some studies on technologies that are labour intensive have shown a negative relationship between off-farm income and adoption. According to Goodwin and Mishra (2004), the pursuit of off-farm income by farmers may undermine their adoption of modern technology by reducing the amount of household labour allocated to farming enterprises. Mazvimavi and Twomlow (2009) argue that farmers who have recent experience with death or illness (HIV/AIDS) in their households are more likely to reduce the intensity of adoption of innovations based on farmers' access to labour and resources. The further away a household is from input and output markets, the smaller the likelihood that they will adopt a new technology (Kassie et al., 2013). Better access to the plot increases the probability of adoption because it influences the availability of the technology, the use of outputs and inputs markets and availability of information and support organizations (Kassie et al., 2013).

5.3 Farmer characteristics, background or socio-demographic factors

Most adoption studies have attempted to measure human capital through farmer's education, age, gender and farming experience (Mignouna et al., 2011; Keelan et al., 2014). The expected effect of farmers' age is unknown. Older farmers have more experience and are better able to assess the characteristics of new technologies than younger farmers, or it could be that older farmers are more risk averse than younger farmers and have a lesser likelihood of adopting new technologies (Adesina and Baidu-Forson, 1995). The gender variable is associated with embedded norms, behaviours and practices in society that encourage or discourage adoption of a particular technology by members of the society (Meinzen-Dick et al., 2004). For example, male-headed farmers are more likely to get information about new technologies and take risky business decisions than female-headed farmers in most African societies (Asfaw and Admassie, 2004). The practice that the production of certain types of crops is the preserve of male members of society can limit the adoption of a particular technology if the crop to be promoted is grown mainly by men. This is because women constitute the majority of rural dwellers in Africa. In some countries, female-headed households are discriminated against by credit institutions and therefore make it impossible for them to finance yield-raising technologies, leading to low adoption rates.

Evidence from Ghana and Nigeria suggests that gender-linked differences in the adoption of modern maize varieties and chemical fertilizer result from gender-linked differences in access to complementary inputs (Ayinde et al., 2013). These findings suggest that ensuring more widespread and equitable adoption of improved technologies requires the introduction of measures that ensure better access for women to complementary inputs, especially land, labour and extension services (Doss and Morris, 2001). Clearly, therefore, an understanding of local cultural practices and preferences is important if all farmers are to benefit from agricultural research (Meinzen-Dick et al., 2004). For example, years of farming experience is expected to be positively related to adoption, because it may influence the ability of the farmer to obtain, process and use relevant information on a new technology (Adesina and Baidu-Forson, 1995; Adesina and Zinnah, 1993). It is expected that a farmer who has more experience also has the right behavioural beliefs about the technology. Education of the farmer has been assumed to have a positive influence on farmers' decision to adopt new technology. The educational level of a farmer increases his ability to obtain, process and use information that is relevant to the adoption of a new technology (Mignouna et al., 2011; Lavison, 2013; Namara et al., 2014). Education enhances the ability of farmers to acquire, synthesize and quickly respond to imbalances, thereby increasing the probability of adoption of a technology (Asfaw and Admassie, 2004). A study on the adoption of improved maize technologies in Ogun state also showed that an increase in the level of education significantly influenced the amount of land a farmer was likely to allocate to improved maize varieties and the appropriateness of fertilizer use in maize production.

5.4 Farm characteristics and related institutional factors

Land tenure arrangements affect the adoption decision, because a farmer is likely to manage owned land more intensely than rented land and to preserve its productivity for future generations (Roberts et al., 2004). Farm size is hypothesized to have a positive impact on the adoption decision (Adesina and Zinnah, 1993). Some technologies are termed scale-dependent because of the importance of farm size in their adoption (Bonabana-Wabbi, 2002). Lavison (2013) showed farm size as an important determinant of technology adoption as it can affect, and in turn be affected by, the other factors influencing adoption. Farm size as a determinant of technology adoption has been found to have mixed effect. Large farm size may have a positive effect on adoption of a certain technology and it may also reveal a negative impact on adoption of another technology such as zero grazing technology. A study by Gabre-Madhin and Haggblade (2001) found that large commercial farmers adopted new high-yielding maize varieties more rapidly than small farm holders. Many studies have also reported a positive relation between farm size and adoption of agricultural technology (Kasenge, 1998; Gabre-Madhin and Haggblade, 2001; Ahmed, 2004; Uaiene et al., 2009; Mignouna et al., 2011). Farmers with large farm size are likely to adopt a new technology as they can afford to devote part of their land to try new technology unlike those with less farm size (Uaiene et al., 2009). Some studies have, however, shown a negative influence of farm size on adoption of improved technology. Farm liquidity and credit are hypothesized to have a positive sign on the adoption decision. Farm liquidity increases the credit worthiness of households and their ability to undertake risky businesses (Asfaw and Admassie, 2004). In addition, access to subsidies facilitates technology adoption and then is hypothesized to have a positive sign on adoption decision (Jara-Rojas et al., 2012). Contact with extension agents is expected to have a positive effect on adoption because exposing farmers to information is expected to stimulate adoption (Kebede et al., 1990). In addition, farmers' confidence in the skill of extension agents is expected to have a positive sign (Kassie et al., 2013). Many authors have reported a positive relationship between extension services and technology adoption. Good examples include adoption of Imazapyr-resistant maize technologies by Mignouna et al. (2011) and adoption of modern agricultural technologies in Ghana by Akudugu et al. (2012), to name a few. This is because exposing farmers to information based on innovation-diffusion theory is expected to stimulate adoption (Uaiene et al., 2009). The extent to which farmers learn from each other and the influence of social network can also play a vital role in accepting and disseminating new technologies to a large population. The main source of information for farmers is other farmers because information is easily available and it is not too costly to utilize it (Gershon et al., 2004). This is confirmed by survey data which showed that farmers cite other farmers as their main source of information regarding agricultural practices (Feder and Slade, 1985; Rees et al., 2000). Institutional factors deal with the extent or degree to which institutions impact on technology adoption by smallholders (Meinzen-Dick et al., 2004). Institutions include all the services to agricultural development, such as finance, insurance and information dissemination. They also include the embedded norms, behaviours and practices in society and the facilities and mechanisms that enhance farmers' access to productive inputs and product markets. In addition, using household-level survey data in the districts of Balaka and Mangochi in Malawi, Girma et al. (2012) identified the determinants of adoption and adoption intensity of improved maize varieties in Malawi and found that access to rural credit, livestock wealth and access to agricultural extension services significantly increased the likelihood of adoption.

5.5 Acquisition of information

As a farmer acquires more information about a new technology by adopting it or partially adopting (trial) or using other sources (workshops, social network, etc.), his perceptions and beliefs are expected to change. The importance of acquiring more information and learning in a dynamic framework was stressed by Abadi and Pannel (1999), who showed in a formal model that information from trialling a new technology involves skill improvement and better decision making. Based on the innovation-diffusion literature, Adesina and Baidu-Forson (1995) hypothesized that participation in workshops is positively related to adoption by exposing farmers to new information. Farmers who are in a network of relation(s) with many previous successful adopters have access to a large information network and, therefore, will be more likely to adopt a new technology (Bamire et al., 2012; Abdoulaye et al., 2012). Farmers may acquire information about new technologies from their peers, and therefore, membership in farmers' groups or associations is hypothesized to be positively associated with adoption (Kassie et al., 2013). Participation in on-farm experimental trials is also hypothesized to be positively related to adoption (Adesina and Baidu-Forson, 1995; Etwire et al., 2013). Aloyce et al. (2000) studied factors influencing the adoption of improved maize seeds and the use of inorganic fertilizer for maize production by farmers in the intermediate and lowland zones of Tanzania. The results indicate that availability of extension services, rainfall, variety characteristics and on-farm field trials were the most important factors that influenced the extent of adopting improved maize seeds and the use of inorganic fertilizer for maize production. Finally, belonging to a social group enhances social capital allowing trust, idea and information exchange (Mignouna et al., 2011). Farmers within a social group learn from each other the benefits and usage of a new technology. Uaiene et al. (2009) suggest that social network effects are important for individual decisions, and that, in the particular context of agricultural innovations, farmers share information and learn from each other. Thus, the dynamics of social networks of exchange of information, knowledge, learning and discussion experiences for building the confidence of adopters needs to be understood and strengthened. Peer effects or learning from other farmers also influence technology adoption. According to Oster and Thorton (2009), in any technology adoption process, peer effects work in three major ways: individuals profit from acting like friends and/or neighbours, individuals gain knowledge of the benefits of the technology from their friends and individuals learn about how to use a new approach from peers. With regard to maize technology adoption, peer effects can lead to economies of scale by lowering transportation costs but can also lead to increased competition and land prices, which can lead to dis-adoption (Carletto et al., 2007).

5.6 Farmers' objectives and goals

A key determinant of sustained adoption is the profitability of agricultural enterprises. Farmers' objective of maximizing profit affects their decision for improved technology adoption. Profit is usually used by researchers as a substitute for the concept of utility as utility is difficult to measure, (Edwards-Jones, 2006). By adding risk attitude, farmers maximize expected utility of profit, rather than the expected profit (Abadi and Pannell, 1999). Thus, if a new technology has a higher expected utility of profit than the old technology, adoption will occur. Kosarek et al. (2001) also found that farmers' decision to adopt hybrid maize was determined by the expected returns (i.e. profitability) of the technology, the availability of hybrid seed and risks associated with the expected outcomes of the new technology. Adoption and expansion of any agricultural activity mostly depend on the profitability and cost of the technology. With small farms, it has been argued that large fixed costs become a constraint to technology adoption (Abara and Singh, 1993), especially if the technology requires a substantial amount of initial set-up cost. Hence, the adoption decisions of farmers can be influenced by the cost and benefit of the technology. Bergevoet et al. (2004) found that goals are important determinants of farmers' behaviour. Further, Greiner et al. (2009), in an exploratory study, hypothesized famers' goals or motivations to be related to the adoption of best management practices. There are many other farmers' objectives and goals identified in the literature as provided in Bergevoet et al. (2004).

5.7 Perceived costs and risks associated with technology

Roberts et al. (2004) used perceived benefits and costs associated with an improved technology as a potential explanatory variable. These variables were measured by asking farmers about their perceptions about profitability of the technology (benefit), the cost associated with the technology (costs) and how they perceive the technology as important to them in the future. For example, Khan et al. (2008) in their study asked farmers of the benefits associated with adoption of 'push–pull' technology for control of cereal stem borers and *Striga* weed in Western Kenya. The benefits mentioned by the farmers were reduced infestation by pests, improved soil fertility, increased maize grain yields, improved fodder and milk productivity. Abadi et al. (2005) used farmers' perceptions of the riskiness of the innovation as another explanatory variable on adoption decision and their findings support that these variables play an important role in adoption, as well as perceived profitability. Lower risk aversion is expected to increase the probability to adopt the technology (Baidu-Forson, 1999).

5.8 Environmental factors

Depending on the type of technology, agro-climatic condition like rainfall is expected to influence technology adoption decisions (Okorie, 1984; Ogunfowora, 1993). Kassie et al. (2013) hypothesized that favourable rainfall positively impacts on decisions to adopt improved seed types and fertilizer use. Similarly, adoption studies conducted in Kenya in 1998 indicated a higher adoption rate at the coast (30%) (Wekesa et al., 2002) but a lower adoption rate in Embu (65%), a district in the moist transitional zone (Ouma et al., 2002), showing variations across agro-climatic zones. Pest and diseases are also expected to influence farmers' decisions and the sign of this variable depends on the type of improved technology (Kassie et al., 2013).

6 Strategies for improving the adoption of improved maize technologies

Improved agricultural technology is an essential strategy for increasing agricultural productivity, achieving food self-sufficiency and alleviating poverty and food insecurity among farmers in Africa. In recent literature, there is a call for change in technology generation and dissemination approaches. The new paradigms are calling for more participatory approaches and moving to more innovation system thinking. These participatory approaches have proven to improve farmers' understanding and uptake of technologies in many cases (Abdoulaye et al., 2012).

For farmers to benefit from improved maize technologies, there is a need to seek ways around the constraints to their adoption. Successful approaches require attention to the market imperfections and other constraints that characterize the contexts in which adoption decisions are made: labour market inefficiencies; input and output inefficiencies; externalities, land market inefficiencies; credit market inefficiencies; risk market inefficiencies; and information inefficiencies. The current chronic shortage of labour during peak agricultural production seasons in Africa provides opportunity for the adoption of improved maize technologies that require less labour. Researchers should, therefore, seek to investigate the reasons why smallholder farmers do the things they do, and attempt to improve on them before displacing the traditional technologies in use by them. Research also has to ensure that improved varieties are generated with traits that are appropriate for the needs and context of smallholders. The tendency for farmers to choose to grow local maize over improved hybrid and open-pollinated varieties is not only an issue of affordability. Farmers have also trait preferences which form the basis of their selection of varieties. An important observation for Malawi is that farmers are interested in varieties that are flint because of their enhanced storability and their higher flour-to-grain extraction rate (Katengeza et al., 2012). In other words, research and extension policies should stress on farmer participation in the research process and on-farm field trials for the purposes of varietal evaluation and demonstration. It is imperative that agricultural training and extension programmes be intensive enough to promote adoption not only of improved maize technologies, such as improved seeds, but also of improved agronomic practices that complement improved seeds and allow improved varieties to express their full potential. African farmers are also widely dispersed and distant from the towns and cities where seed companies often desire to be located. Therefore, to be successful in reaching farmers who will buy the seeds, there has to be a strategic and reliable network by which farmers' access to seed is assured. Dissemination and diffusion of improved varieties largely determines the extent of the demand for quality seeds (Abdoulaye et al., 2009). Efforts made to increase farmers' awareness and adoption of improved maize varieties will increase demand and provide more incentives for the establishment of new seed companies. There is also the need to build the capacity of seed companies and institutions engaged in seed development for production of foundation seeds while support for coordination of foundation seed production should be encouraged; that is, some seed companies could come together to produce foundation seed for all others.

Promotion of various smallholder income sources such as off-farm employment, remittances and livestock production can also lead to higher total household income to finance the purchase of inputs such as fertilizers, improved seed and hired labour. Government support for subsidy programmes for farmers' access to production resources will also, to a large extent, promote adoption of improved technologies. In addition, an increase in farm productivity could influence seed demand in that it is likely to favour a rapid development of the seed delivery system and the seed sector. To build trust and credibility among farmers, the seed producers must ensure that only seeds of high quality, with high-yield potential, are sold to farmers. In addition, the transfer of genetic materials between public and private sectors should be improved to allow seed companies have easy access to suitable and adapted varieties. The implementation of the harmonized regional seed laws and regulations in Africa should also be quickly embarked upon to allow for rapid regional spillovers of varieties released in one country to similar agro-ecologies in other countries.

Promoting the adoption of improved maize technologies requires that research must, apart from simply increasing productivity, consider those variables that reduce the farmers' vulnerability to loss of income, bad health, natural disasters and other factors. In addition, an understanding of local cultural practices and preferences is important if farmers, particularly smallholders, are to benefit from improved technologies developed through research in Africa. Furthermore, technologies that are built on assets which resource-poor farmers already have are more likely to be adopted. In some countries, households headed by women are discriminated against by the local communities and/or credit institutions. There is clearly a case for improving current smallholder credit systems to ensure that a wider spectrum of smallholders are able to access credit, especially households headed by women. The risk of farmers' exposure to external shocks can be lessened to some extent through the provision of crop insurance. Other steps that may be taken to encourage the adoption of technologies that increase agricultural productivity and reduce land degradation include improving the relative price of maize grain (price of grain divide by price of fertilizers) , improving access to credit and/or waiving some of the taxes levied on agricultural inputs.

The promotion of greater research-extension linkages will also improve technology adoption. Since seed production is a highly specialized operation that draws heavily on science and technology, it is always desirable for the proximity of seed businesses to an agricultural research institute/university. The technical services of breeders, pathologists, entomologists and other scientists will thus be made readily available to organizations, as and when required. Stronger partnerships between agricultural researchers and other agents of change, including local organizations, farmers, community leaders, NGOs, national policy makers and donors, are also important in stimulating technology adoption for increased agricultural productivity. There is need for synergies to be created between government departments, non-governmental organizations, researchers, donors and local communities in implementing programmes that promote smallholder farmers' adoption of the technologies that can increase maize production and agricultural productivity. In general, measures that can be taken to increase adoption of improved maize technologies include lowering fertilizer costs, lowering the price of other inputs and raising agricultural product prices, improving smallholder farmers' access to finance for agricultural development, adopting a 'package' approach to provision of improved maize technologies and development and rehabilitation of infrastructure for agricultural inputs and product markets.

7 Conclusion

Technology is key in any national strategy that aims to increase maize productivity. Over the decades, improved technologies have been developed for practically every stage of the production process, including technologies for tilling the soil, planting seeds, fertilizer application (organic and inorganic), irrigating the land, weeding (application of agro-chemicals like herbicides), protecting them from pests and weeds, harvesting and general agronomic practices (e.g. plant spacing). However, adoption of these improved technologies is taking place across Africa at different rates. This chapter reviewed past studies in understanding farmers' decisions on the factors that influence or hinder adoption of the improved technologies as an essential element in planning and executing technology-related programmes for meeting the challenges of food production in Africa.

The perception of farmers towards improved technologies is a key precondition for adoption to occur. Adoption of new technology will be determined by its suitability to the farmers, the profitability and risk associated with it and the institutional capability to communicate to farmers about the technology. Many individual farm and household socio-economic and demographic characteristics affect technology adoption. Some households are more likely to be risk takers and early adopters of new technology than others. Other factors that have been shown to determine adoption of improved maize technology include farmer-specific factors, economic factors, acquisition of information and environmental and institutional factors. From the review, determinants of improved technology adoption do not always have the same effect on adoption; rather the effect varies depending on the type of technology being introduced and where they are being introduced.

Constraints to improved maize technology adoption were identified to include availability of cultivable land, input and output prices, weather/rainfall uncertainties and production risks. Other constraints include infrastructure, incentives, liquidity and complementary factors such as soil quality which impedes adoption and retention of the improved technology. Gender, however, cuts across the constraints and affects the strategies for and the distributional consequences of overcoming each of the adoption constraints.

Promoting the adoption of improved technologies for maize production, therefore, requires addressing multiple constraints that are related. Policy makers and developers of new or improved technologies would, therefore, need to carefully assess and understand farmers' needs, constraints and the ability to adopt a given technology. This will enable them to come up with an appropriate technology that will suit farmers. The intensification of adequate and effective research–extension–farmer linkages and the possible establishment of extension liaison services by international agricultural research centres for easy and better dissemination of their research are also imperative.

8 Where to look for further information

This chapter attempted to discuss the constraints hindering adoption of improved technologies for maize cultivation in Africa. It certainly has not been exhaustive and further information can be obtained in the numerous peer-reviewed journal articles published on the same topic, in addition to those cited in the reference list. Moreover, on the specific topic of maize adoption, more information can be obtained from the following links:

- http://impact.cigiar.org/pdf/52.pdf
- http://repository.cimmyt.org/xmluvbitstream/handle/10883/964/66324.pdf
- http://www.jatropha.pro/PDF%20bestanden/NorthernTnz%20maize.pdf
- http://ageeconsearch.umn.edu/bitstream/56109/2/coastalkenya.pdf
- http://www.iiste.org/journals/index.php/RHSS/article/viewfile/17564/17762
- http://www.yale.edu/macmillan/faculty/papers/7.pdf
- http://www.sciencedirect.com/science/article/pii/s0306919214001808
- http://impact.cigiar.org/pdf/52.pdf
- http://repository.cimmyt.org/xmluvbitstream/handle/10883/964/66324.pdf

- http://www.jatropha.pro/PDF%20bestanden/NorthernTnz%20maize.pdf
- http://ageeconsearch.umn.edu/bitstream/56109/2/coastalkenya.pdf
- http://www.iiste.org/journals/index.php/RHSS/article/viewfile/17564/17762
- http://www.yale.edu/macmillan/faculty/papers/7.pdf
- http://www.sciencedirect.com/science/article/pii/s0306919214001808

9 References

Abadi, G. A. K. and Pannell, D. J. (1999). A conceptual framework of adoption of an agricultural innovation. *Agricultural Economics* 21: 145–54.

Abadi, G. A. K., Pannell, D. J. and Burton, M. P. (2005). Risk, uncertainty, and learning in adoption of a crop innovation. *Agricultural Economics* 33: 1–9.

Abara, I. O. C. and Singh, S. (1993). Ethics and biases in technology adoption: the small farm argument. *Technological Forecasting and Social Change* 43: 289–300.

Abdoulaye, T. and Lowenberg-DeBoer, J. (2000). Intensification of Sahelian farming systems: evidence from Niger. *Agricultural Systems* 64: 67–81.

Abdoulaye, T. and Sanders, J. H. (2005). Stages and determinants of fertilizer use in semiarid African agriculture: the Niger experience. *Agricultural Economics* 32: 167–79.

Abdoulaye, T. and Sanders, J. H. (2006). New technologies, marketing strategies and public policy for traditional food crops: Millet in Niger. *Agricultural Systems* 90: 272–92.

Abdoulaye, T., Amaza, P. S., Olanrewaju, A., Ellis-Jones, J. (2012). Promoting the adoption of innovations through participatory approaches: example from northern Nigeria. *Tropicultura* 30(3): 155–60, ISSN 0771-3312.

Abdoulaye, T. and Anyebe, O. (2013). Promoting the use of drought tolerant maize in Nigeria. *R & D Review* 10: 22–3, ISSN 2071-3681.

Abdoulaye, T., Bamire, A. S., Wiredu, A. N., Baco, M. N. and Fofana, M. (2012). Project community surveys, characterization of maize-producing communities in Benin, Ghana, Mali and Nigeria: West Africa Regional Synthesis Report. Drought Tolerant Maize for Africa (DTMA), 18p.

Abdoulaye, T., Sanogo, D., Langyintuo, A., Bamire, A. S. and Olanrewaju, A. (2009). *Assessing the Constraints Affecting Production and Deployment of Maize Seed in DTMA Countries of West Africa*. Ibadan, Nigeria: IITA, 40p.

Adegbola, P. and Gardebroek, C. (2007). The effect of information sources on technology adoption and modification decisions. *Agricultural Economics* 37: 55–65.

Adesina, A. A. and Baidu-Forson, J. (1995). Farmers' perceptions and adoption of new agricultural technology: evidence from analysis in Burkina Faso and Guinea, W. Africa. *Agricultural Economics* 13: 1–9.

Adesina, A. A. and Zinnah, M. M. (1993). Technology characteristics, farmers' perceptions and adoption decisions: a Tobit model application in Sierra Leone. *Agricultural Economics* 9: 297–311.

Adesina, A. A. and Sanders, J. H. (1991). Peasant farmer behavior and cereal technologies: stochastic programming analysis in Niger. *Agricultural Economics* 5: 21–38.

Ahmed, A. (2004). Making technology work for the poor: strategies and policies for African sustainable development. *International Journal of Technology, Policy and Management* 4(1): 1–17, ISSN 1468-4322.

Ajayi, M. T. and Oloruntoba, A. (2007). Assessment of factors affecting farmers' adoption and utilisation of major agricultural technologies developed by International Institute of Tropical Agriculture (IITA). *Journal of Agriculture, Forestry and the Social Sciences* 5(1): 20–8.

Akinbode, W. O. and Bamire, A. S. (2015). *Discontinued Use Decision of Improved Maize Varieties in Osun State*, Nigeria. *Journal of Development and Agricultural Economics* 7(3): 85–91.

Akinola, A. A. Alene, A. D., Adeyemo, R., Sanogo, D. and Olanrewaju, A. S. (2009). Impact of balanced nutrient management systems technologies in the Northern Guinea savanna of Nigeria. *Journal of Food, Agriculture and Environment* 7(2): 496–504.

Akinola, A. A., Alene, A. D., Adeyemo, R., Sanogo, D. and Olanrewaju, A. S. (2009). Economic impacts of soil fertility management research in West Africa. *African Journal of Agricultural and Resource Economics* 3(2): 159–75.

Akinola, A. A., Alene, A. D., Adeyemo, R., Sanogo, D., Olanrewaju, A. S., Nwoke, C., Nziguheba, G. and Diels, J. (2010). Determinants of adoption and intensity of use of balanced in the Northern Guinea Savanna of Nigeria. *Quarterly Journal of International Agriculture* 49(1): 25–45, Berlin.

Akudugu, M., Guo, E. and Dadzie, S. (2012). Adoption of modern agricultural production technologies by Farm Households in Ghana: what factors influence their decisions? *Journal of Biology, Agriculture and Healthcare* 2(3): 12–20.

Alene, A. and Mwalughali, J. (2012). 'The Effectiveness of Crop Improvement Programs in Sub-Saharan Africa from the Perspectives of Varietal Output and Adoption: The Case of Cassava, Cowpea, Maize, and Soybean'. Draft Technical Report for Measuring and Assessing the Impacts of the Diffusion of Improved Crop Varieties in Africa (DIVA) Project, International Institute of Tropical Agriculture (IITA), Ibadan, Nigeria.

Aloyce, R. M. K., Verkuijl, H. and Mwangi, W. (April 2000). Factors affecting adoption of improved maize seeds and use of inorganic Fertilizer for maize production in the intermediate and lowland zones of Tanzania. *Journal of Agricultural and Applied Economics* 32(01): 35–47.

Asfaw, A. and Admassie, A. (2004). The role of education on the adoption of chemical fertiliser under different socioeconomic environments in Ethiopia. *Agricultural Economics* 30: 215–28.

Asfaw, S., Shiferaw, B., Simtowe, F., Haile, M. G. (2012). Agricultural technology adoption, seed access constraints and commercialization in Ethiopia. *Journal of Development and Agricultural Economics* 3(9): 436–47.

Austin, E., Willock, J., Deary, I., Gibson, G., Dent, J., Edwards-Jones, G., Morgan, O., Grieve, R. and Sutherland, A. (1998). Empirical models of farmer behaviour using psychological, social and economic variables. Part I: linear modeling. *Agricultural Systems* 58: 203–24.

Ayinde, O. E., Abdoulaye, T., Olaoye, G. and Akangbe, J. A. (2013). Gender and innovation in agriculture: a case study of farmers' varietal preference of drought tolerant maize in Southern Guinea Savannah Region of Nigeria. *Albanian Journal of Agricultural Sciences* 12(4): 617–25.

Baco, M. N., Abdoulaye, T., Sanogo, D. and Languitou, A. (2010). Caracterisation des ménages producteurs de mais en zone de savane seche au Benin. Rapport pays – Enquete-menage – Benin. Ibadan, Nigeria: IITA, 38p.

Badu-Apraku, B., Asuboah, R. A., Fakorede, B. and Asafo-Adjei, B. (2014). *Strategies for Sustainable Maize Seed Production in West and Central Africa*. Ibadan, Nigeria: IITA, 140p.

Baidu-Forson, J. (1999). Factors influencing adoption of land-enhancing technology in the Sahel: lessons from a case study in Niger. *Agricultural Economics* 20: 231–9.

Bamire, S. A. and Ocran, V. K. (1995). Traditional seed production, distribution and marketing systems. A paper presented at the Crop Management Research Training (CMRT) – Seed Maize Production Techniques, 14–25 August 1995. Kumasi, Ghana.

Bamire, A. S. (1999). Factors Influencing the Adoption of Fertilizer Technology in Osun State of Nigeria. A PhD. Thesis, Department of Agricultural Economics, Faculty of Agriculture, Obafemi Awolowo University, Ile-Ife, Ibadan, Nigeria.

Bamire, A. S. (1999). Socio-economic determinants of soil conservation practices in Nigeria. *Indian Journal of Soil and Water Conservation* 27(3): 246–53.

Bamire, A. S. (2007). Gender, land tenure systems and the adoption of intensification technologies among smallholder maize farmers in the agroecological zones of Nigeria. Report submitted to the United Nations University Institute for Natural Resources in Africa (UNU-INRA), P.M.B., K.I.A., Accra, Ghana.

Bamire, A. S., Abaidoo, R., Jemo, N., Abdoulaye, T., Yusuf, A. A. and Nwoke, C. (2012). Profitability analysis of commercial chemical and biological crop products among farm households in agro-ecological zones of West Africa. *African Journal of Agricultural Research* 7(23): 3385–94.

Bamire, A. S., Abdoulaye, T. and Sanogo, D. (2009). Community assessment report for the drought tolerant maize for Africa (DTMA) project in the dry savanna of Nigeria. Submitted to the International Institute of Tropical Agriculture, Ibadan, Nigeria.

Bamire, A. S., Abdoulaye, T., Amaza, P., Tegbaru, A., Alene, A. D. and Kamara, A. Y. (2010). Impact of promoting sustainable agriculture in Borno (PROSAB) program on adoption of improved crop varieties in Borno State of Nigeria. *Journal of Food, Agriculture & Environment* 8(3 and 4): 391–8.

Bamire, A. S., Abdoulaye, T. and Sanogo, D. (2012). Determinants of improved maize adoption in Nigeria: are they the same for poor and well endowed households? *Ife Journal of Agriculture* 25: 83–100.

Bamire, A. S., Abdoulaye, T., Sanogo, D. and Langyintuo, A. (2010). Characterization of maize producing households in the dry savanna of Nigeria. Country Report – Nigeria, IITA, Ibadan, Nigeria, 46p.

Bamire, A. S., Adejobi, A. O., Akinola, A. A. and Olagbaju, O. (2007). Socioeconomic factors influencing the choice of soil amendment techniques among maize farmers in Oyo State, Nigeria. In U. Haruna, S. A. Jubril, Y. P. Mancha and M. Nasiru (eds), *Proceedings of the 9th Annual Conference of the Nigerian Association of Agricultural Economists, on Consolidation of Growth and Development of Agricultural Sector*, held in ATBU, Bauchi, Bauchi State, Nigeria, 5–8 November 2007, 433–41.

Bamire, A. S. and Manyong, V. M. (2003). Profitability of intensification technologies among smallholder maize farmers in the forest-savanna transition zone of Nigeria. *Agriculture, Ecosystems & Environment* 100: 111–18.

Bamire, A. S., Fabiyi, Y. L. and Manyong, V. M. (2002). Adoption patterns of fertilizer among small-scale farmers in two ecologies of Southwest Nigeria: a Tobit analysis. *Australian Journal of Agricultural Research* 53: 901–10.

Batz, F. J., Peters, K. and Janssen, W. (1999). The influence of technology characteristics on the rate and speed of adoption. *Agricultural Economics* 21: 121–30.

Bedru, B. and Wegary, D. (2014). Determinants of smallholder farmers' hybrid maize adoption in the drought prone Central Rift Valley of Ethiopia. *African Journal of Agricultural Research* 9(17): 1334–43.

Bergevoet, R. H. M., Ondersteijn, C. J. M., Saatkamp, H. W., van Woerkum, C. M. J., Huirne, R. B. M. (2004). Entrepreneurial behaviour of dutch dairy farmers under a milk quota system: goals, objectives and attitudes. *Agricultural Systems* 80: 1–21.

Bezu, S. and Holden, S. (2008). Can food for work encourage agricultural production. *Food Policy* 33: 541–9.

Binswanger, H. P. and von Braun, J. (1991). Technological change and commercialization in agriculture: the effect on the poor. *World Bank Research Observer* 6(1): 57–80.

Binswanger-Mkhize, H. P. (2009). Challenges and opportunities for African agriculture and food security: High food prices, climate change, population growth, and HIV and AIDs. Expert Meeting on How to feed the World in 2050, Food and Agriculture Organization of the United Nations Economic and Social Development Department, 24–25 June 2009.

Bonabana-Wabbi, J. (2002). Assessing Factors Affecting Adoption of Agricultural Technologies: The Case of Integrated Pest Management (IPM) in Kumi District, Msc. Thesis Eastern Uganda.

Carletto, C., Kirk, A. and Winters, P. (2007). Non-traditional exports, traditional constraints: the adoption and diffusion of cash crops among smallholders in Guatemala. World Bank Policy Research Working Paper No. 4347.

CIMMYT (2015). Seeds of Transformation, News from Africa, International Maize and Wheat Improvement Center (CIMMYT), September, 2015.

Conny, J., Almekinders, M. and Louwaars, N. P. (2002). The importance of the farmers' seed systems in a functional national seed sector. *Journal of New Seeds* 4(1 and 2): 15–33.

De Groote, H., Doss, C., Lyimo, S. D. and Mwangi, W. (2003). Adoption of maize technologies in East Africa – what happened to Africa's emerging maize revolution? Paper presented at the FASID Forum V, 'Green Revolution in Asia and its Transferability to Africa', Tokyo, December 8–10 2003.

De Janvry, A., Dustan, A. and Sadoulet, E. (2011). Recent advances in impact analysis methods for ex-post impact assessments of agricultural technology: options for the CGIAR. CGIAR Independent Science and Partnership Council.

De Janvry, A. and Sadoulet, E. (2002). World poverty and the role of agricultural technology: direct and indirect effects. *Journal of Development Studies* 38(4): 1–26.

Deininger, K. and Ayalew Ali, D. (2008). Do overlapping land rights reduce agricultural investment? evidence from Uganda. *American Journal of Agricultural Economics* 90(4): 869–82.

Diiro, G. (2013). Impact of Off-farm Income on Technology Adoption Intensity and Productivity: Evidence from Rural Maize Farmers in Uganda. International Food Policy Research Institute, Working Paper 11.

Doss, C. R. (2006). Analyzing technology adoption using microstudies: limitations, challenges, and opportunities for improvement. *Agricultural Economics* 34: 207–19.

Doss, C. R. and Morris, M. L. (2001). How does gender affect the adoption of agricultural innovations? The case of improved maize technology in Ghana. *Agricultural Economics* 25(1): 27–39.

Duflo, E., Kremer, M. and Robinson, J. (2008). How high are rates of return to fertilizer? evidence from field experiments in Kenya. *American Economic Review* 98(2): 482–8.

Edwards-Jones, G (2006). Modelling farmer decision-making: concepts, progress and challenges. *Animal Science* 82: 783–90.

Ekboir, J., Boa, K. and Dankyi, A. A. (2002). *Impacts of No-Till Technologies in Ghana*. Mexico, DF: International Maize and Wheat Improvement Center (CIMMYT).

Ellis, F. and Freeman, H. and Ade, H. (2004). Rural livelihoods and poverty reduction strategies in four African Countries. *Journal of Development Studies* 40(4):1–30.

Etwire, M., Abdoulaye, T., Obeng-Antwi, K., Bua, A., Kanton, R., Asumadu, H., Abdulai, M. S., Haruna, A. and Etwire, J. (2013). On-farm evaluation of maize varieties in the transitional and savannah zones of Ghana: determinants of farmer preferences. *Journal of Development and Agricultural Economics* 5(6): 255–62, ISSN 2006-9774.

FAOSTAT (2015). Food and Agricultural Organisation of the United Nation. FAOSTAT Database available on line http://faostat fao.org/site/567/default.aspx (accessed 20 August 2015).

Feder, G., Just, R. E. and Zilberman, D. (1985). Adoption of agricultural innovations in developing countries: a survey. *Economic Development and Cultural Change* 33: 255–98.

Feder, G. and Umali, D. L. (1993). The adoption of agricultural innovations: a review. *Technological Forecasting and Social Change* 43: 215–39.

Feder, G. and Slade, R. (1985). The role of public policy in the diffusion of improved agricultural technology. *American Journal of Agricultural Economics* 67(2): 423–8.

Fishbein, M. and Ajzen, I. (1991). *Theory of Planned Behaviour*. New York: Academic Press, Inc.

Fishbein, M. and Ajzen, I. (2010). *Predicting and Changing Behavior: The Reasoned Action Approach*. New York: Psychology Press.

Fofana, M. T., Abdoulaye, T., Ntji Coulibaly, Sanogo, D. and La Rovere, R. (2010). Characterization of maize producing communities in the Sikasso region of Republic of Mali, DTMA Project Country Report – Community Survey, 17.

Gabre-Madhin, Z. and Haggblade, S. (June 2001). *Success in African Agriculture: Results of an Expert Survey*. Washington DC: International Food Policy Research Institute.

Gianessi, L. and Williams, A. (2011). Overlooking the obvious: the opportunity for herbicides in Africa. *Outlooks on Pest Management* October: 211–15.

Goodwin, B. and Mishra, A. (2004). Farming efficiency and the determinants of multiple job holding by farm operators. *American Journal of Agricultural Economics* 86(3): 722–9.

Government of Malawi (2004). Behavioural surveillance survey (BSS) report, Lilongwe, Malawi.

Greiner, R., Patterson, L. and Miller, O. (2009). Motivations, risk perceptions and adoption of conservation practices by farmers. *Agricultural Systems* 99: 86–104.

Guang-Hai, Q., Qi-Yu, D., Yan, T., Shu-Geng, W. and Shi-Huang, Z. (2007). Nutritional evaluation and utilization of quality protein maize (QPM) in animal feed. In Protein sources for the animal industry, FAO Corporate document repository, Agriculture and Consumer Protection, Italy, Rome. http://www.fao.org/docrep/007/y5019e/y5019e01.htm#Top of Page (accessed 6 February 2016).

Hall, B. and Khan, B. (2002). *Adoption of New Technology*. Berkeley: New Economy Handbook. University of California.

Hansson, H., Ferguson, R. and Olofsson, C. (2012). Psychological constructs underlying farmers' decisions to diversify or specialise their businesses – an application of theory of planned behaviour. *Journal of Agricultural Economics* 63: 465–82.

Hassan, R., Mekuria, M. and Mwangi, W. (2001). *Maize Breeding Research in Eastern and Southern Africa: Current Status and Impacts of Past Investments Made by the Public and Private Sectors ,1966–97*, Mexico, DF: CIMMYT.

Hayami, Y. and Ruttan, V. W. (1985). *Agricultural Development: An International Perspective*. Revised edition. Baltimore MD, USA: Johns Hopkins University Press.

Idrisa, Y. L., Abdoulaye, T., Mohammed, S. T. and Ibrahim, A. A. (2014). Analysis of Drought-Tolerant maize adoption and its effect on food security among farmers in the Sudan Savanna of Northeastern Nigeria. *Asian Journal of Agricultural Extension, Economics & Sociology* 3(6): 496–504; Article no. AJAEES.2014.6.003 SCIENCEDOMAIN international. www.sciencedomain.org.

IFPRI (2011). *Dataset on Farm Budgets Collected for Analyzing Profitability of Maize, Rice, and Soybean Production in Ghana*. Accra, Ghana: International Food Policy Research Institute.

James, C. (2003). *Global Review of Commercialized Transgenic Crops: 2002 Feature: Bt Maize* (ISAAA Brief No. 29). Ithaca, NY: International Service for the Acquisition of Agri-Biotech Applications.

Joao, A. R. B., Foletto, L. and Xavier, V. T. (2015). An interdisciplinary framework to study farmers' decisions on adoption of innovation: insights from expected utility theory and theory of planned behaviour. *African Journal of Agricultural Research* 10(29): 2814–25.

Jara-Rojas, R., Bravo-Ureta, B. E. and Díaz, J. (2012). Adoption of water conservation practices: a socioeconomic analysis of small-scale farmers in Central Chile. *Agricultural Systems* 110: 54–62.

Justice, A. T. and Abdoulaye, T. (2012). Climate change and agricultural technology adoption: the case of drought tolerant maize in rural Nigeria. *Mitigation Adaptation Strategy Global Change* 17: 277–92.

Kalinda, G. T. and Kuntashula, E. (2014). Adoption of improved maize seed varieties in Southern Zambia. *Asian Journal of Agricultural Sciences* 6(1): 33–9, ISSN: 2041-3882; e-ISSN: 2041-3890 © Maxwell Scientific Organization.

Karanjaa, D. D., Renkow, M. and Crawford, E. W. (2003). Welfare effects of maize technologies in marginal and high potential regions of Kenya. *Agricultural Economics* 29: 331–41.

Kasenge, V. (1998). Socio-economic factors influencing the level of Soil Management Practices on Fragile Land. In A. J. Shayo-Ngowi, G. Ley and F. B. R Rwehumbiza (eds), *Proceedings of the 16th Conference of Soil Science Society of East Africa*, 13–19 December 1998, Tanga, Tanzania, 102–12.

Kassie, G. T., La Rovere, R., Mwangi, W., Erenstein, O., Langyintuo, A., and Sonder, K. (2011). Drought Risk and Maize Production in Southern Africa – Lessons for Ethiopia. In *Adapting to Crisis: Ethiopian Agriculture in the 21st Century. Proceedings of the 13th Annual Conference of the Agricultural Economics of Ethiopia*. November 2011, Addis Ababa, Ethiopia, 99–124.

Kassie, M., Jaleta, M., Shiferaw, B., Mmbando, F. and Mekuria, M. (2013). Adoption of interrelated sustainable agricultural practices in smallholder systems: evidence from rural Tanzania. *Technological Forecasting and Social Change* 80: 525–40.

Katengeza, S. P., Mangisoni1, J. H., Girma, T. K., Sutcliffe, C., Langyintuo, A., La Rovere, R. and Mwangi, W. (2012). Drivers of improved maize variety adoption in drought prone areas of Malawi. *Journal of Development and Agricultural Economics* 4(14): 393–403, December 2012. Available online at http://www.academicjournals.org/JDAE.

Kebede, Y., Gunjal, K. and Coffin, G. (1990). Adoption of new technologies in Ethiopian agriculture: the case of Tegulet-Bulga district Shoa province. *Agricultural Economics* 4: 27–43.

Keelan, C., Thorne, F., Flanagan, P. and Newman, C. (2014). Predicted willingness of Irish farmers to adopt GM technology. *The Journal of Agrobiotechnology Management and Economics* 12(3): 394–403.

Kijima, Y., Otsuka, K. and Sserunkuuma, D. (2011). An inquiry into constraints on a green revolution in Sub-Saharan Africa: the case of NERICA rice in Uganda. *World Development* 39(1): 77–86.

Khan, Z. R., Amudavi, D. M., Midega, C. A. O., Wanyama, J. M. and Pickett, J. A. (2008). Farmers' perceptions of a 'push–pull' technology for control of cereal stemborers and Striga weed in western Kenya. *Crop Protection* 27: 976–87.

Koppel, B. M. (ed.) (1994). *Induced Innovation Theory and International Agricultural Development: A Reassessment*. Baltimore: Johns Hopkins University Press.

Kormawa, P. M., Chiano, J. N. and Manyong, V. M. (2000). Cowpea Demand and Supply Pattern in West Africa: the case of Nigeria. Paper presented at the world cowpea conference III, organized by the Institute of Tropical Agriculture, Ibadan, Nigeria, 4–7 September. Accessed from, http.//www.iita.org/info/cowpea.

Kosarek, J. L., Garcia, P. and Morris, M. L. (2001). Factors explaining the diffusion of hybrid maize in Latin America and the Caribbean region. *Agricultural Economics* 26(3): 267–80.

Koundouri, P., Nauges, C. and Tzouvelekas, V. (2006). Technology adoption under production uncertainty: theory and application to irrigation technology. *American Journal of Agricultural Economics* 88(3): 657–70.

Kumar, N. and Quisumbing, A. R. (2010). Access, Adoption, and Diffusion Understanding the Long-term Impacts of Improved Vegetable and Fish Technologies in Bangladesh, Poverty, Health, and Nutrition Division. IFPRI Discussion Paper 00995 June 2010.

La Rovere, R., Kostandini, G., Abdoulaye, T., Dixon, J., Mwangi, W., Guo, Z. and Bänziger, M. (2010). *Potential Impact of Investments in Drought Tolerant Maize in Africa*. Addis Ababa, Ethiopia: CIMMYT, 28.

Lambrecht, I., Vanlauwe, B. and Maertens, M. (2015). Integrated soil fertility management: from concept to practice in eastern DR Congo. Bioeconomics Working Paper Series Working Paper 2014/5, Division of Bioeconomics, Department of Earth and Environmental Sciences, University of Leuven, Belgium. http://ees.kuleuven.be/bioecon/.

Langyintuo, A. S., Mwangi, W., Diallo, A. O., MacRobert, J., Dixon, J. and Banziger, M. (2010). Challenges of the maize seed industry in eastern and southern Africa: a compelling case for private–public intervention to promote growth. *Food Policy* 35(4): 323–31.

Läpple, D. and Kelley, H. (2013). Understanding the uptake of organic farming: accounting for heterogeneities among Irish farmers. *Ecological Economics* 88: 11–19.

Lavison, R. (2013). Factors Influencing the Adoption of Organic Fertilizers in Vegetable Production in Accra, Msc Thesis, Accra, Ghana.

Lopes, H. (2010). Adoption of Improved Maize and Common Bean Varieties in Mozambique an unpublished thesis submitted to Michigan State University, In partial fulfillment of the requirements for the degree of Master of Science, Agricultural, Food, and Resource Economics.

Louwaars, M. P. and Marrewijk, G. A. M. (1999). *Seed Supply Systems in Developing Countries*. Wageningen: CTA.

Magyembe, J. M. (1997). Herbicide use in Uganda. In E. Adipala, G. Tusiime and P. Okori (eds), Proceedings of the 16th Biennial Weed Science Society for Eastern Africa, Kampala, Uganda: Technical Centre for Agricultural and Rural Co-operation, September, http://books.google.com/books/about/Proceedings_of_the_ 16th_Biennial_ Weed_Sc.html?id=vuwfAQAAIAAJ, 165–9.

Makokha, S., Kimani, S., Mwangi, W., Verkuijl, H. and Musembi, F. (2001). *Determinants of Fertilizer and Manure use in Maize Production in Kiambu District, Kenya*. Mexico, DF: CIMMYT and KARI.

Martínez-García, C. G., Dorward, P. and Rehman, T. (2013). Factors influencing adoption of improved grassland management by small-scale dairy farmers in central Mexico and the implications for future research on smallholder adoption in developing countries. *Livestock Science* 152: 228–38.

Mathenge, M., Smale, M. and Olwande, J. (2014). The impact of maize hybrids on income, poverty, and inequality among smallholder farmers in Kenya. *Food Policy* 44: 262–71.

Mazvimavi, K. and Twomlow, S. (2009). Socioeconomic and institutional factors influencing adoption of conservation farming by vulnerable households in Zimbabwe. *Agricultural Systems* 101: 20–9.

McNamara, K. T., Wetzstein, M. E. and Douce, G. K. (1991). Factors affecting peanut producer adoption of integrated pest management. *European Review of Agricultural Economics* 13: 129–39.

Meinzen-Dick, R., Adato, M., Haddad, L. and Hazell, P. (2004). *Science and Poverty: An Interdisciplinary Assessment of the Impact of Agricultural Research*. Washington, DC: International Food Policy Research Institute, 31p.

Mendola, M. (2007). Agricultural technology adoption and poverty reduction: a propensity-score matching analysis for rural Bangladesh. *Food Policy* 32(3): 372–93.

Mensah-Bonsu, A., Sarpong, D. B., Al-Hassan, R., Asuming-Brempong, S., Egyir, I., Kuwornu, J. and Osei-Asare, Y. (2011). Technology Adoption and Land and Water Management Practices among Maize Farmers in Ghana. Paper presented at the International Conference of Agricultural Production, 'Increasing Agricultural Productivity and Enhancing Food Security in Africa: New Challenges and Opportunities', Addis Ababa, Ethiopia, 1–3 November.

Mignouna, B., Manyong, M., Rusike, J., Mutabazi, S. and Senkondo, M. (2011). Determinants of adopting Imazapyr-Resistant maize technology and its impact on household income in western Kenya. *AgBioforum* 14(3): 158–163.

Morris, M. L., Tripp, R. and Dankyi, A. A. (1999). Adoption and impacts of improved maize production technology: a case study of the Ghana grains development project. Economics Program Paper No 99-01. Mexico, DF: CIMMYT.

Moshi, A. (1997). *Economic Impact of Maize Research in Tanzania. Southern African Center for Cooperation in Agricultural and Natural Resources Research & Training*. Gaborone, Botswana, 103p.

Muzari, W., Gatsi, W. and Muvhunzi, S. (2012). The impacts of technology adoption on smallholder agricultural productivity in sub-Saharan Africa: a review. *Journal of Sustainable Development* 5(8): 69–77.

Namara, R. E., Hope, L., Owusu, S. E., De Fraiture, C. and Owusu, D. (2014). Adoption patterns and constraints pertaining to small-scale water lifting technologies in Ghana. *Agricultural Water Management* 131: 194–203.

Natsa, R. T. (2014). Counting Gains of Agricultural Transformation Agenda. Leadership. http://leadership.ng/reporter/ruth-tene-natsa/ (accessed 25 February 2016).

Negatu, W. and Parikh, A. (1999). The impact of perception and other factors on the adoption of agricultural technology in the Moret and Jiru Woreda (district) of Ethiopia. *Agricultural Economics* 21: 205–16.

Nowak, P. J. (1987). The adoption of conservation technologies: economic and diffusion explanations. *Rural Sociology* 42: 208–20, doi: 10.3945/an.110.000182.

Nuss, E. T. and Tanumihardjo, S. A. (2011). Quality protein maize for Africa: closing the protein inadequacy gap in vulnerable populations. *Advanced Nutrition* 2: 217–24, doi: 10.3945/an.110.000182.

Ogunfowora, B. (1993). Analysis of fertilizer supply and demand in Nigeria. In N. B. Migindadi, D. O. A. Phillip and P. Jayaraman (eds), *Alternative Pricing and Distribution Systems for Fertilizers in Nigeria*. FACU Study series No. 93/1, Ibadan: FACU.

Ogunsumi, I. O., Farinde, A. J. and Ogbosuka, G. E. (2010). Comparative analysis of extension services of agricultural programme in Edo and Osun states of Nigeria. *American Journal of Social and Management Sciences* 1(2): 186–90.

Öhlmér, B., Olson, K. and Brehmer, B. (1998). Understanding farmers' decision making processes and improving managerial assistance. *Agricultural Economics* 18: 273–90.

Okorie, A. (1984). Recent experiences in farm input supply and distribution in Nigeria. In M. Feldman and F. S. Idachaba (eds), *Crop Marketing and Input Distribution in Nigeria*. Ibadan: FACU.

Oster, E. and Thornton, R. (2009). Determinants of technology adoption: private value and peer effects in menstrual cup take-up mimeo University of Chicago. *Journal of the European Economic Association* 10(6): 1263–93.

Ouma, J., Murithi, F., Mwangi, W., Verkuijl, H., Gethi, M. and De Groote, H. (2002) *Adoption of Maize Seed and Fertilizer Technologies in Embu District, Kenya*. Mexico, DF: CIMMYT (International Maize and Wheat Improvement Center).

Overfield, D., Murithi, F. M., Muthamia, J. N., Ouma, J. O., Birungi, K. F., Maina, J. M. and Kibata, G. N. (2001). Analysis of the Constraints to Adoption of Herbicides by Smallholder Maize Growers in Kenya and Uganda. In The BCPC Conference Weeds 2001: Proceedings of an International Conference Held at the Brighton Hilton Metropole Hotel, UK, 12–15 November 2001, Volume 1, UK: British Crop Protection Council, 907–12.

Pannell, D. J., Marshall, G. R., Barr, N., Curtis, A., Vanclay, F. and Wilkinson, R. (2006). Understanding and promoting adoption of conservation practices by rural landholders. *Australian Journal of Experimental Agriculture* 46: 1407–24.

Pingali, P. L. (2001). *CIMMYT 1999-2000 World Maize Facts and Trends. Meeting World Maize Needs: Technological Opportunities and Priorities for the Public Sector*. Mexico, DF: CIMMYT.

Prokopy, L., Floress, K., Klotthor-Weinkauf, D. and Baumgart-Getz, A. (2008). Determinants of agricultural best management practice adoption: evidence from the literature. *Journal of Soil and Water Conservation* 63: 300–11.

Quiñones, E. and Diao, X. (2011). Assessing Crop Production and Input Use Patterns in Ghana – What Can We Learn from the Ghana Living Standards Survey (GLSS5)? Ghana Strategy Support Program Working Paper 0024. Washington, DC: International Food Policy Research Institute.

Ravallion, M. and Chen, S. (2004). How have the world's poorest fared since the early 1980s? *World Bank Research Observer* 19(2): 141–70.

Rees, D., Momanyi, M., Wekundah, J., Ndungu, F., Odondi, J., Oyure, A. O. and Joldersma, R. (2000). *Agricultural Knowledge and Information Systems in Kenya: Implications for Technology Dissemination and Development*. London: Overseas Development Institute.

Reardon, T., Stamoulis, K. and Pingali, P. (2007). Rural nonfarm employment in developing countries in an era of globalization. *Agricultural Economics* 37: 173–83.

Regasa, C., Dankyi, A., Acheampong, P., Wiredu, A. N., Chapoto, A., Asamoah, M. and Tripp, R. (2013). Patterns of adoption of improved maize technologies in Ghana. Ghana Strategy Support Programme, IFPRI Working Paper 36, July 2013.

Reimer, A. P., Weinkauf, D. K. and Prokopy, L. S. (2012). The influence of perceptions of practice characteristics: an examination of agricultural best management practice adoption in two Indiana watersheds. *Journal of Rural Studies* 28: 118–28.

Roberts, R. K., English, B. C., Larson, J. A., Cochran, R. L., Goodman, W. R., Larkin, S. L., Marra, M. C., Martin, S. W., Shurley, W. D. and Reeves, J. M. (2004). Adoption of site-specific information and variable-rate technologies in cotton precision farming. *Australian Journal of Agricultural Economics* 36: 143–58.

Rogers, E. (2003). *Diffusion of Innovations*. New York: Free Press.

Rosegrant, M. W., Ringler, C., Msangi, S., Zhu, T., Sulser, T., Valmonte-Santos, R. and Wood, S. (2007). Agriculture and food security in Asia: the role of agricultural Res. knowledge in a changing environment. *Journal of Semi-Arid Tropical Agricultural Research* 4: 1–35.

Schultz, T. W. (1964). *Transforming Traditional Agriculture*. New Haven, CT: Yale University Press.

Tripp, R. and Mensah-Bonsu, A. (2013). Ghana's Commercial Seed Sector: New Incentives or Continued Complacency? Ghana Strategy Support Program Working Paper 32. Washington, DC: International Food Policy Research Institute.

Uaiene, R., Arndt, C. and Masters, W. (2009). Determinants of Agricultural Technology Adoption in Mozambique. Discussion papers No. 67E.

Uganda Bureau of Statistics (2010). Uganda National Household Survey 2009/10: Socio-Economic Module Abridged Report, Kampala: Uganda Bureau of Statistics.

Wekesa, E., Mwangi, W., Verkuijl, H., Danda, K. and De Groote, H. (2003). *Adoption of Maize Technologies in the Coastal Lowlands of Kenya*. Mexico, DF: CIMMYT.

Willock, J., Deary, I. J., Edwards-Jones, G., Gibson, G. J., McGregor, M. J., Sutherland, A., Dent, J. B., Morgan, O. and Grieve, R. (1999). The role of attitudes and objectives in farmer decision making: business and environmentally-oriented behaviour in Scotland. *Journal of Agricultural Economics* 50: 286–303.

World Bank (2006). World Bank Annual Report.worldbank.org/INTANNREP2K6/Resources/2838485-1158333614345/AR06_final_LO_RES.pdf.

Wu, J. J. and Babcock, B. A. (1998). The choice of tillage, rotation, and soil testing practices: economic and environmental implications. *American Journal of Agricultural Economics* 80: 494–511.

Wubeneh, N. G. and Sanders, J. H. (2006). Farm-level adoption of sorghum technologies in Tigray, Ethiopia. *Agricultural Systems* 91: 122–34.

Chapter 12

Supporting smallholders in maize cultivation: using an agricultural innovation systems approach

Mariana Wongtschowski and Remco Mur, Royal Tropical Institute (KIT), The Netherlands; and Carolina Camacho, The International Maize and Wheat Improvement Center (CIMMYT), Mexico

1 Introduction: agricultural innovation systems

Over the years, paradigms of agricultural innovation have shifted considerably. Perspectives have moved from linear transfer-of-technology models in the 1960s to 'farmer first' and farming systems research (1980s and 1990s) to the more recent perspective of agricultural innovation systems (AIS).

An innovation system can be defined as a network of organizations, enterprises and individuals focused on bringing new products, new processes and new forms of organization into use, together with the institutions and policies that affect their behaviour and performance. The innovation systems concept embraces not only the science suppliers but also the totality and interaction of actors involved in innovation. It extends beyond the creation of knowledge to encompass the factors affecting demand for and use of knowledge in novel and useful ways (Hall et al., 2006: 16). Table 1 outlines a number of principles related to the AIS concept.

With the AIS concept comes a switch from a focus on 'technology' to 'innovation'. Technology, on the one hand, emphasizes new 'hardware' (seed varieties, ploughs, water management techniques, etc.). Innovation, on the other hand, has knowledge, organizational

http://dx.doi.org/10.19103/AS.2016.0001.19

and institutional elements. Organizational innovations look at new ways of organizing groups or production, whereas institutional innovation reflects new or revised partnerships, policies and behaviour.

The agriculture innovation systems framework maintains that improved interaction and joint work between stakeholders result in better information exchange, and more ideas and opportunities for sector improvement. This, in turn, leads to further collaboration and more opportunities for more and faster innovation.

An important implication for the role of research and researchers is that whereas research was long upheld as the unique initiator or driver of the process of agricultural innovation, research is now considered as one of the providers of knowledge and information, alongside farmers and other actors in the wider system. Innovation is regarded as a process of co-learning and co-creation (Hawkins et al., 2009; Hall et al., 2006). Innovation and experimentation also take place without the involvement of researchers (Biggs, 1990; Chambers, 1983). Research, as an actor, is able to add much value to the process of innovation and increase the likelihood of impact at scale. It is embedded in the system. Some refer to 'Agricultural Research in Development' (ARinD) in contrast to the more well-known term of Agricultural Research for Development (AR4D). The WorldFish Centre seems to have been the first to coin the term 'ARinD' in this context, back in 2013. The terminology is meant to bring in the perspective of doing research in a dynamic context, in

Table 1 Main Principles of the agriculture innovation system (Nederlof et al., 2011)

1	Focus on innovation rather than production. Innovation is understood to be the application of knowledge (of all types) to achieve desired social or economic outcomes.
2	Interaction and learning. Innovation is an interactive process through which knowledge acquisition (from different sources) and learning take place. Patterns of interaction between different knowledge sources form a central component of an organization's or sector's capacity to innovate.
3	Linkages for accessing knowledge and learning. The relationships that sustain the acquisition of knowledge and permit interactive learning are critical and can take many forms. They can be partnerships, commercial transactions (a contract or licence), networks or platforms.
4	There are new actors and new roles in the innovation process. The concept recognizes that (1) there is an important role for a broad spectrum of actors outside government; (2) the actors' relative importance changes during the innovation process; (3) as circumstances change and actors learn, roles can evolve and (4) actors can play multiple roles.
5	Attitudes, practices and interaction of behavioural patterns determine the propensity to innovate. The common attitudes, routines, practices, rules or laws that regulate the interactions between individuals and groups largely determine the propensity of actors and organizations to innovate.
6	Policies are important in innovation. Policy support for innovation is the outcome of a set of policies that work together. In evaluating the effectiveness of policies on enabling innovation, one must be sensitive to a wide range of policies and seek ways to coordinate them.
7	The demand side must be included in the innovation process. The innovation systems concept recognizes the importance of including stakeholders and developing behavioural patterns that make organizations and policies sensitive to stakeholder agenda or demands.
8	Changing to cope with change. The classic response of more successful innovation systems, when faced with external shocks, is to reconfigure linkages or networks of partners.
9	Building on 'sticky' information. Innovation can be based on different kinds of knowledge possessed by different actors: local, context-specific (farmers or local entrepreneurs) and generic knowledge (researcher, etc.). Information is sticky when it is local and specific to the owner and not easily available to others.

which research needs to collaborate with actors working directly on development issues, finding its added value in that process, supporting, understanding and accelerating the change processes desired (Douthwaite et al., 2015).

Gildemacher and Mur (2013) created a conceptual model on agricultural innovation and the mechanisms for reaching impact at scale (see Fig. 1). It provides a useful framework to discuss the role of agricultural research in different moments of an innovation process.

In efforts to catalyse agricultural innovation, the model distinguishes three moments, leading to three types of results (Gildemacher and Wongtschowski, 2015):

1. Opportunity and needs identification, leading to the 'identification of entry points for change'.
2. 'Experimentation', leading to tried and tested promising new practices.
3. 'Bringing into routine use', potentially leading to impact at scale.

This chapter discusses the role of research in innovation processes; drawing on cases from research efforts on maize-based food systems. The next three sections (artificially) define the 'moments of an innovation process', as defined above and then discuss the contribution of research at each stage from an innovation systems perspective. This chapter ends with a discussion on the implications of this analysis to the way research is carried out and organized.

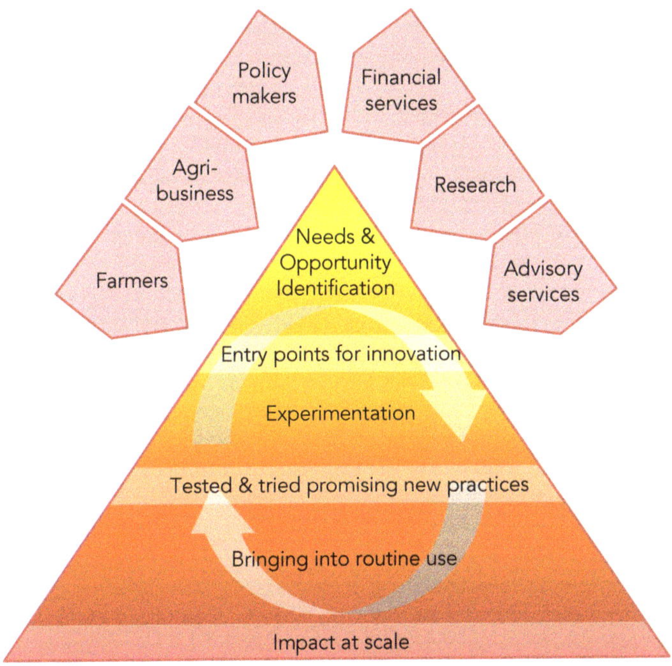

Figure 1 The different moments of an innovation process (Gildemacher and Wongtschowski, 2015).

2 Entry points for change

Opportunities and needs can be identified by a number of stakeholders. Farmers, needless to say, have a key role in raising those diverse needs (and opportunities), alongside development actors working directly with farmers, such as NGOs and extension workers. Researchers working in the field, with a good understanding of the local farming systems, can spot problems and opportunities from a different point of view; complementary to that of other actors. And, last but not least, research contributes new methods and products that provide new ideas, technologies and practices that contribute to a better understanding of, or ways to tackle, a particular problem.

It is not uncommon, nevertheless, for research to assume what the problems and opportunities are in a particular reality; without further consultation of local actors or a good understanding of farming systems and the context they are couched in. For example, Nelen et al. (2015) point out that targeted agricultural interventions are too often narrowly focused on increasing productivity, and fail to consider the (lack of) access to land and other productive assets.

Some of the International Maize and Wheat Improvement Center (CIMMYT)-led projects have started to experiment with ways to face these shortcomings. MasAgro and Buena Milpa provide interesting examples of such efforts.

MasAgro is a Mexican Government-funded project that engages research, development and knowledge transfer to support farmers in Mexico to sustainably increase their productivity of maize and wheat and post-harvest handling by offering them training, technical support on sustainable technologies (e.g. conservation agriculture, integrated pest management, fertility management, post-harvest technologies) and high-yielding maize and wheat seeds. CIMMYT initiated and coordinates the project. MasAgro has more than 150 partners from the public and private sector who contribute with technical support, training and research, amongst others. One of its four components is 'MasAgro Productor'; tasked to bring research (and its results) closer to farmers.

MasAgro Productor uses 'Experimental Platforms' to research new practices and technologies. They serve simultaneously as knowledge exchange and training platforms. MasAgro Productor has recently started promoting 'Experimental Platform Committees', which are farmers groups that discuss with researchers and technicians what the platform should experiment with, review experiments' results and discuss them. The assumption is that the committees is one way to 'reconnect' experimental platforms and farmers, which will lead to more relevant research and ownership by farmers, who then are more inclined to spread the news/info on the Platform and its results. However, there are other, more effective, mechanisms in which research takes into account other actors' demands such as hub meetings where different actors meet to plan activities, including research. The idea behind these mechanisms is to engage farmers in a more structured way, to get their direct input for research priority setting, as well as their feedback on research findings.

Another such example is the Buena Milpa project, also led by CIMMYT, and implemented in Guatemala. The project, focused on reducing poverty, malnutrition and vulnerability on climate change, was designed to provide space and time for researchers and local actors to engage in a meaningful dialogue about what are the local needs and opportunities. The idea was to avoid pushing for 'entry points for action' as researchers, but rather it jointly identifies regional and local needs and demands through meetings with stakeholders.

The project is still in an early stage, but initial activities point out the need to support the ongoing work of local partners on issues such as soil conservation and biodiversity management, with researchers playing an important role in better understanding how the different elements of farming systems interact and how biodiversity management can be improved locally.

3 Local experimentation

This second 'moment' in the innovation process refers to experimenting to tackle problems and take advantage of opportunities; therefore on the basis of the earlier-defined 'entry points for change'. Experimentation is here done under local conditions – so the assumption is that off-farm research has provided the necessary knowledge basis.

The off-farm research that precedes (local) experimentation tackles the 'technological' aspect – but a broader take is fundamental in this phase. A new practice (planting a variety with shorter cycle, for example) needs to be adapted further at local level, and actors need to re-organize themselves around it (e.g. having input suppliers to sell it, and government-delivered subsidized fertilizer to be delivered on time). Planting a new variety – like almost all new practices – has social, gender, economic and environmental implications that must be re-determined locally (Beuchelt and Badstue, 2013).

An important contribution of research in this phase is the ability to design experimentation in such a way that differences between options become clearer (i.e. the results of applying certain options can be distinguished from other variables). Researchers can support farmers and other stakeholders in listing criteria for judging options, and analyse the data when these criteria are applied. An important difference is made between joint experimentation and 'demonstration plots'. The latter are designed often to 'prove a point', and not for learning together with farmers.

In addition, process documentation (done usually by social scientists) is important to provide feedback to research, through better understanding the farming systems, gender issues and institutional constraints faced by farmers. It can also 'provide input for the dimension of "bringing into routine use," beyond the environment in which it is being tested. Analysis of the process of experimentation, and the context in which "experimentation" was successful, is important to consider how it can be successful on a wider scale outside the pilot context' (Gildemacher and Wongtschowski, 2015). The importance of social sciences in understanding these processes and working alongside technical scientists is often underestimated.

In MasAgro, this is currently been tackled by a sound social/anthropological research on the functioning of the (above-mentioned) experimental platforms and their relation with the work done in farmers' field, in which experimentation also occurs. CIMMYT's socio-economic programme is putting considerable effort in understanding how MasAgro, as a project, has been adapted in the different regions/hubs where it operates; what has worked where, and why. In this effort, it has collected evidence on how the adoption of a new practice involves not only technological challenges but mainly institutional ones. The programme is also gathering learnings from the different actors showing that researchers' involvement to experimentation should be from an interdisciplinary perspective.

Another way of organizing feedback to research is through innovation platforms (IPs). CIMMYT's Sustainable Intensification of Maize and Legume Systems for Food Security in Eastern and Southern Africa (SIMLESA) program adopted IPs as one of the ways to facilitate the interaction between stakeholders in its project areas. In Western Kenya, the major groundnut-producing area in the country, CIMMYT initiated an IP in Bungoma South and introduced and experimented with new varieties of groundnuts. Different varieties were tested and compared by the stakeholders, including farmers and processors. Initially, farmers were very enthusiastic about a variety that was particularly suitable for making peanut butter. The platform stakeholders decided to start multiplying this variety and to promote it to fellow farmers within and beyond the IP. However, the processing of the seed proved problematic. Processing it manually turned into drudgery and existing shelling machines damaged the groundnut, leading to significant losses. Hence, the platform stakeholders approached researchers to help them to develop more appropriate shelling machines, suitable for shelling these specific seeds and for processing larger quantities.

4 From new innovation to routine practice

'Bringing into routine use' is referred to as the process of making a certain process, new idea or practice used by a large number of farmers (or other stakeholders). Researchers cannot – alone – bring a new (set of) ideas to scale. Their role in this step is more humble. Partnerships with private, public, non-governmental and farmers' organizations play an important role. The focus of research is on understanding the complex and building evidence.

CSISA (Cereal Systems Initiative for South Asia), a CIMMYT-led project being implemented in Asia, seeks to achieve impact at scale through working with intermediary stakeholders operating at different scales. The project aims at intensification of agricultural systems by promoting combinations of technologies and practices, including the introduction of short-duration rice varieties, early planting of long-duration wheat varieties, improved germplasm and zero-tillage/direct seeding for rice and wheat, and rice transplanting machinery. Dissemination of new scientific knowledge and 'promising' practices is done through training events to extension workers and through printed and electronic media. The programme puts strong emphasis on training farmers to become service providers (i.e. providing access to tractors equipped with CA-friendly tools) and influencing the government agenda to support such service providers through subsidies on the equipment and on their services. CSISA staff attend district-level planning meetings in which staff of the Department of Agriculture converge to discuss the programmes to be implemented in the next season. CSISA staff seize the opportunity to discuss the advantages of early sowing and zero tillage, indirectly influencing the government's agenda.

As mentioned above, AIS thinking implies that a re-adaptation of practices/technologies must take place when they are applied in new settings. The role of researchers is to follow, analyse and understand these re-adaptation efforts and draw lessons from them. Analysis and lessons from these efforts are important to shape future research agendas by understanding the limitations of the current proposed practices/technologies. It also provides new entry points for action as new (often institutional and organizational) barriers for development are better understood. Unfortunately, this step is often misrepresented as 'spreading the word', with little or no attention to process analysis and discussions of the implications of its conclusions. This lack of problematization of the research agenda has as much to do with the way projects are organized as with donor requirements.

The Sustainable Intensification of Maize Legume Systems for the eastern Province of Zambia project (SIMLEZA), implemented by CIMMYT and IITA (International Institute for Tropical Agriculture) between 2011 and 2015, combined on-farm research on new maize and legume varieties, conservation agriculture and soya agronomy, with the targeting of such new technologies and the household-level training in soya processing. SIMLEZA set up project-based IPs in these three districts as an outscaling mechanism – for communicating project findings to a wider audience. In 2014, the project started to strengthen its work on IPs in one district; Lundazi. The effort showed that the IP was well placed to understand the larger developments in the area and trigger better coordination and joint action between stakeholders. It has, nevertheless, had little effect on shaping SIMLEZA's own research agenda – which was seen by researchers as pre-set and with little room for new ideas.

5 Roles and capacities required from agricultural research

The shift from linear transfer-of-technology models in the 1960s to 'farmer first' and farming systems research (1980s and 1990s) towards AIS perspectives implies new roles for agricultural research. If research is to play the roles described in the previous section, it needs capacities, skills, means and institutional backup to do so. When analysing the embedding research programmes in AIS thinking, Schut et al. (2016) conclude that '*system innovations will not only require reframing of paradigms, but also changes in organisational mandates and incentives, procedures and funding*'.

Specifically for research project, an AIS perspective requires an in-depth understanding of the agricultural system it is embedded in – both at farm level and at higher (landscape) levels. It requires knowledge of current and previous work in the area and a critical reflection of the contribution of research: both international research institutes and national ones.

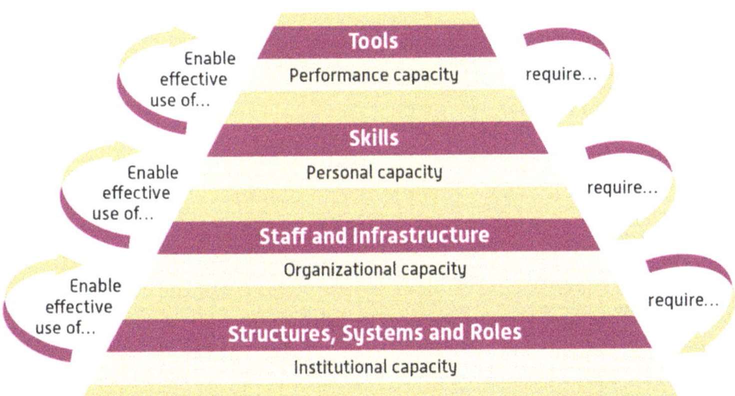

Figure 2 Capacity pyramid.
Source: Potter and Brough, 2004.

Table 2 Innovation 'moments' and the consequent roles and capacities required of agricultural research

Innovation 'moments'	Roles of research	Required individual capacities, attitudes	Required organizational capacities	Required institutional capacities and linkages
Identify entry points for action	Helping people to (continuously) reflect on and analyse their situation, problems and opportunities from different perspectives Eliciting critical uncertainties and translating these into researchable issues for different disciplines and across disciplines	Interact with farmers and other stakeholders Systems analysis Ability to translate complex problems into 'researchable issues'	Interdisciplinary teams, approaching complex problematic situations from an interdisciplinary angle Projects designed around problems/ opportunities and not around disciplinary fields of work	A common objective or problematic situation as starting point, stakeholders' agendas are leading Trust and rapport with stakeholders Trust and rapport with donors to allow flexibility
Experimentation	Experiment with a variety of options Do on-farm, participatory research Providing relevant knowledge and information Facilitate assessment of the trade-offs and feedback, adjust research agenda	Facilitate participatory approaches to analyse and experiment Experiment with organizational and institutional arrangements Experiment with options, addressing the needs of different categories of stakeholders	Resources to facilitate and engage in local experimentation Link different disciplines and types of research (on-station, on-farm, farmer-led)	Provide incentives for working in an interdisciplinary manner Consider multi-stakeholder bodies/ mechanisms as spaces for experimentation, not for dissemination
Bringing into routine use	Create opportunities for people to assess research results Study the contribution of different interaction mechanisms to agricultural development and building a system's capacity to innovate Integrate feedback in the research agenda Ensure effective partnerships and institutional arrangements that can support upscaling Understand local adaptation (technical, social, organizational, institutional) Provide evidence through research and prepare evidence-based policy papers	Researcher able to facilitate feedback sessions Ability to document successful and not-successful innovation Ability to facilitate local adaptation of successful innovations Ability to synthesize research findings (evidence) into policy messages	Working in interdisciplinary research teams, process-orientation Strong interaction between policy researchers (socio-economic) and other content-related researchers (technical)	Mobilize local stakeholders (networks) to support upscaling Maintain good linkages to policy makers at different levels Mobilize different stakeholders to support lobby and advocacy/create synergetic actions Relationship with donors must recognize research as a 'neutral' and self-critical provider of information

It requires an understanding of (complex) problems from an interdisciplinary perspective – understanding and asking oneself whose problem it is, how it comes about and what it is linked to. More often than not, this leads to the need to tackle it from an interdisciplinary perspective. Finally, it requires projects and donors alike to be open to learn from failures, mistakes and successes.

Potter and Brough (2004) distinguish four levels of capacity needs for systemic change: the performance capacity (tools and resources in place), personal capacity (referring to individual skills and attitudes), organizational capacity (infrastructure, collaboration between staff, capacities of all staff) and institutional level (policies, relations with other stakeholders). These have characteristics of their own but are interrelated (see Fig. 2). Table 2 summarizes the different requirements of research(ers) at individual, organizational and institutional levels for the different 'innovation moments'.

Lessons from the above-mentioned CIMMYT-initiated projects revealed that a number of capacities are key for R4D from an AIS perspective.

5.1 Personal capacities

New roles of research as described above demand different capacities of individual researchers. These capacities are primarily related to the interaction with other stakeholders and other research in the different stages of the (iterative) research cycle, including the identification of entry points for change, experimentation, feedback and scaling.

A second capacity is related to the ability of researchers to engage in systems analysis and approach a problem from a holistic and interdisciplinary perspective. That also implies the ability of the researcher to translate complex problems into relevant interdisciplinary and disciplinary research agendas and to synthesize and translate back the research finding as part solutions that fit the wider system. It includes the ability to experiment with different options and with organizational and institutional arrangements as part of the solution. An understanding of gender issues and diversity among farmers beyond their agro-ecological characteristics is key.

The capacity to facilitate stakeholder interaction throughout the process is a third major capacity required from researchers. This capacity can be embedded in research or can be outsourced, for example, to public agricultural extension or an NGO operating in the area.

'Facilitation capacity' also refers to the ability to facilitate systems analysis in a participatory manner, to facilitate participatory approaches to experimentation, addressing the needs of different socio-economic categories of stakeholder. In addition, it refers to capacity to organize feedback, both related to the experiments and the scaling process. The latter relates to the ability to understand why farmers adopt certain new practices, but even more important, why others do not adopt: the ability to analyse and document successful and not-successful innovation. Finally, researchers need the ability to synthesize research findings (evidence) into policy messages.

5.2 Organizational capacity

In order to facilitate researchers' engagement in R4D projects from an AIS perspective, there is a need to organize research projects differently. Organizational capacities that are required include the establishment of interdisciplinary teams; approaching complex problematic situations from an interdisciplinary angle. This includes strong interaction between socio-economic researchers and technical researchers (breeders, agronomists).

Projects, for example, can be organized around challenges or opportunities faced by farmers and other stakeholders, and not disciplinary fields of work. Working in a more participatory manner, with more interaction with farmers and other stakeholders, requires resources to facilitate and engage in local experimentation and feedback. The emphasis is more on the process. It requires time and financial resources. It also requires linkages between different types of research (on-station, on-farm, farmer-led).

5.3 Institutional capacity

Research from a systems perspective demands less linear and more iterative, flexible open-ended research projects. It is important to take a common (development-related) problem as starting point, not the available (set of) technologies or practices. Stakeholders' realities and agendas are leading. Trust and rapport with farmers and other local stakeholders is needed for meaningful stakeholder involvement in agenda setting, experimentation and feedback. Equally important is good trust and rapport with donors, which is needed to allow flexibility in project design and implementation. Donors must recognize research as the self-critical provider of information.

Multi-stakeholder bodies or mechanisms such as IPs are to be considered as spaces for experimentation and the generation of feedback, not for dissemination. However, the mobilization of local stakeholders and their networks (something IPs can support with) is key to support upscaling. Policy makers are part of the innovation system and are key for developing the appropriate policies. Thus, maintaining good linkages to policy makers at different levels is of utmost importance. At the same time, the mobilization of different stakeholders to support lobbying and advocacy and to create synergetic actions to create an enabling environment for scaling is key to successful research for development projects.

A final element is the need to create incentives for working in an interdisciplinary manner. Currently, interdisciplinary research is not being rewarded in most research organizations. The existing opportunities for interdisciplinary scientific publications are limited. Platforms for this type of publication need to be created.

6 Conclusions

The need for systems approaches is derived from the increasing acknowledgement of the complexity of the problems that farmers and other stakeholders face in agricultural development. In the last decades, many research organizations have made efforts to adopt elements of systems approaches to research for development. However, the shift towards AIS perspectives implies new roles for capacities of agricultural research.

However, little acknowledgement and attention has been paid to establish the required capacities that allow for these new roles related to the implementation of more interdisciplinary and holistic ways of research for development. Very few organizations and projects have successfully adopted and institutionalized AIS perspective as the main paradigm for research for development. At best, elements of AIS have been built into more traditional linear research projects. A good example is the attempts to establish IPs for scaling.

To make a difference, more attention is required for establishing the conditions to allow for AIS as a leading paradigm for agricultural research for development. Research for development projects need to pay specific attention to the development of individual, organizational and institutional capacities. In other words, they need an explicit focus on establishing the capacity to innovate.

7 Where to look for further information

More information on the collaboration between KIT and the CRP Maize can be found at:

http://www.kit.nl/sed/project/maize-crp-innovation-systems-improving-impact-research/
http://maize.org/agricultural-innovation-systems-in-maize/
https://www.youtube.com/watch?v=0wZWlkBaE2U

Publications that touch upon issues discussed here include:

Gildemacher, P. and Mur, R. 2013. *Bringing New Ideas into Practice; Experiments with Agricultural Innovation. Learning form Research into use in Africa*. KIT Publishers, Amsterdam. Found at: http://213ou636sh0ptphd141fqei1.wpengine.netdna-cdn.com/sed/wp-content/uploads/publications/2054_bringing_new_ideas_into_practice.pdf

Gildemacher, P. R. and Wongtschowski, M. 2015. *Catalysing Innovation: From Theory to Action*. KIT Working Papers Series 2015: 1. Found at: http://213ou636sh0ptphd141fqei1.wpengine.netdna-cdn.com/sed/wp-content/uploads/sites/2/2015/06/WPS1_2015_online.pdf

8 References

Beuchelt, T. D. and Badstue, L. 2013. Gender, nutrition and climate smart food production: opportunities and trade-offs. *Food Security*, 5 (5): 709–21.

Biggs, S. D. 1990. A multiple source of innovation model of agricultural-research and technology promotion. *World Development*, 18: 1481–99.

Chambers, R. 1983. *Rural Development: Putting the Last First*. Longman, London.

Douthwaite, B., Apgar, J. M., Schwarz, A., McDougall, C., Attwood, S., Senaratna Sellamuttu, S. and Clayton, T., eds. 2015. *Research in Development: Learning from the CGIAR Research Program on Aquatic Agricultural Systems*. Penang, Malaysia: CGIAR Research Program on Aquatic Agricultural Systems. Working Paper: AAS-2015-16.

Gildemacher, P. and Mur, R. 2013. *Bringing New Ideas into Practice; Experiments with Agricultural Innovation. Learning form Research into use in Africa*. KIT Publishers, Amsterdam.

Gildemacher, P. R. and Wongtschowski, M. 2015. *Catalysing Innovation: From Theory to Action*. KIT Working Papers Series 2015: 1.

Hall, A., Janssen, W., Pehu, E. and Rajalahti, R. 2006. *Enhancing Agricultural Innovation: How to Go beyond the Strengthening of Research Systems*. World Bank, Washington.

Hawkins, R., Heemskerk, W., Booth, R., Daane, J., Maatman, A. and Adekunle, A. 2009. *Integrated Agricultural Research for Development (IAR4D)*. Forum for Agricultural Research in Africa, Accra.

Nederlof, S., Wongtschowski, M. and van der Lee, F., eds. 2011. *Putting Heads Together. Agricultural Innovation Platforms in Practice*. Bulletin 396, KIT Publishers.

Nelen, J., de Steenhuijssen Piters, B. and Posthumus, H. 2015. *Enhancing Rural Labour Productivity: Reaching the Rural Poor.* KIT Working Paper Series 2015: 8.

Potter, C. and Brough, R. 2004. Systemic capacity building: A hierarchy of needs. *Health Policy and Planning* 19(5): 336–45. Oxford University Press, Oxford.

Schut, M., Klerkx, L., Sartas, M., Lamers, D., Campbell, M. M. C, Ogbonna, I., Kaushik, P., Atta-Krah, K. and Leeuwis, C. 2016. Innovation Platforms: experiences with their institutional embedding in agricultural research for development. *Experimental Agriculture*, Available on CJO2015. doi: 10.1017/S001447971500023X.

Chapter 13

Women and maize cultivation: increasing productivity through gender analysis

Cheryl Doss[1], *Yale University, USA*

1 Introduction: the importance of considering the impact of gender in maize production

When maize production is seen as a technical process, in which breeders seek to develop varieties that produce higher yields under a range of agro-ecological conditions, it is difficult to understand why it might be important to pay attention to women farmers. However, agricultural researchers are often frustrated that seemingly promising technologies are not taken up. Moreover, even when they are adopted, they rarely deliver the expected transformational impacts. There may be many possible reasons, but at least one is that scientists are not necessarily working to alleviate the constraints that are binding. The underlying challenge is that agricultural science functions best when the research targets are narrowly defined and clearly identified. However, the systems in which farmers operate are seldom simple. Farmers are managing complex portfolios of activities amid a bewildering array of constraints. Moreover, there is such huge variation in the conditions

1. With thanks to Meredith Byrne for excellent research assistance.

http://dx.doi.org/10.19103/AS.2016.0001.20

that farmers face that the nature of the agricultural challenges may vary even from one plot to another on the same farm.

Maize scientists are being pushed to move beyond the traditional breeding goals of simply increasing yields. Instead, they are expected to play a role in increasing incomes, decreasing poverty and increasing the welfare of farmers and consumers. While the scientific process of breeding maize that produces higher levels of output under specific conditions may not need to consider gender issues, the processes of setting priorities and developing the research agenda need to include gender in order to develop technologies that can increase the net profits – both cash and non-cash – for farmers.

Men and women farmers choose which crops to grow and which varieties of particular crops to plant. They consider the inputs that are available to them individually as farmers and to their households. These inputs may include fertilizer and other purchased inputs, and labour, including that of themselves, their family and those that they can hire. In addition, land is required and the property rights that farmers have over the land may affect their choices. Most smallholder farmers produce for both home consumption and the market, and thus choose to grow both what will provide their own food security and what will sell in the market.

Maize production is embedded in all of these human processes; therefore, we need to consider both men and women, as producers and consumers, and how they are embedded in households, communities and markets. Gender plays a key role in terms of how individuals interact at all these levels.

Evidence suggests that women farmers are less likely to use improved technologies for maize production, such as improved seeds and inorganic fertilizer, than are men. This tends to be true whether we consider women individually as farmers (Morris et al., 1999), women as heads of agricultural households (Bezu et al., 2014; Doss et al., 2003) or both (Doss and Morris, 2001; Fisher and Kandiwa, 2014; Namonje-Kapembwa and Chapoto, 2016).

In order to improve the well-being of women farmers, we need to understand not only women as farmers but also the gender relations in which the decisions by men and women farmers are embedded. We need to understand both women as farmers of their own maize plots and women involved in household maize production. It is important to note that a discussion of gender issues in maize production is not simply one about women's roles on the farm. It extends to a deeper understanding of men's and women's roles and responsibilities in a broader context. Analyses of how gender influences agricultural production must take into account how social norms, roles, responsibilities and access to resources affect both men and women.

In particular, researchers must understand the opportunities and constraints that men and women face within maize production systems. Some programmes should specifically target women farmers, who often face additional constraints in comparison with men. In addition, women need to be incorporated into the processes of structural transformation of the agricultural sector, ensuring that new opportunities are available for both women and men, both within and beyond maize production. The conditions are not static; in fact, it is precisely the conditions of production that agricultural research is working to change.

There is an extensive literature on women farmers and agricultural production, some of which focuses on maize. This chapter, however, is not a systematic review of the literature. Numerous such reviews are available and many are referenced throughout this chapter (see e.g. Food and Agriculture Organization, 2011; Quisumbing et al., 2014; World Bank, Food and Agriculture Organization, and International Fund for Agricultural Development, 2009). Instead, this chapter seeks to highlight the key ways in which gender analysis could impact the design, focus, and implementation of maize research, thereby ensuring that women participate and benefit from growth in the maize sector.

While some of the points in this chapter would be relevant for any type of agricultural research, there are some specific challenges to maize research. Maize farming takes place under very heterogeneous conditions, especially when compared to wheat and rice. In developing countries, the irrigated lowlands of South and Southeast Asia reflect fairly similar conditions for rice cultivation, while the Indo-Gangetic plain has many agroecologically similar areas for wheat cultivation. However, there is no comparable homogenous area for maize cultivation in the developing world. This means that understanding the local contexts is particularly important for maize. Especially in Africa, not only is the agroecology of maize production areas varied but also the access to markets, land tenure systems and social norms vary widely. Finally, maize is a particularly dynamic crop; farmers are shifting to maize cultivation in areas that formerly grew millets and sorghums. Thus, the constraints faced by farmers are likely to matter at a much more granular scale.

This chapter will begin by situating men and women farmers within households and discussing why the conceptualization of households matters for agricultural research. Section 2 analyses the ways in which gender impacts agricultural production systems, including access to resources and decisions about inputs. Because labour is a key input, and one particularly impacted by technological innovations, it is addressed in Section 4. A discussion of the gendered roles in maize processing and storage is followed by an analysis of the varietal choice. While the focus on this chapter is on maize farming, urban consumers shape the demand for maize, and Section 7 considers their preferences. Finally, since agricultural programmes are increasingly asked to consider how they affect women's empowerment, a brief discussion of the Women's Empowerment in Agriculture Index highlights a tool that is being used within the sector for monitoring and evaluation. The final section discusses the implications for maize scientists.

2 Men and women within farming households

Many discussions of gender and agriculture compare the differences between male- and female-headed households. This is almost always an unhelpful approach for understanding production and consumption decisions. Thus, it is useful to begin by considering a range of approaches to thinking about men and women farmers within smallholder farming households.

Smallholder famers live in households and analyses of smallholder farming have typically focused at the household level. Most agricultural research and programmes have assumed that the household head is the primary farmer and is the one responsible for making decisions about agricultural production. This leads to targeting and analysis at the household level, categorizing households by whether they are headed by a man or a woman. Typically, if there is man living in the household, he is considered the head.[2] One result is that men have been targeted for agricultural services, such as extension, credit or membership in producer organizations. Moreover, the voices of men farmers are the ones that have primarily shaped the agricultural research agenda.

2. There is increasingly a shift in changing the language to refer to 'dual headed households' for those that are headed by a couple and referring to sole male-headed and sole female-headed for those households headed by a person who is not married. This addresses the issue that the more traditional language of male- and female-headed households confounds the issue of headship with that of the structure of the household.

In response to concerns about focusing on the household head, some analyses have focused on the primary farmer. This is preferable, because it does not assume that the head is also the primary farmer. Other analyses consider the agricultural holder, the person who has management responsibility for an agricultural holding. In practice, this approach usually identifies one person as responsible for the agricultural enterprises of a household. And, in practice, the primary farmer or holder usually is the household head, although in some circumstances, such as when the household head has migrated, it may be the spouse or another adult.

These definitions lead to classifications of households based on the one person who is considered the head, holder or farm manager. When a household is labelled female-headed or female-managed, it usually indicates that no adult male is present. This lack of a man in the household may result in the household having less labour available or less access to other inputs such as information or credit. But only considering the sex of the head or manager renders invisible all of the others who live in those household, especially women who live in households thus defined as male-headed, even when they are intimately involved with farming. These women's roles in deciding what and how to produce is ignored.

Research coming out of Africa, particularly parts of West Africa, has drawn attention to agricultural systems in which women farm their own plots of land, even within male-headed households. This shifts the unit of analysis from the household to a plot of land. Multiple farmers may thus be identified within a single household. Men and women within the same household may grow similar or different crops on their plots; maize may be grown by either or by both. New plot-level data available through the Living Standards Measurement Study – Integrated Surveys on Agriculture (LSMS-ISA) for six African countries has facilitated this level of analysis (see e.g. Backiny-Yetna and McGee, 2015; Oseni et al., 2015; Slavchevska, 2015; Aguilar et al., 2015; Kilic et al., 2015. The extent to which men and women farm independently varies across these countries).

Typically, in these systems, men are able to mobilize women's labour for their plot, but women have a harder time mobilizing men's labour on their plots. One implication of this was identified in Burkina Faso where Udry (1996) found inefficiency across plots within the same household. Reallocating some fertilizer and male labour from men's plots to women's plots, within the same household, would have increased total household production.

Yet, in many households, men and women work together on all or most of the household's plots of land, growing crops for both home consumption and for the market. There has been little research on how the women in these households influence the decisions about what to plant and how much to sell when they are not identified as the primary farmer or a plot manager. Women may have some voice in terms of what is grown, but we are missing their voices in such analyses. One recent study finds that joint management of maize plots by husband and wife is associated with higher fertilizer use (Marenya et al., 2015). This may be particularly relevant for maize, where women often work on household plots.

There is an extensive literature, primarily in economics, regarding what gives women decision-making power within households, often referred to as the bargaining power to influence decisions (for a review of this literature, see Doss (2013)). Land ownership is often used as a proxy for bargaining power, yet it is rare to consider agricultural decisions as the outcomes in the bargaining process. In one of the few studies that specifically considers agricultural decision-making, women's land ownership and their involvement in fieldwork is associated with women's greater involvement in decision-making on the farm in Ecuador (Twyman et al., 2015).

Thus, to understand maize production systems, it is important to consider the household and to understand who is involved in making the decisions, providing the labour and controlling the output. This will ensure that interventions reflect the priorities and constraints of those who are involved and making decisions, rather than the one person identified as responsible for all agriculture within the household or on a plot.

3 Men and women in maize production systems

Agricultural scientists frequently focus on ways to improve yields, given varied agro-ecological environments that have droughts and floods, pests and weeds, biotic and abiotic stresses. But farmers face problems that are even more complex. In particular, farmers are not simply maximizing yields. Even in completely commercialized settings, we know that farmers are not maximizing yields but are instead maximizing profits. Clearly, higher yields are a component of this objective, but farmers are concerned with costs as well as revenues. Similarly, smallholder farmers who are producing both for themselves and for the markets are concerned about profits. Yet, not all of their inputs and outputs are traded within markets. They are using household labour and consuming some of the outputs themselves. In addition, they are not simply maximizing profit – even broadly defined – of maize, but are doing so across the full spectrum of agricultural and non-agricultural activities. Thus, the choices that farmers make and the technologies that will ultimately benefit them are influenced by the costs of all their inputs, including those (like family labour) that are not traded in the market. These costs and constraints may differ for men and women farmers.

Land is the most important resource for agricultural production. Throughout the world, women are less likely to be landowners compared to men, and – when they do own land – their plots tend to be smaller and of poorer quality (see the following on the gendered patterns of agricultural land ownership in Africa and Asia: Doss et al., 2015; Kieran et al., 2015). For agricultural production, ownership is only one dimension. We might also be concerned with the access to and control over land. Someone who is not a formal owner may still have rights to use the land.

Especially in areas where land is under customary tenure, rather than being formally titled, women often access land through their husbands or fathers. Thus, their access to land is dependent upon their relationships to men. When a woman accesses land through her husband, she risks losing it when the marriage ends, either through divorce or through the death of her spouse. This gives her an additional layer of tenure insecurity.

It is widely accepted that farmers invest more in their land when they have secure tenure.[3] The investments relevant for maize production may include improving soil fertility or water management. For many women, the additional tenure insecurity may affect the investments that they are willing and able to make. In turn, this may affect varietal choice and the returns to purchased inputs. If women lack tenure security, they may have less interest in managing long-term soil fertility. This may, in turn, affect the profitability of fertilizer and other inputs.

3. Secure tenure does not necessarily mean that the land is titled. People may have secure rights under a customary system and, if the legal system is not adequate, titled land is not necessarily secure.

Rental markets provide another way to access land for farming. The extent of rental markets varies widely across locations. In principle, rental markets should be good ways for women to access land; renting land requires less capital than purchasing land. Yet, there are often social norms that limit women's access to rental markets; women may not be considered real farmers and thus, landowners are reluctant to rent to them. These attitudes may also affect the rental rates that women have to pay.

In addition to land, inputs and information are also key to agricultural production (for a detailed review of women's access to non-land inputs, see Peterman et al., 2010). (Labour inputs will be discussed in the following section.) The most widely studied input is inorganic fertilizer. Most studies of fertilizer use find that women farmers are less likely to use fertilizer and use it less intensively than men. However, once researchers move beyond simple descriptive statistics of men's and women's fertilizer use and analyse these patterns controlling for other farm and farmer characteristics, such as land size and education, the differences are often reduced or eliminated. Thus, it appears not to be the case that women are inherently less likely to use fertilizer; instead, the characteristics of women's farms or the women themselves account for their decisions to use less fertilizer.

In practice, the prices that men and women farmers may face for inputs may also be different. The issue is not necessarily that women are charged more by dealers. However, the effective price at the farm gate may be higher for women. For instance, the price per unit of fertilizer may be higher when it is purchased in small quantities. Women may also have less access to fertilizer purchases through producer cooperatives or other lower cost suppliers. In addition, the cost of moving fertilizer and other inputs from the market to the farm may be higher for women in some contexts. Women farmers are less likely to own a bicycle (and cultural norms in some places restrict women's use of bicycles) or motorized transport.

Thus, when considering the profitability of using fertilizer, men and women find themselves in different situations. A recent study examining farmers' adoption of improved maize varieties in Kenya finds great heterogeneity across farm households with regard to the profitability of fertilizer use (Suri, 2011). The author controls for household composition, but not specifically the sex of the farmers, but the analysis lends credence to the argument that the profitability varies widely across farms.

One suggestion has been to promote smaller input packages that are well suited to farmers who are operating on a small scale and have limited access to cash or credit (Quisumbing and Pandolfelli, 2010); this would be one way to increase women's access to inputs such as seeds or fertilizer.

A vast amount of literature has also shown that women are less well served compared to men by agricultural extension programmes and have less access, in general, to agricultural credit programmes, although in recent years, extension and credit programmes have made greater efforts to include women in a number of countries (for a historical perspective, see Saito and Weidemann (1990); for a recent discussion of gender and extension, see Manfre et al. (2013); for a discussions of the constraints facing women farmers, see Doss (2001); Quisumbing and Pandolfelli (2010)). To the extent that women lack information and credit, it is likely to affect their input choices and, in turn, their selection of crop varieties.

Finally, it is worth noting that if women do not control the marketing of their crops or the resulting revenue, it may reduce their returns to farming, functioning like a tax on their output. This may influence their willingness to purchase inputs and to exert effort in managing and farming their plots.

4 Gender divisions in labour and access to labour for maize production

The traditional patterns of the gender division of labour have been well documented in a variety of settings. Similar to other crops, for maize, they generally conform to the pattern that men are responsible for clearing land and ploughing. Particularly in Africa, women are frequently responsible for planting, weeding and harvesting. There are differences across locations as to who is responsible for fertilizer or pesticide application. Similarly, harvesting and marketing may be done by men or women.

While these are the typical patterns, there are variations. There are places where women are involved with ploughing, for example, even though it is generally not the norm. In addition, the patterns may change over time. As men migrate out of the rural areas, their wives often take on additional agricultural tasks (for a discussion of gender, migration and agriculture in Asia, see Mueller et al., 2015). Moreover, with increasing commercialization within the agricultural sector, men and women may take on new or different roles.

Although it is often assumed that smallholder farms have an excess of labour available, in practice, they often face labour shortages, especially at peak seasons. Because men and women may engage in different agricultural activities, they may face different labour constraints. A recent analysis of six countries in Africa highlighted the fact that women farmers face challenges in accessing labour for their plots. They often live in smaller households, with fewer adult men, and have more difficulty compared to men in accessing male labour (World Bank and ONE Campaign, 2014). And while women would be interested in hiring men's labour if they had the resources, the times during which women require men's labour are precisely the times when men are working on their own farms.

In addition, women have significant other responsibilities, including much of the household work and childcare. These limit the amount of time and effort that they have available for farm work. Many studies have documented the time poverty that women face (Arora, 2015; Blackden and Wodon, 2006). Thus, technologies that reduce their labour burdens, both in agriculture and in household work, could have a beneficial impact for women farmers.

New technologies may change labour requirements along numerous dimensions: the total amount of labour needed, the timing of labour inputs or the type of labour required. Given the gendered patterns of labour, this suggests that new technologies will have different impacts of men's and women's labour inputs and thus whether women accrue any of the advantages is very context specific.

Many labour-saving technologies for maize production, such as tractors for land preparation, primarily reduce the labour burden for men's activities. If nothing else changes, then using tractors would reduce men's labour input and have little impact on women's labour. However, if using tractors for land preparation results in larger areas under cultivation for the household, women's labour burden could increase, because they now have a larger area to plant and weed. The net welfare impacts for women will depend on the extent to which they have control over the increased production. On the other hand, when women are farming their own plots, they have limited access to male labour; therefore, the mechanization of men's labour has the potential to benefit them, if they have the resources to hire tractors.

Technologies that are land saving – in other words, those that increase the intensity of production on a plot of land – may also be labour increasing. They may increase production per unit of land, but not necessarily per unit of labour. As such technologies are developed, it may be important also to develop technologies that reduce the labour burdens in other aspects of maize production or processing.

Some heavily promoted practices, such as conservation agriculture, may assume that farmers have abundant and underutilized labour that can be mobilized – or at least that family labour has little or no cost. However, this is not necessarily the case. The labour demand for these practices may come at a high cost in terms of other household activities, including farm and non-farm businesses and home production.

A number of technologies that were designed for women have had very low levels of adoption. Studies have tended to suggest two recurring reasons. One is that women often do not have the resources to purchase the required items. This may be the case for improved seeds, inorganic fertilizer or mechanization. A second reason is that while some of the technologies were designed for women, they were not designed to meet women's needs and thus did not address key constraints. Many processing and storage technologies fall into this category. Thus, any technologies designed as labour saving for women must have women's input into the design and must be affordable in the local context.

In areas where poor women are more likely to be landless agricultural labourers than smallholder farmers, some forms of labour-saving technology may displace them, rendering them worse off if no other forms of wage labour are available. In the long run, we would want to ensure that these labour-saving technologies are complemented with opportunities for women in other sectors that offer higher-quality wage work.

5 Post-harvest processing and storage

In smallholder systems, women are often responsible for much of the post-harvest activities, particularly processing and storage (Quisumbing and Pandolfelli, 2010). Women's work in processing of maize on the farm has been largely ignored by researchers. Early efforts to develop better technologies were not successful, and this area of research was largely dropped. Although some of the issues were technical, particularly finding processing technologies that reduced women's time but had low investment costs, other challenges were social, with issues of who had control over the equipment and how it was maintained. Struggles between men and women in the community often ensued when new technologies were created and often women were not able to benefit (see example cited in Doss (2001)).

One of the challenges in technologies to save women's labour inputs into processing is that the opportunity cost of women's time is low. Although many women are time-poor, as noted above, the activities that they would do if their time was freed up from processing cannot cover the cost of the technology. It might be the case if that if women spent less time in processing maize, they could spend more time on child health and nutrition. But although that is a valuable activity, it is not economically remunerated and does not allow them the cash to purchase the equipment.

Although the evidence on the extent of post-harvest losses is varied, in some places, they are quite high. The risk of loss in storage may affect farmer decisions. Farmers may prefer not to store grain on farm and thus sell it immediately after harvest when prices are low. This may also affect farmer's level of participation in the market.

There are two types of technologies that could reduce losses in storage. The first is maize varieties that are well suited for on-farm storage. This would involve incorporating good storage characteristics in the breeding process. In addition, better methods of storage could also help to reduce losses. Keeping grain dry and free from mould and insects could reduce losses and benefit both men and women farmers. Yet, new technologies to improve storage could have differential impacts on men and women due to their different roles. Some new technologies, such as metal silos, may be viewed as being within men's domain, rather than that of women. Thus, they have the potential to shift the stored grain from being under women's control to that of men (Swiss Agency for Development and Cooperation, 2015). It is not clear what impact this would have in the long run on production decisions made by women. This suggests that women be involved in the design of improved storage technologies, both so that they suit the needs of women and so that women are able to retain some control over them (labour-saving policies for women are discussed in Grassi et al. (2015)).

6 Crop and varietal choice

The choices of which crops to grow – beginning with whether or not to grow maize and if so, which varieties to grow – are influenced by the agro-environment, the markets and availability of seeds and other inputs, the preferences of the farmers and the profitability of different crops and varieties. Farmers may choose to produce for home consumption or for the market or both. As the sections above indicate, men and women farmers may face different constraints and thus different implicit prices which will affect their production decisions and choice of variety (see de Brauw (2015) for a discussion of crop choice among male and female farmers in Northern Mozambique).

Much of the research on varietal preference focuses on the choices that are made by producers, and one of the key factors that affect trait preferences is whether the maize is being grown for home consumption or for the market. Studies of smallholder farmers typically interview farmers who eat the maize that they grow, even if they also sell small or large quantities. As markets increasingly commercialize, this may change the preferences of smallholder producers.

When growing primarily for the market, generally the characteristics preferred are those typical of cash crop production, such as higher yields, early maturation and tolerance to drought. For example, in Western Kenya, where farmers are primarily growing maize for sale, one study found no gender differences in varietal preferences (Odendo et al. 2001). To the extent that often more of women's production is for home consumption and more of men's production is marketed, we would expect to see some differences in varietal preference.

Because many women farmers are both producers and consumers of maize, their varietal choices will be based on both production and consumption characteristics. The preferred characteristics for maize to be consumed at home may include taste, nutrition and ease of cooking. And thus, we find that women farmer's often express preferences for varieties that have good taste and nutrition characteristics. For example, in Eastern Ethiopia, women farmers preferred varieties that combined sweetness (when consumed green), earliness, big ears and more usable flour (Mulatu and Zelleke, 2002).

When asked about preferences, women farmers often rank higher nutrition as an important consideration. The challenge is that although women may state that they prefer the varieties that are more nutritious or fortified, they do not always know the nutrition content of different varieties. For example, in Ghana, women's systematic lack of access to extension officers results in lower levels of awareness to the nutrition benefits of improved varieties when compared to their counterparts (Morris et al., 1999).

Because women are the ones responsible for food preparation, they often express preferences for varieties with good cooking characteristics. Depending on the method of preparation, these may include the ease of grinding grain into flour and the ability to form dough. Thus, in Mexico, women farmers choose the local maize because of its culinary superiority (Preibisch et al., 2002). Moreover, in parts of East Africa, women prefer grain that is flint rather than dent because the flint grain yields better flour when pounded with a mortar and pestle (Mulatu and Zelleke, 2002).

Some of these preferences may be changing because of the structural changes in the economy. With a shift towards cash crop economies and higher levels of urbanization, women's roles and responsibilities are changing. With greater responsibility outside the home, varietal preferences have also evolved. In areas of Mexico where rural women are purchasing tortillas and processed maize flour, they are less concerned whether their household grows local varieties of maize with better culinary properties (Hellin et al., 2010). In addition, with the move towards a cash economy, non-yield attributes also appear to be less important. The move towards commercialized production often pressures farmers to grow more uniform products, moving away from some of the traditional varieties (Chambers and Momsen, 2007).

While some of the literature considers the preferences over traits expressed by farmers, another body of literature considers whether or not farmers have adopted improved varieties of maize. Many studies suggest that women farmers are less likely to adopt improved varieties compared to men, although the size of the gap reduces or disappears when the analysis controls for other farmer and farm characteristics, such as farm size and access to extension services. As discussed above, women farmers tend to have less access to these complementary inputs. One study in Ghana finds that after controlling for these complementary inputs, there are no differences in the adoption of improved varieties of maize between men and women farmers in households with a male head, but women farmers in households with a sole female head are less likely to adopt these varieties. This suggests that there are differences between the women in these two types of households that are not captured by other observable characteristics that affect adoption (Doss and Morris, 2001).

The challenge is that most studies analysing the adoption of improved varieties either define improved varieties quite broadly or do not clearly define it. Thus, it often includes not only new material but also seeds that have been recycled multiple times and are no longer recognizable as the original improved varieties (for a discussion of these issues, see Doss (2007)). Thus, the gender implications are less clear.

Women often play an additional role in varietal choice through their role in seed selection. Since many households save maize seeds, rather than purchasing fresh hybrid seed each season, the management of seed is vitally important. Seed quality and germination are significantly related to handling, but there may also be genetic and agronomic factors that would affect the quality of saved seed.

In addition, there have been suggestions that women farmers are more risk averse, and thus prefer varieties that involve less risky investments, often familiar, tested local varieties. It is hard to identify whether this is due to inherent higher risk aversion among women farmers, or that women farmers face multiple other constraints that results in their preferring less risky varieties. A recent study uses both experimental games and household surveys to analyse the risk preferences of men and women within the same maize-growing households. They find that women's risk preferences affect the choice of maize seeds in both male- and female-headed households. When wives were risk averse or loss averse, households were less likely to adopt high-yielding hybrid varieties (Love et al., 2014).

7 Women as urban consumers

The section above discusses how women farmers express preferences over varietal traits, including food preparation, consumption and nutrition characteristics. Yet, many maize consumers are now in urban areas and are purchasing maize rather than growing it themselves. While they share some concerns with the producers who consume their own maize, they also face some different issues. They also express preferences for taste and nutrition as well as cooking and storage characteristics. In addition, urban consumers who are purchasing grain express preferences for higher-quality maize and concerns about food safety issues.

Taste and nutrition are primary consumption considerations. The specific preferred taste characteristics of urban consumers vary across countries. Similar to those who produce for home consumption, urban consumers usually report a preference for higher nutritive value for maize, all else equal. For example, in Kenya, when urban consumers were given a choice between freshness, nutritive value, taste, price, texture and brand, both women and men ranked nutritive value as the most important attribute, although the results were stronger for women (Kengo and de Groot, 2015). In urban Mozambique, both men and women preferred the taste and texture of a less nutrient dense variety of white maize, but were willing to trade for a bio-fortified variety to obtain the associated nutritional benefits (Stevens and Winter-Nelson, 2008). Information on the nutrition characteristics of purchased maize may be lacking. For urban consumers, moves towards consistent branding of maize products may resolve this issue.

Urbanization and rising incomes will change the demand for maize. Urban consumers tend to place a high value on convenience, which often results in their switching from maize to wheat or rice. In Nairobi, for example, the trend is for urban consumers to move away from maize and rice to wheat (Muyanga et al., n.d.). In addition, the income elasticity of demand for maize is low, indicating that as consumers have higher incomes, they shift from consuming maize to other foods.

Poorer urban consumers often continue to rely on maize. They are the group that especially benefits from cheaper maize. To the extent that women are overrepresented among poor urban consumers and are the ones responsible for household food purchases, they particularly benefit when prices are low.

8 Agricultural research to empower women

Agricultural research is now expected to not only improve production and livelihoods, but the United States Agency for International Development, (USAID), among other donors, is now expecting that agricultural programmes will also empower women in the agricultural sector. The Women's Empowerment in Agriculture Index (WEAI) is a tool developed to monitor the changes in women's empowerment and is currently required for all of USAID's Feed the

Future programmes (Alkire et al., 2013). The WEAI was developed by USAID, International Food Policy Research Institute and the Oxford Poverty and Human Development Initiative (for more information on the WEAI, see http://www.ifpri.org/topic/weai-resource-center). The tool is being widely adopted and adapted for use in a range of agricultural programmes.

The WEAI considers five domains of empowerment:

- *Production*: Sole or joint decision-making over food and cash crop farming, livestock and fisheries, as well as autonomy in agricultural production
- *Resources*: Ownership, access to, and decision-making power over productive resources such as land, livestock, agricultural equipment, consumer durables and credit
- *Income*: Sole or joint control over income and expenditures
- *Leadership*: Membership in economic or social groups and comfort in speaking in public
- *Time*: Allocation of time to productive and domestic tasks and satisfaction with the available time for leisure activities

Women are considered to be empowered if they attain adequacy in at least four of the five domains (or they can attain a weighted adequacy in 80% of the indicators to be considered empowered). The index was designed to monitor agricultural programming to ensure that it has a positive impact on women's empowerment. Agricultural programmes that exclude women and centralize production and decision-making within the hands of men may result in lower WEAI scores. On the other hand, programmes that support women's participation in agricultural decision-making and lower women's labour burden will increase the scores.

In order for programmes to successfully empower women, these considerations will have to be included at the design stage of the project. It will also require the voices of women farmers in this design and implementation.

9 Conclusions

In smallholder agriculture systems, both men and women farmers are seeking to maximize their profits, both cash and non-cash, across a range of activities, including but not limited to maize production. The costs and benefits may be expressly monetized or may be within the non-market household sphere, such as family labour and household meals from home-grown maize. Men and women farmers implicitly face different prices for land, inputs and labour. Women typically have less access to information and credit, which further affects the relative prices for women and men.

In addition, women farmers may have different priorities for the variety of maize produced. Because they are primarily responsible for preparing the family meals, they are often concerned about taste, cooking quality and nutrition of the maize grown for home consumption. Thus, they may have higher willingness to trade off other characteristics for these. These differences influence the choices that men and women make regarding maize production.

Thus, men and women farmers may make many different decisions about what to plant, what inputs to use, how to allocate their labour and which technologies to use.

Understanding these decisions within the context of their households and the constraints that they face is critical to identifying potential ways to increase agricultural productivity and improve the lives of rural households.

The literature on women as maize farmers has a number of gaps. We continue to know relatively little about women who are farming on household plots rather than their own individual plots. To what extent, and under what conditions, are they involved in production decisions? What programmes and policies targeted at women in maize-growing households may increase output and household welfare? In addition, the literature is quite uneven across locations. Research in Africa often focuses on technology adoption and the constraints that women face; research in Mexico and Central America often emphasizes women's roles in seed selection and maintaining genetic diversity.

There are numerous areas where agricultural scientists can work to strengthen women's position within maize production systems. This would involve developing varieties that meet the needs of smallholder, resource-constrained farmers, many of whom are women. It would also include developing technologies that reduce the labour burdens for women farmers. Low-cost technologies that reduce women's time burden in processing have the potential to have a significant impact on women's lives. Given the numbers of technologies that have not been widely adopted by women, it may be useful to incorporate women's input into the design of the technologies to ensure that they are appropriate in that particular context. Involving women from the initial stages could also ensure that women were able to reap the benefits from these technologies.

Women farmers use a range of improved technologies on their maize farms, including improved seed and fertilizer. They do so when the conditions are right: when it is profitable, where profit is broadly defined to include the non-marketed costs and benefits; when the inputs are available and accessible; and when at least some of the benefits accrue to the women themselves. Developing a broader range of such technologies that are accessible to women and are profitable under the conditions in which they farm has the potential to benefit women farmers and increase productivity in the maize sector.

10 Where to look for further information

References to key summaries of the literature are included throughout this chapter. For discussions of gender and agriculture more generally, see Food and Agriculture Organization (2011), Quisumbing et al. (2014), World Bank, Food and Agriculture Organization and International Fund for Agricultural Development (2009) and Qisumbing and Pandolfelli (2010). Grassi et al. (2015) discuss issues of time constraints for rural women. Approaches to decreasing the gender gap in extension services are presented in Manfre et al. (2013).

11 References

Aguilar, A., Carranza, E., Goldstein, M., Kilic, T., and Oseni, G. (2015). Decomposition of gender differentials in agricultural productivity in Ethiopia. *Agricultural Economics*, *46*(3), 311–34. doi:10.1111/agec.12167.

Alkire, S., Meinzen-Dick, R., Peterman, A., Quisumbing, A., Seymour, G., and Vaz, A. (2013). The women's empowerment in agriculture index. *World Development*, *52*, 71–91. doi:10.1016/j.worlddev.2013.06.007.

Arora, D. (2015). Gender differences in time-poverty in rural Mozambique. *Review of Social Economy*, *73*(2), 196–221. doi:10.1080/00346764.2015.1035909.

Backiny-Yetna, P., and McGee, K. (2015). *Gender Differentials and Agricultural Productivity in Niger.* Policy Research Working Paper, No. 7199. Washington DC: World Bank.

Bezu, S., Kassie, G. T., Shiferaw, B., and Ricker-Gilbert, J. (2014). Impact of improved maize adoption on welfare of farm households in Malawi: a panel data analysis. *World Development*, *59*, 120–31. doi:10.1016/j.worlddev.2014.01.023.

Blackden, C. M., and Wodon, Q. (eds) (2006). *Gender, Time Use, and Poverty in Sub-Saharan Africa. World Bank Working Paper*, No. 73. Washington DC : World Bank.

Chambers, K. J., and Momsen, J. H. (2007). From the kitchen and the field: gender and maize diversity in the Baja region of Mexico. *Singapore Journal of Tropical Geography*, *28*(1), 39–56. doi:10.1111/j.1467-9493.2006.00275.x.

de Brauw, A. (2015). Gender, control, and crop choice in northern Mozambique. *Agricultural Economics*, *46*(3), 435–48. doi:10.1111/agec.12172.

Doss, C. (2013). Intrahousehold bargaining and resource allocation in developing countries. *World Bank Research Observer*, *28*, 52–78. doi:10.1093/wbro/lkt001.

Doss, C., Kovarik, C., Peterman, A., Quisumbing, A., and van den Bold, M. (2015). Gender inequalities in ownership and control of land in Africa: myth and reality. *Agricultural Economics*, *46*, 403–44. doi:10.1111/agec.12171.

Doss, C. R. (2001). Designing agricultural technology for African women farmers: lessons from 25 years of experience. *World Development*, *29*(12), 2075–92. doi:10.1016/S0305-750X(01)00088-2.

Doss, C. R. (2007). Analyzing technology adoption using microstudies:limitations, challenges, and opportunities for improvement. *Agricultural Economics*, *34*, 207–19.

Doss, C. R., and Morris, M. L. (2001). How does gender affect the adoption of agricultural innovations? the case of improved maize technology in Ghana. *Agricultural Economics*, *25*(1), 27–39. doi:10.1016/S0169-5150(00)00096-7.

Doss, C. R., Mwangi, W., Verkuijl, H., and De Groote, H. (2003). *Adoption of Maize and Wheat Technologies in Eastern Africa : A Synthesis of the Findings of 22 Case Studies.* CIMMYT Economics Working Paper 03-01. Mexico, D.F.: CIMMYT.

Fisher, M., and Kandiwa, V. (2014). Can agricultural input subsidies reduce the gender gap in modern maize adoption? evidence from Malawi. *Food Policy*, *45*, 101–11. doi:10.1016/j.foodpol.2014.01.007.

Food and Agriculture Organization. (2011). *State of Food and Agriculture. Women and Agriculture: Closing the Gender Gap for Development.* Rome: FAO.

Grassi, F., Landberg, J., and Huyer, S. (2015). *Running Out of Time: The Reduction of Women's Work Burden in Agricultural Production.* Rome: Food and Agriculture Organization of the United Nations.

Hellin, J., Keleman, A., and Bellon, M. (2010). Maize diversity and gender: research from Mexico. *Gender & Development*, *18*(3), 427–37. doi:10.1080/13552074.2010.521989.

Kengo, D. M., and de Grooto H. (2015). Consumer preferences raise hope for increasing effective demand for maize products and maize research in Kenya, Finance. *Economics and Applied Research Journal*, *2*(2), 1–12.

Kieran, C., Sproule, K., Doss, C., Quisumbing, A., and Kim, S. M. (2015). Examining gender inequalities in land rights indicators in Asia. *Agricultural Economics*, *46*, n/a–n/a. doi:10.1111/agec.12202.

Kilic, T., Palacios-López, A., and Goldstein, M. (2015). Caught in a productivity trap: a distributional perspective on gender differences in Malawian agriculture. *World Development*, *70*, 416–73. doi:10.1016/j.worlddev.2014.06.017.

Love, A., Magnan, N., and Colson, G. J. (2014). Male and female risk preferences and maize technology adoption in Kenya. In *Presented at the American Agricultural Economics Association*. Minneapolis, MN, 27–29 July 2014.

Manfre, C., Rubin, D., Allen, A., Summerfield, G., Colverson, K., and Akeredolu, M. (2013). *Reducing the Gender Gap in Agricultural Extension and Advisory Services*. Washington, DC: Measuring Extension and Advisory Services Discussion Paper 2.

Marenya, P., Kassie, M., and Tostao, E. (2015). Fertilizer use on individually and jointly managed crop plots in Mozambique. *Journal of Gender, Agriculture and Food Security*, *1*(2), 62–83.

Morris, M. L., Tripp, R., and Dankyi, A. A. (1999). *Adoption and Impacts of Improved Maize Production Technology: A Case Study of the Ghana Grains Development Project*. Mexico, DF: CIMMYT.

Mueller, V., Kovarik, C., Sproule, K., and Quisumbing, A. R. (2015). *Migration, Gender, and Farming Systems in Asia: Evidence, Data, and Knowledge Gaps*. IFPRI Discussion Paper No. 01458. Washington DC. Retrieved from http://ebrary.ifpri.org/cdm/ref/collection/p15738coll2/id/129391.

Mulatu, E., and Zelleke, H. (2002). Farmers' highland maize (Zea mays L.) selection criteria: implication for maize breeding for the Hararghe highlands of eastern Ethiopia. *Euphytica*, *127*(1), 11–30. doi:10.1023/A:1019939721444.

Muyanga, M., Jayne, T. S., Argwings-Kodhek, G., and Ariga, J. (n.d.). *Staple Food Consumption Patterns in Urban Kenya: Trends and Policy Implications*. Tegemeo Institute of Agricultural Policy and Development, Egerton University, Working Paper No. 19. Nairobi.

Namonje-Kapembwa, T., and Chapoto, A. (2016). *Improved Agricultural Technology Adoption in Zambia: Are Women Farmers Being Left Behind* ? Working Paper, No. 106. Lusaka, Zambia: Indaba Agricultural Policy Research Institute.

Odendo, M., De Groote, H., and Odongo, O. M. (2001). Assessment of Farmers' Preferences and Constraints To Maize Production in Moist Midaltitude Zone of Western Kenya. In *African Crop Science Conference Proceedings*, 769–75. African Crop Science Society. Volume 5.

Oseni, G., Corral, P., Goldstein, M., and Winters, P. (2015). Explaining gender differentials in agricultural production in Nigeria. *Agricultural Economics*, *46*(3), 285–310. doi:10.1111/agec.12166.

Peterman, A., Behrman, J. A. and Quisumbing, A. R. (2014). A review of empirical evidence on gender differences in nonland agricultural inputs, technology and services in developing countries. In A. R. Quisumbing, R. Meinzen-Dick, T. L. Raney, A. Croppenstedt, J. Behrman and A. Peterman (eds), *Gender in Agriculture: Closing the Knowledge Gap*, pp. 145–86. New York: Springer.

Preibisch, K. L., Herrejon, G. R., and Wiggins, S. L. (2002). Defending food security in a free-market economy: the gendered dimensions of restructuring in Rural Mexico. *Human Organization*, *61*(1), 68–79.

Quisumbing, A. R., and Pandolfelli, L. (2010). Promising approaches to address the needs of poor female farmers: resources, constraints, and interventions. *World Development*, *38*(4), 581–92. doi:10.1016/j.worlddev.2009.10.006.

Quisumbing, A. R., Meinzen-Dick, R., Raney, T. L., Croppenstedt, A., Behrman, J., and Peterman, A. (2014). *Gender in Agriculture: Closing the Knowledge Gap*.

Saito, K. A., and Weidemann, C. J. (1990). *Agricultural Extension for Women Farmers in Africa*. Women in Development Working Paper WPS 398. Washington, DC: World Bank.

Slavchevska, V. (2015). Gender differences in agricultural productivity: the case of Tanzania. *Agricultural Economics*, *46*(3), 335–55. doi:10.1111/agec.12168.

Stevens, R., and Winter-Nelson, A. (2008). Consumer acceptance of provitamin A-biofortified maize in Maputo, Mozambique. *Food Policy*, *33*(4), 341–51. doi:10.1016/j.foodpol.2007.12.003.

Suri, T. (2011). Selection and comparative advantage in technology adoption. *Econometrica*, *79*(1), 159–209. doi:10.3982/ECTA7749.

Swiss Agency for Development and Cooperation. (2015). *Gender Analysis of Maize Post-Harvest Management in Zambia : A Case Study of Chipata and Katete Districts*. Swiss Cooperation Office in Southern Africa.

Twyman, J., Useche, P., and Deere, C. D. (2015). Gendered perceptions of land ownership and agricultural decision-making in ecuador: who are the farm managers? *Land Economics*, *91*(3), 479–500. doi:10.3368/le.91.3.479.

Udry, C. (1996). Gender, agricultural production, and the theory of the household. *Journal of Political Economy*, *104*(5), 1010–46. doi:10.1086/262050.

World Bank, Food and Agriculture Organization, and International Fund for Agricultural Development (2009). *The Gender and Agriculture Sourcebook*. Washington DC: World Bank.

World Bank and ONE Campaign. (2014). *Leveling the Field: Improving Opportunities for Women Farmers in Africa*. Washington, DC: World Bank.

Index

Lightning Source UK Ltd.
Milton Keynes UK
UKHW020921070819
347547UK00001B/11/P

9 781786 760081